Algebra

Volume 1

Algebra

Volume 1

P. M. Cohn

Bedford College
University of London

JOHN WILEY & SONS

London · New York · Sydney · Toronto

Library of Congress catalog card number 73-2780

ISBN 0 471 16430 5 Cloth bound
ISBN 0 471 16431 3 Paper bound

Printed in Great Britain at the Aberdeen University Press, Aberdeen

For Juliet and Ursula

Preface

Although algebra has a long history, it has undergone some quite striking changes in the past few decades. Not least among these is the way the subject has entered into the development of other branches of mathematics, over and above its new applications elsewhere. Its changing role is reflected in the importance of algebra in the curricula, as well as in the many excellent textbooks that now exist. Most of these are designed for undergraduates at North American universities and are either (a) a very broad introduction to linear algebra, with a little groups and rings, for general students taking mathematics, or (b) a course for graduates, or junior–senior students majoring in mathematics, who have already taken a course of type (a). The pattern in Britain is a little different: the honours student specializing in mathematics takes algebra for two or three years (depending on his ultimate interests) and his need is for a textbook which combines (a) and (b) above and is somewhere between them in level. The object of the present work is to provide such a book: the present first volume includes most of the algebra taught in the first two years to undergraduates at British universities; this will be followed by a second volume covering the third year (and some graduate) topics. Hopefully, this first volume will also prove useful to North American students at the Junior, Senior and Graduate levels, both as a reappraisal of some familiar topics and as an introduction to more advanced topics that will lead into the planned second volume of this work.

The actual prerequisites are quite small: students coming to this book will normally have met calculus and some analytic geometry, complex numbers and a little elementary algebra (binomial theorem, quadratic equations, etc.). In any case, some of the topics will be familiar ideas in a new form. There is no doubt that the chief difficulty for the student is the abstractness of the subject so some pains have been taken to motivate the ideas introduced. Connexions between different parts of the subject have been stressed and, on occasion, important applications are briefly discussed. There are numerous exercises, ranging from routine problems to further developments or alternative proofs of results in the text. Some of the harder ones are marked by an asterisk.

The central ideas are *group* and *ring*: once the example of the integers (and the integers mod m) has been described in Ch. **2**, the basic properties of groups and rings are developed in Chapters **3,9** and Chapters **6,10** respectively. The other main topic is *linear algebra.* Ch. **4** describes vector spaces (without mention of a metric or determinants) and this is followed by a brief account of methods of solving linear equations (Ch. **5**). Determinants are introduced in Ch. **7**; although they have not been needed so far, they provide an important invariant and have regained in recent years some of their theoretical importance. Ch. **8** deals with metric questions (quadratic forms, Euclidean spaces) and Ch. **11** discusses the various normal forms for matrices. In the first chapter the all-important ideas of *set* and *mapping* are briefly described, as well as some notions from formal logic. Although not explicitly used in what follows, the latter has had an important influence and some degree of awareness may preserve the reader from pitfalls. There is no separate chapter on categories in this first volume, but the basic terms are introduced to help systematize the results obtained.

It is clearly impossible to be in any sense comprehensive; even if it were possible, this would not be the place. Any selection of material must necessarily be governed by personal taste; my aim has been, if possible, to sustain the reader's interest, while introducing him to the ideas which are important and useful in present-day mathematics. A more detailed idea of the contents may be gleaned from the table of contents and the index. Vol. 2 is planned to deal with: cardinal numbers, lattices, categories, Galois theory, real fields and valuations, tensor products and exterior algebra, ideal theory and Artinian rings.

A book written at this level clearly owes a great deal to other people besides the author (who merely acts as collector, or rather, selector of the material). From the many inspiring lectures I have heard, let me single out the most recent: a delightful series by Paul Halmos in St. Andrews, at the time I was collecting exercises. I am also grateful to the Senate of London University for allowing me to draw on past examination papers. The manuscript was read by Walter Ledermann, and by Warren J. Dicks, who also read proofs; my thanks go to both of them for their helpful comments. Finally I should like to thank the publishers for the efficient way they carried out (and often anticipated) my wishes.

Bedford College P. M. COHN
July 1973

Contents

9 Further group theory

10 Rings and modules

11 Normal forms for matrices

Appendixes

Table of interdependence of chapters (Leitfaden)

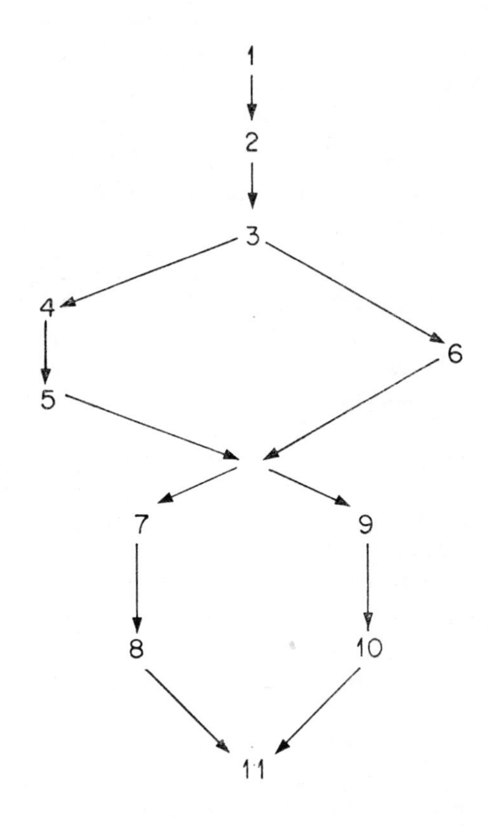

1

Sets and mappings

1.1 The need for logic

Mathematics is the study of relations between certain ideal objects such as numbers, functions and geometrical figures. These objects are not to be regarded as real, but rather as abstract models of physical situations. As examples of the relations that can hold, consider the following assertions that can be made about the natural numbers:

(a) *every even number is the sum of two odd numbers,*

(b) *every odd number is the sum of two even numbers,*

(c) *every even number greater than 2 is the sum of two primes.*

Of these assertions, (a) is true, (b) is false, while for (c) it is not known whether this is true or false ((c) was conjectured by Goldbach in 1742 and has so far resisted all attempts to prove or disprove it).

If our mathematical system is to serve as a model of reality we must know how to recognize true assertions, at least in principle (even though in practice some may be hard to prove). When the object of discussion is intuitively familiar to us—as in the case of the natural numbers—we take certain assertions recognized to be true as our axioms and try to derive all other assertions from them. Once that is done, we can forget the intuitive interpretation and regard our objects as abstract entities subject to the given axioms. When we come to apply our system to a concrete case, we need to find an interpretation for each notion introduced and verify that each axiom holds in the interpretation; we are then able to conclude that all the assertions derived from the axioms also hold. This underlines the need to keep the axiom system as small as possible.

The advantage of this axiomatic method of study is that we can examine the effect on our system of varying the axioms and that the proofs become more transparent the more abstract the system. On the other hand, it takes a little time to familiarize oneself with the abstract notions; here the (more or less concrete) model on which it was based will help, although it is not strictly necessary and certainly no part of the theory. Studying these abstract notions is rather like learning a new language; as in that case we shall find that as our knowledge widens we recognize more landmarks; this makes learning very much easier.

But there is an important respect in which the process differs from learning a language: we shall need to reason about the new concepts and this will require careful attention to the logical interrelation of statements. Of course it is true that even in everyday affairs we can spurn logic only at our peril, but there the patent absurdity of our conclusion usually forces us to abandon a faulty line of reasoning. By contrast, when we pursue an abstract line of thought, involving unfamiliar concepts, we may reach conclusions by logical reasoning, but we will no longer be able to check these conclusions by commonsense. It is therefore important to be fully aware of the rules of logic we need and to realize that these rules can be applied without regard for the actual meaning of the statements on which they are used. For this reason we begin by describing very briefly some concepts and notations from logic.

Propositional logic describes ways in which true statements (also called *assertions* or *propositions*) can be combined to produce other true statements. E.g., if it is asserted that 'Jack was running' and 'Jill was singing', then we may conclude that

'Jack was running and Jill was singing'. (1)

On the other hand, if Jack was not running then statement (1) is false irrespective of what Jill was doing. By enumerating further possibilities we can thus give a precise description of the way the word 'and' is used to link assertions. In order to do this concisely, let 'A' stand for an assertion, such as 'Jack was running', and 'B' for a second assertion, not necessarily different from A. Then we can form the expression 'A and B', also written '$A \wedge B$' and called the *conjunction* of A and B, and make a table which indicates in which cases $A \wedge B$ is true, using 'T' for 'true' and 'F' for 'false':

A	T	T	F	F
B	T	F	T	F
$A \wedge B$	T	F	F	F

This is called the *truth-table* for conjunction. It shows that $A \wedge B$ is true when A is true and B is true, and false in all other cases. For our purposes we may assume that each statement is either true or false; the relevant value T or F is called the *truth-value* of the statement. Since there are two possible truth-values for A and two for B, we have $2 \,.\, 2 = 4$ possibilities in all, which are listed in the above table.

A second way in which assertions can be combined is by using 'or': 'John went to the cinema last night, or to the theatre'. This is a true statement if in fact John last night went to the cinema, and also true if he went to the theatre; the possibility that he went to both is not really envisaged, but if he did, the statement would still be regarded as true. This causes some ambiguity in everyday life: if A and B are both true, is the statement 'A or B' to be regarded as true? The situation is usually cleared up by the context (but not always, cf. 'This summer Jane will go to Italy or Norway'). In mathematics the expression 'A or B' is always taken to mean 'A or B or both'; it is written '$A \vee B$' and is called the *disjunction* of A and B. Its truth-table is

A	T	T	F	F
B	T	F	T	F
$A \vee B$	T	T	T	F

A typical use of disjunction in mathematics is the sentence: 'If a and b are two real numbers whose product is zero: $ab = 0$, then $a = 0$ or $b = 0$'. Clearly we must not exclude the case where $a = 0$ and $b = 0$.

With every statement we can associate its opposite or *negation* by inserting 'not' in the appropriate place. Thus 'Max is the biggest liar' has the negation 'Max is not the biggest liar'. Generally, if A is any statement, then its negation is 'not A', also written '$\neg A$', and it is true precisely when A is false. Its truth-table is

A	T	F
$\neg A$	F	T

Here there are only two possibilities because only one statement is involved.

The notion of *implication* is particularly important for us and its use in mathematics differs in some ways from everyday usage, though the underlying meaning is of course the same. Thus 'A implies B' or 'if A, then B', written '$A \Rightarrow B$' means for us: 'either A is false or B is true'. It is expressed in the truth-table

A	T	T	F	F
B	T	F	T	F
$A \Rightarrow B$	T	F	T	T

For example, a mathematical proof might contain the line: 'If $n > 5$, then $n > 3$'. A parallel use in everyday English would be: 'If this book was influenced by Shakespeare it must have been written after *The Canterbury Tales*'.

We note from the truth-table for $\Rightarrow$ that '$A \Rightarrow B$' holds for any B whenever A is false; in other words, a false statement implies anything. This may seem strange at first, but it has its counterpart in ordinary usage where we underline the absurdity of an assertion by drawing an even more absurd conclusion ('If Jones wins the election I'll eat my hat').

We also note that implication can be defined in terms of the other connectives by the rule

'$A \Rightarrow B$' stands for '$(\neg A) \vee B$'.

More generally, if we are given a rule for forming a proposition $P(A, B)$ whose truth-value depends only on those of A and B, then P can be defined in terms of $\neg$, $\vee$ alone. E.g., if P is given by the table

A	T	T	F	F
B	T	F	T	T
P	F	T	T	T

then $P(A, B) = (\neg A) \vee (\neg B)$.

Another connective of frequent use is the *bi-implication* or *equivalence*: '$A \Leftrightarrow B$'. This is defined as '$(A \Rightarrow B) \wedge (B \Rightarrow A)$'; thus '$A \Leftrightarrow B$' is true precisely when A, B are both true or both false.

Some composite statements are true for all assignments of truth-values, e.g., $A \vee (\neg A)$. Such statements are called *tautologies*. To check whether a given assertion is a tautology we can again use truth-tables. E.g., consider $[A \wedge (A \Rightarrow B)] \Rightarrow B$.

A	B	$A \Rightarrow B$	$A \wedge (A \Rightarrow B)$	$[A \wedge (A \Rightarrow B)] \Rightarrow B$
T	T	T	T	T
T	F	F	F	T
F	T	T	F	T
F	F	T	F	T

This is seen to be a tautology because only Ts occur in the last column. There is a quicker method, based on the fact that in 3 out of 4 cases $(P \Rightarrow Q)$ has the value T. Thus assume that for some assignment of truth-values, $[A \wedge (A \Rightarrow B)] \Rightarrow B$ is F. Then by the truth-table for $\Rightarrow$, B is F and $A \wedge (A \Rightarrow B)$ is T. By the truth-table for $\wedge$ it follows that A is T and $A \Rightarrow B$ is T. But we already found B to be F, hence $A \Rightarrow B$ is F, a contradiction.

Usually the simple statements discussed above are not enough to deal with mathematical situations. We need, besides the propositions, also *propositional functions* or *predicates*. Consider for example:

(a) *x is an odd number* (x ranges over the natural numbers),

(b) *x forgot his hat this morning* (x ranges over humans),

(c) *x is married to y* (x and y range over humans),

(d) *x is greater than y* (x, y range over natural numbers).

Unlike a proposition, a propositional function is no longer true or false but only becomes so when particular values are substituted for the variables ('2 is an odd number'; 'Mary's baby forgot his hat this morning').

In practice one often wants to say that some assertion involving x, say $P(x)$, holds for *all* x (in the universe of discourse). We write this as

$(\forall x)P(x)$, which stands for: For all x, $P(x)$ holds.

We say that the variable x is *bound* by the *universal quantifier* $\forall$. To express that $P(x)$ holds for *some* x, we write

$$(\exists x)P(x), \text{ which stands for: There exists } x \text{ such that } P(x);$$

here x is bound by the *existential quantifier* $\exists$. Of course when all the variables occurring in a propositional function have been bound (by universal or existential quantifiers), we have an assertion of the type considered before.

For example, in the domain of natural numbers, $(\forall x)(\forall y)(x+y = y+x)$ expresses the fact that for any x and y the sum $x+y$ is independent of the order of the terms. Similarly, $(\forall x)(\exists y)(x < y)$ states that for every x, there is a y greater than x, i.e. there is no last number. Note particularly that if we apply the quantifiers in the opposite order we get the propostion $(\exists y)(\forall x)(x < y)$, which states that there exists a y greater than all x. Clearly this is false whereas the other was true; thus one must pay attention to the order in which the quantifiers are applied.

We note that a bound variable may always be renamed without changing the meaning, thus $(\forall x)P(x)$ means exactly the same as $(\forall y)P(y)$; for this reason the bound variable is often described as a 'dummy' variable. One sometimes makes use of this freedom to rename bound variables, to avoid a clash of notation. E.g., instead of $(a = x+y) \wedge (\forall x)(a \neq 2x)$, one would write $(a = x+y) \wedge (\forall z)(a \neq 2z)$. Both forms mean the same, but the second is less likely to cause misunderstanding.

The universal and existential quantifiers are related by the following equivalences, which enable us to define either in terms of the other (this means that we can choose either of them as the basic one and define the other in terms of it):

$$\neg(\exists x)\neg P(x) \Leftrightarrow (\forall x)P(x),$$

$$\neg(\forall x)\neg P(x) \Leftrightarrow (\exists x)P(x).$$

With the help of these formulae it is easy to write out the negation of any formula with quantifiers, e.g.

$$\neg[(\forall x)(\exists y)(\forall z)F(x, y, z)] \Leftrightarrow (\exists x)(\forall y)(\exists z)[\neg F(x, y, z)].$$

As an illustration, consider the assertion that every number has an immediate successor. This is expressed by the formula

$$(\forall x)(\exists y)(\forall z)[(x < y) \wedge \{(x < z) \Rightarrow (y \leqslant z)\}].$$

The negation is

$$(\exists x)(\forall y)(\exists z)[\neg(x < y) \vee \{(x < z) \wedge \neg(y \leqslant z)\}]$$

or, if we write $x \geqslant y$ for $\neg(x < y)$ and $y > z$ for $\neg(y \leqslant z)$,

$$(\exists x)(\forall y)(\exists z)[(x \geqslant y) \vee \{(x < z) \wedge (y > z)\}];$$

this says that there is some x which does not have an immediate successor.

In any mathematical theory one has axioms from which the assertions of the theory (the theorems) are derived by logical deduction ('proofs'), using also the logical theorems, i.e. the tautologies. It is not necessary, nor indeed appropriate, to describe in detail the form such a proof would take. The customary presentation of proofs, logically informal though mathematically rigorous, is best assimilated by studying examples. But it may be helpful to end this section with a word or two on the chief methods of proof.

A direct proof (or step in a proof) usually takes the form: 'A' is true and '$A \Rightarrow B$' is true, hence 'B' is true. This is called *modus ponens* (which was the term used in Scholastic Logic).

It is important to distinguish between '$A \Rightarrow B$' on the one hand and 'A, hence B' on the other. The distinction may seem pedantic in cases where A is true, but to ignore it can easily give rise to confusion. E.g. compare the following two statements about positive real numbers x and y:

(a) $x > y \Rightarrow (x^2 > xy \text{ and } xy > y^2) \Rightarrow x^2 > y^2$,

(b) $x > y$, hence $x^2 > xy$ and $xy > y^2$, therefore $x^2 > y^2$.

Here (a) is a conditional statement which tells us nothing unless we are told that $x > y$ to begin with. In any case, it is ambiguous; it is of the form $A \Rightarrow B \Rightarrow C$ and this can mean either $(A \Rightarrow B) \Rightarrow C$, or $A \Rightarrow (B \Rightarrow C)$. In fact, it usually means neither but is offered by some slipshod writers who intend (b) (which is unambiguous).

However, there is a legitimate use of the expression $A \Rightarrow B \Rightarrow C$. When a theorem asserts that a number of statements, say A, B, C, are equivalent, we frequently indicate beforehand the order in which we prove the parts of the theorem. E.g., we may prove in turn $A \Rightarrow B$, $B \Rightarrow C$, $C \Rightarrow A$; clearly this will establish the claim that A, B, C are equivalent. This is often shortened to '$A \Rightarrow B \Rightarrow C \Rightarrow A$.'

An indirect proof usually takes one of the following forms. In order to prove '$A \Rightarrow B$' we may prove '$\neg B \Rightarrow \neg A$' (called the *contrapositive*), but *not* '$B \Rightarrow A$', for the latter is not generally equivalent to '$A \Rightarrow B$', cf. Ex. (3). E.g., suppose we wish to prove '$(\forall x)(x^2$ is even $\Rightarrow x$ is even)'. It is not correct to argue: 'If x is even, then x^2 is even, hence the result', but we can argue: 'If x is odd, then x^2 is odd, hence the result'.

Another form of indirect proof is by contradiction, also called *reductio ad absurdum*. In order to prove A we show that '$(\neg A) \Rightarrow F$', i.e. we show that (not A) leads to a contradiction. Thus to prove $\sqrt{2}$ irrational, let us assume the contrary, i.e. $2 = (m/n)^2$, where m/n is a rational number. If we take m/n in its lowest terms, m and n cannot both be even and, on multiplying up, we have $m^2 = 2n^2$, i.e. m^2 is even, hence m is even, say $m = 2h$; now $m^2 = 4h^2 = 2n^2$, hence $n^2 = 2h^2$ and so n^2 is even, therefore n is also even, a contradiction. This proves $\sqrt{2}$ to be irrational. This proof goes back to the school of Pythagoras.

Finally there is the proof by *counter-example*. Many statements are of the form $(\forall x)P(x)$; if we want to disprove this we must prove its negation, i.e. $(\exists x) \neg P(x)$, and this is done by finding a c such that $\neg P(c)$. E.g., it may be that Goldbach's conjecture is false; to establish this one would have to find a counter-example, i.e. an even number greater than 2, which cannot be written as the sum of two primes.

In mathematics we often have a non-constructive existence proof. This may sound strange at first sight, but on reflection we see that this is no different from everyday life: an assembly of 500 persons must include two with the same birthday, though closer examination is needed to find such a pair.

Frequently a theorem is in the form of an implication or an equivalence. We note a number of equivalent ways of saying this:

$A \Rightarrow B$: A holds only if B holds, or A is sufficient for B.

$A \Leftarrow B$ (which stands for $B \Rightarrow A$): A holds if B holds, or A holds whenever B holds, or A is necessary for B.

$A \Leftrightarrow B$: A holds if and only if B holds, or A is necessary and sufficient for B.

In particular, the phrase 'if and only if' occurs frequently and it is sometimes abbreviated by 'iff'.

It is also useful to have a sign to indicate the end of a proof (instead of saying each time 'this is the end of the proof'). We shall follow current usage by employing the sign ■ at the end of a proof, or at the end of a theorem, either because the proof precedes it or, when the proof is omitted, because it is so easy that it can be supplied by the reader.

Exercises

(1) Establish the following equivalences: (i) $A \wedge A \Leftrightarrow A$, (ii) $A \vee A \Leftrightarrow A$, (iii) $A \wedge B \Leftrightarrow B \wedge A$, (iv) $A \vee B \Leftrightarrow B \vee A$, (v) $(A \wedge B) \wedge C \Leftrightarrow A \wedge (B \wedge C)$, (vi) $(A \vee B) \vee C \Leftrightarrow A \vee (B \vee C)$, (vii) $A \wedge (B \vee C) \Leftrightarrow [(A \wedge B) \vee (A \wedge C)]$, (viii) $A \vee (B \wedge C) \Leftrightarrow [(A \vee B) \wedge (A \vee C)]$.

(2) Establish the following tautologies: (i) $A \Rightarrow (B \Rightarrow A)$, (ii) $[A \Rightarrow (B \Rightarrow C)] \Rightarrow [(A \Rightarrow B) \Rightarrow (A \Rightarrow C)]$, (iii) $(A \Rightarrow B) \Rightarrow [(B \Rightarrow C) \Rightarrow (A \Rightarrow C)]$, (iv) $\neg(A \wedge B) \Leftrightarrow (\neg A) \vee (\neg B)$, (v) $\neg(A \vee B) \Leftrightarrow (\neg A) \wedge (\neg B)$.

(3) Show that $A \Rightarrow B$ is equivalent to $\neg B \Rightarrow \neg A$, but that of the assertions $A \Rightarrow B$, $B \Rightarrow A$, neither implies the other.

(4) Show that from any tautology expressed in terms of $\vee$, $\wedge$ and $\neg$ another tautology can be derived by interchanging $\vee$ and $\wedge$ and inserting $\neg$ in suitable places. Can you formulate a general rule?

(5) Define $A \mid B$ to mean $\neg(A \wedge B)$ (this is the Sheffer stroke function). Show how to express $\neg A$ and $A \vee B$ in terms of $\mid$; do the same for the other connectives introduced: $\wedge$, $\Rightarrow$, $\Leftrightarrow$.

(6) $P(A, B, C)$ is defined to be true if precisely one of A, B, C is true. Express P in terms of $\wedge$, $\vee$, $\neg$, and hence in terms of $|$.

(7) Express the following assertions in words, where x, y, z range over the natural numbers (including 0), and replace those that are false by their negations: (i) $(\forall x)(\exists y)(x = y+y)$, (ii) $(\forall x)(\forall y)(\exists z)(y \geqslant x \Rightarrow y = x+z)$, (iii) $(\exists x)(\exists y)((x \neq y) \wedge (x^y = y^x))$, (iv) $(\exists x)(\forall y)(\exists z)[(y > x) \Rightarrow (y = xz)]$.

(8) Express the following assertion in symbols alone: Between any two distinct real numbers there is another real number.

(9) What is wrong with the following argument? Any soap is better than no soap; but no soap is better than Wonder-Bubble Soap, hence any soap is better than Wonder-Bubble Soap.

1.2 Sets

Many of the objects we shall study are themselves collections of other objects. These collections or *sets* may be finite or infinite; later we shall meet sets with additional structure, but for the moment we shall look at abstract sets and the ways in which they can be combined to form new sets.

By a *set* we understand, then, any collection of objects. For example, the following are sets: (i) all the stars visible from my house at 9 pm tonight, (ii) all one-legged magicians, (iii) all odd numbers. We see that in some cases it may be difficult to check which objects belong to the set (e.g. (i) above) or whether the set has any members at all ((ii) above). All that matters is that the definition is sufficiently clear for us to be able to tell (in principle at least) whether any given object is or is not a member of the set. We usually denote sets by capitals and their members, also called their *elements*, by lower case letters. However, it will not always be possible to keep to this convention, especially when we are dealing with sets whose members are themselves sets. If S is a set, we write $x \in S$ to indicate that x is a member of S; in the contrary case we write $x \notin S$.

A set, in this sense, is no more and no less than the totality of its members; no considerations of order or multiplicity enter. Thus: Adam and Eve; Eve and Adam; Adam, Eve and the mother of Cain, all describe the same set. In a more precise form this can be stated by saying that S and T denote the same set: $S = T$, if and only if, for all x, $x \in S \Leftrightarrow x \in T$.

Every set encountered in the real world is *finite*; by this we mean that its members can (at least in principle) be counted, using the natural numbers, and this process stops at a certain number. Otherwise the set is infinite, e.g., the set of all odd numbers is infinite. It is this occurrence of infinite sets in mathematics that requires rather careful analysis. Although no critical situations will arise in these pages, it should be kept in mind that too free a use of set-theoretic notions can easily lead to contradictions. The best known of these is Russell's paradox: A set may be a member of itself, e.g., the Union of all Registered Charities may be a Registered Charity. Now consider the set M of those sets that are not members of themselves. Is M a member of itself, i.e. is $M \in M$? If $M \in M$, then $M \notin M$ (by the definition of M), while if $M \notin M$, then $M \in M$. Thus we reach a contradiction in either case. The paradox is resolved by restricting the ways in which sets can be formed, so that it becomes inadmissible to consider 'the set of all those sets that are not members of themselves'. There are several ways of doing this, but they need not concern us here; they will not play a role in the rather simple set-theoretical arguments we shall meet.

Let S and T be sets. If every member of T also belongs to S, we say that T is a *subset* of S and write $T \subseteq S$ or also $S \supseteq T$. E.g., the odd numbers form a subset of the set of all numbers. Any set S is a subset of itself; this is called an *improper* subset of S in contrast to the *proper* subsets of S which are different from S itself. We write $T \subset S$ or $S \supset T$ to indicate that T is a proper subset of S.

Frequently we describe a subset of S by means of a propositional function, thus $\{x \in S \mid P(x)\}$ denotes the subset of S consisting of those (and only those) elements x for which $P(x)$ holds. E.g., if the set of all natural numbers is denoted by $\mathbf{N}$, then the subset of odd numbers may be denoted by $\{x \in \mathbf{N} \mid x \text{ is odd}\}$. On the other hand, forms like $\{x \mid P(x)\}$, in which the domain over which x ranges is left unspecified, are best avoided.

Let S and T be any sets, then the elements which belong to both S and T form a set which is called the *intersection* of S and T and is denoted by $S \cap T$. E.g., if S is the set of all one-legged creatures and T the set of magicians, then $S \cap T$ is the set of all one-legged magicians. It may happen that $S \cap T$ has no members at all; this means that $S \cap T$ is the *empty set*. By definition this is the set with no members; it is generally denoted by $\varnothing$. Two sets whose intersection is the empty set are said to be *disjoint*.

From two sets S, T we can form another set, the *union*, written $S \cup T$, which consists of all the members of S or T. E.g., a public library may admit as borrower anyone who is either (i) a householder in the district or (ii) a resident of at least 3 years' standing. Denoting the sets of persons named in (i), (ii) by A, B respectively, we see that the set of people eligible as borrowers is $A \cup B$.

In most cases, the sets under consideration in any given case will all be subsets of some given set U, the 'universe of discourse'. Thus U might be the set of natural numbers, or of triangles in the plane etc. In this situation we can, for any set S, form its *complement*, i.e. the set of all members of U that are not in S; it is denoted by S'. Thus if S is the set of all odd numbers, then its complement (in the set of all natural numbers) is the set of all even numbers. This example makes it clear why the complement, to be useful, has to be taken within a given set as universe. We also note the following brief way of describing intersection, union and complement, which brings out a certain analogy with the rules for combining propositions.

$$S \cap T = \{x \in U \mid x \in S \wedge x \in T\},$$
$$S \cup T = \{x \in U \mid x \in S \vee x \in T\},$$
$$S' = \{x \in U \mid x \notin S\}.$$

Here U is the universe containing all the objects under discussion.

Although many of our sets are infinite, we shall also be dealing with finite sets. In particular, with any object x we can associate the set $\{x\}$ whose only member is x. It is important to distinguish between the set $\{x\}$ and the object x (which may itself be a set). E.g., the set **N** of natural numbers is infinite, but the set $\{\mathbf{N}\}$, whose only member is **N**, is finite. If S is any finite set, its members can (by the definition of finite set) be labelled or indexed by the integers from 1 to n, for some integer n. Thus if the elements of S are $x_1, \ldots, x_n$, then $S = \{x_1, \ldots, x_n\}$; if distinct elements have received distinct labels, i.e. if $x_i \neq x_j$ for $i \neq j$, then S consists of exactly n elements. But it is usually more convenient not to impose this restriction, so that there may be repetitions among $x_1, \ldots, x_n$. If we wish to consider the objects $x_1, \ldots, x_n$ in the order given, we use parentheses: $(x_1, \ldots, x_n)$, and call the result a *sequence* or, more particularly, an *n-tuple*; e.g., a roster selecting pilots for flying duties is of this form. By contrast, the set $\{x_1, \ldots, x_n\}$ where the order is immaterial, is written with curly brackets (braces). A set which has been indexed in some way by the numbers from 1 to n is also called a *family*; more generally even infinite sets can be indexed if we use an infinite indexing set. E.g. if Δ_{ABC} denotes the plane triangle with vertices A, B, C, this provides an indexing of all triangles in the plane by triples of points and we may speak of the family $\{\Delta_{ABC}\}$ of triangles obtained in this way.

From any two objects x and y we can form the sequence (x, y); it is called an *ordered pair*, and of course is different from (y, x), unless $x = y$. If S and T are any sets, we denote by $S \times T$ the set of all ordered pairs (x, y) with $x \in S$ and $y \in T$. When $T = S$, we also write S^2 in place of $S \times S$. The set $S \times T$ is called the *Cartesian product* of S and T, after R. Descartes who showed how to describe points of the plane by the Cartesian product of the real line with itself.

Examples. (i) At a dance, let S be the set of gentlemen and T the set of ladies, then $S \times T$ is the set of possible couples. (ii) If $S = \{0, 4, 6\}$, $T = \{1, 4\}$, then $S \times T = \{(0, 1), (0, 4), (4, 1), (4, 4), (6, 1), (6, 4)\}$. (iii) If $S = T = \mathbf{R}$, the set of real numbers, then $\mathbf{R}^2$ is the set of pairs of real numbers and these pairs may be used to represent points in the plane.

More generally, from n sets $S_1, \ldots, S_n$ we can form the product $S_1 \times S_2 \times \cdots \times S_n$ whose elements are all the sequences $(x_1, \ldots, x_n)$, in which $x_i \in S_i$

$(i = 1, \ldots, n)$; when $S_1 = \cdots = S_n = S$, say, one also writes S^n in place of $S \times S \times \cdots \times S$ and S^n is called the nth *Cartesian power* of S.

Exercises

(1) Prove the following formulae for subsets of a set U: (i) $A \cap A = A$, (ii) $A \cup A = A$, (iii) $A \cap B = B \cap A$, (iv) $A \cup B = B \cup A$, (v) $(A \cap B) \cap C = A \cap (B \cap C)$, (vi) $(A \cup B) \cup C = A \cup (B \cup C)$, (vii) $A \cap (B \cup C) = (A \cap B) \cup (A \cap C)$, (viii) $A \cup (B \cap C) = (A \cup B) \cap (A \cup C)$. (Compare with Ex. (1), **1.1.**)

(2) Illustrate the formulae of Ex. (1) by taking A, B, C to be the set of all quadrilaterals, all regular polygons and all polygons large enough to cover a penny (not necessarily respectively).

(3) If A has α elements and B has β elements, find the number of elements in $A \times B$. If, moreover, A and B are disjoint, find the number of elements in $A \cup B$. What is this number when $A \cap B$ has δ elements?

(4) How many subsets are there in a set of n elements? (Do not forget to include $\varnothing$ and the set itself.)

(5) Give examples of sets such that (i) all and (ii) none of their members are also subsets.

(6) Every number can be defined by a sentence in English and since English has a finite vocabulary, the number of numbers definable by a sentence of at most twenty words, say, is finite. So it makes sense to speak of 'the least number which cannot be defined by a sentence of at most twenty words'. Does it?

1.3 Mappings

Let S and T be sets; any subset of $S \times T$ is called a *correspondence* from S to T. E.g., let S be the set of all points in the plane and T the set of lines; the relation of *incidence* (the point P is incident with the line l if P lies on l) defines a correspondence from points to lines. To each point P correspond all the lines through P and to each line l correspond all the points on l. This is an example of a 'many–many correspondence', where to each element of S correspond many elements of T and vice versa. In general, to each element of S there may correspond many, one or no elements of T.

An important special case of a correspondence is that of a *bijective* or *one–one* correspondence, also called a *bijection*. Here there corresponds to each $s \in S$ just one $t \in T$ and to each $t \in T$ just one $s \in S$. The following are some examples of bijections: (i) At a gathering of married couples there is a bijection between the set of men and the set of women: to each person there corresponds precisely one spouse of the other sex. (ii) The real numbers may be represented on a line (the x-axis, say, in coordinate geometry) so that to each real number corresponds just one point on the line and to each point on the line corresponds one real number. (iii) The correspondence $x \leftrightarrow 2x$

defines a bijection between real numbers; on the other hand (iv) the correspondence $x \to x^2$ does not, because x and $-x$ have the same square, for any real x.

The last example makes it clear that the notion of bijection is too restrictive to account for such simple functions as x^2. But the notion of function, or mapping, in various guises, plays a basic role in mathematics. For this reason the next definition is fundamental in all that follows.

A *mapping* from S to T is a correspondence between S and T such that to each $x \in S$ there corresponds exactly one $y \in T$. If f is the mapping, one writes $f\colon S \to T$ or $S \xrightarrow{f} T$ and calls S the *domain* and T the *range* of f. The unique element $y \in T$ that corresponds to $x \in S$ is called the *image* of x and is written $f(x)$ or f_x or more often xf. We also write $x \mapsto y$ to indicate the correspondence between x and its image y. Often the set of all images, namely $\{y \in T \mid y = xf \text{ for some } x \in S\}$ is also called the *image* of the mapping f and is written Sf or $\operatorname{im} f$; in practice this double use of the term 'image' does not lead to confusion.

Examples of mappings. (i) With each newborn baby associate its weight in grams to the nearest gram. This is a mapping from the set of newborn babies to the natural numbers. (ii) Let S be the United Kingdom and T a map of the United Kingdom. There is a 'mapping' which associates with each place in the country a point on the map. (iii) and (iv) The examples (iii) and (iv) of correspondences considered earlier define mappings from $\mathbf{R}$ to itself, namely $x \mapsto 2x$ and $x \mapsto x^2$ respectively. (v) If $S = (x_\alpha)$ is a family indexed by a set A, then we have a mapping $\alpha \mapsto x_\alpha$ from A to S. (vi) Given a Cartesian product $P = S \times T$, we can define mappings from P to S and T by the rules $(x, y) \mapsto x$ and $(x, y) \mapsto y$; they are called the *projections* on the factors S and T. Similarly, in an n-fold product $P = S_1 \times \cdots \times S_n$ we have for each $i = 1, \ldots, n$ a projection $\varepsilon_i\colon P \to S_i$ given by $(x_1, \ldots, x_n) \mapsto x_i$.

Clearly a bijection is a particular type of mapping. On closer examination we see that two properties are required for a mapping to be bijective; it is useful to consider them separately. A mapping $f\colon S \to T$ is said to be *injective* or an *injection* or *one–one* if distinct elements of S have distinct images, i.e. $s \neq s'$ implies $sf \neq s'f$. The mapping is called *surjective* or a *surjection* or *onto* T if every element of T is an image, i.e. if $Sf = T$. Thus a mapping is bijective precisely if it is injective and surjective. In the above examples, (i) is neither surjective nor injective (at least if we take enough babies), while (ii) and (iii) are bijective. The mapping $x \mapsto x^2$ of $\mathbf{R}$ into itself considered in (iv) is neither injective nor surjective, but on the set $\mathbf{C}$ of complex numbers it defines a mapping from $\mathbf{C}$ to itself which is surjective, though not injective (the surjectivity is just an expression of the fact that every complex number has a square root).

Let $f\colon S \to T, g\colon T \to U$ be any mappings, then we can compose them to get a mapping $h\colon S \to U$, given by

$$xh = (xf)g \qquad \text{for all } x \in S. \tag{1}$$

A graphic way of expressing this equation is shown in the accompanying diagram. Starting from an element $x \in S$, we reach the same element of U whether we go via T, $x \mapsto xf \mapsto (xf)g$ or direct, $x \mapsto xh$; we express this by saying that the triangle shown *commutes*.

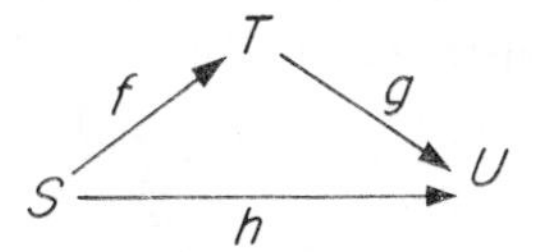

The mapping h defined by (1) is called the *composite* or *product* of f and g and is denoted by fg; in this notation (1) reads

$$x(fg) = (xf)g. \tag{2}$$

As a rule one omits the parentheses and denotes either side of (2) by xfg. We observe that fg is defined only when the range of f is contained in the domain of g. Further we note that if we had written mappings on the left, (2) would read: $(fg)x = g(fx)$. It is to avoid this reversal of factors that we put mappings on the right of their arguments.†

As an example, let $f, g\colon \mathbf{N} \to \mathbf{N}$ be given by $xf = x+1$, $xg = x^2$, then $xfg = (x+1)^2$, $xgf = x^2+1$. We see that $fg \neq gf$, so attention must be paid to the order in which the mappings are composed.

When S is a finite set, f and g may be given explicitly. Let us indicate each mapping by writing down the elements of S as a sequence and under each element write its image. Thus if $S = \{1, 2, 3\}$, and f is given by $\begin{pmatrix}1 & 2 & 3\\ 2 & 3 & 1\end{pmatrix}$, while g is $\begin{pmatrix}1 & 2 & 3\\ 1 & 3 & 2\end{pmatrix}$, then fg is $\begin{pmatrix}1 & 2 & 3\\ 3 & 2 & 1\end{pmatrix}$, and gf is $\begin{pmatrix}1 & 2 & 3\\ 2 & 1 & 3\end{pmatrix}$. Again $fg \neq gf$; on the other hand if h is $\begin{pmatrix}1 & 2 & 3\\ 3 & 1 & 2\end{pmatrix}$ then $fh = hf$.

An important rule in composing mappings is the *associative law*: For any mappings $f\colon S \to T$, $g\colon T \to U$, $h\colon U \to V$ we have

$$(fg)h = f(gh). \tag{3}$$

Observe that both sides of (3) are defined, by what was assumed about f, g, h. More generally, the equation (3) holds for any mappings f, g, h such that both sides of (3) are defined.

To prove (3) we apply each side to an element x of S, remembering (2): $x[(fg)h] = [x(fg)]h = ((xf)g)h$ and $x[f(gh)] = (xf)(gh) = ((xf)g)h$; now a comparison gives (3).

† Another possibility (often used) is to denote the composition of f and g by gf. With this convention, and writing mappings on the left, (2) reads $(gf)x = g(fx)$ and so the order of the factors is the same on both sides. But this would oblige us to read products like gf from right to left and we shall not use this convention.

With every set S we can associate the *identity mapping* 1_S which maps each element of S to itself: $x \mapsto x$. Clearly this is always a bijection. Further, for any $f: S \to T$ and $h: U \to S$ we have $1_S f = f$, $h1_S = h$.

Let S be any set and T a subset, then there is a mapping ι from T to S, defined by $x\iota = x$ for all $x \in T$; this is called the *inclusion mapping* of T in S. Although ι and 1_S have the same effect wherever they are defined (namely on T), they must be carefully distinguished; e.g., whereas 1_S is bijective, ι is injective, but not surjective (except when $T = S$ and so $\iota = 1_S$). In fact ι may be obtained from 1_S by restricting the domain to T; this is often expressed by writing $\iota = 1_S \mid T$. Generally, if $f: X \to Y$ is any mapping and X' is a subset of X, then the *restriction* of f to X', denoted by $f \mid X'$, is defined as the mapping $f': X' \to Y$ given by $xf' = xf$ for all $x \in X'$. We observe that this restriction may be written as $f' = \iota f$, where ι is the inclusion of X' in X.

Using the composition of mappings we can describe bijections. In the first place we have

LEMMA 1 *If $f: S \to T$, $g: T \to S$ are any mappings such that*

$$fg = 1_S, \tag{4}$$

then f is injective and g is surjective.

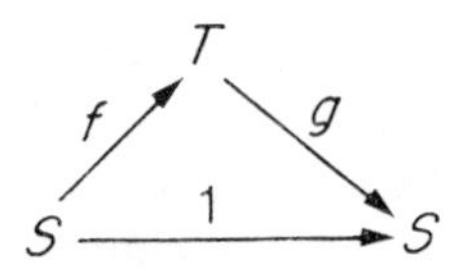

For let $x, y \in S$ and $xf = yf$, then $x = xfg = yfg = y$, hence f is injective Given any $x \in S$, $x = xfg = (xf)g$ and this shows g to be surjective. ∎

Suppose that $f: S \to T$, $g: T \to S$ satisfy

$$fg = 1_S \qquad gf = 1_T, \tag{5}$$

then f is both injective and surjective, by the lemma, and so is a bijection. Conversely, if $f: S \to T$ is a bijection, then we can always find a unique mapping $g: T \to S$ to satisfy (5). For, given $u \in T$, we know there exists just one $x \in S$ such that $xf = u$. Put $ug = x$, then this defines g on T and $ugf = u$, $xfg = x$, therefore (5) holds. This proves

THEOREM 2 *A mapping $f: S \to T$ is a bijection if and only if there is a mapping $g: T \to S$ to satisfy* (5). ∎

There can be at most one mapping g to satisfy (5), for any given f. For assume that (5) holds and that $g': T \to S$ is another mapping such that $fg' = 1_S$, $g'f = 1_T$, then by the associative law, $g' = g'1_S = g'fg = 1_T g = g$. The unique mapping g satisfying (5) is called the *inverse* of the bijection f and is written f^{-1}. In this notation (5) reads $ff^{-1} = 1_S$, $f^{-1}f = 1_T$.

The distinction between finite and infinite sets is an important one which will be taken up in greater detail in Vol. 2. Here we shall only note one useful property of finite sets (actually it can be used to characterize them):

LEMMA 3 *An injective mapping from a finite set to itself is also surjective.*

For let $f: S \to S$ be injective and take $a \in S$; we must find $b \in S$ such that

$$a = bf. \tag{6}$$

Consider the effect of performing f repeatedly. Let us write f^2 for ff and generally abbreviate $ff \ldots f$ (with n factors) as f^n. In the series of elements $a, af, af^2, \ldots$ there must be repetitions, because S is finite, so assume that

$$af^r = af^s, \tag{7}$$

where $r > s$ say. Since f is injective, $xf = yf$ implies $x = y$, so we may cancel f in (7). If we do this s times, we get $af^{r-s} = a$, i.e. (6) holds with $b = af^{r-s-1}$. ■

Later we shall meet many applications of this lemma. It is not really a surprising result and, to newcomers at least, not as surprising as the fact that it no longer holds for infinite sets. To give an illustration, if in a club each member succeeds in borrowing £1 from another member but no two have borrowed from the same person, then everyone has also had to lend £1 (by the lemma) so no one is any better off. But suppose that we have an infinite club, with members $A_1, A_2, \ldots$ indexed by the positive integers † (where it is assumed that $A_m \neq A_n$ for $m \neq n$). If now for each n, A_n borrows £1 from A_{n+1}, then A_1 is £1 better off, while all the others come out even.

The failure of Lemma 3 for infinite sets makes it seem difficult at first sight to extend the notion of counting and cardinality (or 'number of elements') to infinite sets. These difficulties were overcome by Cantor who laid the foundations of set theory in the 1870s. This does not concern us directly as we shall (in this volume) use the notion of cardinality only for finite sets. With every finite set S we associate a natural number $|S|$, the number of its elements (sometimes called the *cardinal* of S). In a complete account one would have to show that this is uniquely defined, i.e. that different ways of counting S give the same answer. This will be proved when we come to the axiomatic development of numbers in Vol. 2.

Exercises

(1) S is a set of four elements. Find (i) the number of mappings of S into itself, (ii) the number of bijections of S to itself. (Hint. Try sets of two and three elements first.)

(2) If $f: A \to B$ and $g: B \to C$ are both injective (or both surjective), show that fg is so too. If fg is injective (or surjective) what can be said about f and g?

† It should be pointed out that there are infinite sets that cannot be indexed by the integers (the uncountable sets). This is proved in books on analysis or set theory (see also Vol. 2).

(3) If f is any bijection and f^{-1} its inverse, show that the domain and range of f^{-1} are the range and domain respectively of f. Show also that f^{-1} is again bijective and $(f^{-1})^{-1} = f$.

(4) If $f: A \to B$, $g: B \to C$, $h: C \to A$ are three mappings such that $fgh = 1_A$, $ghf = 1_B$ and $hfg = 1_C$, show that each of f, g, h is a bijection and find their inverses.

(5) Let $fg = 1_S$; if f is surjective, or if g is injective, show that both f and g are bijective and inverse to each other.

(6) Let $fg = 1_S$; we say that g is a *right inverse* of f and that f is a *left inverse* of g. Show that f is bijective iff it has a single right inverse, or also iff it has a single left inverse. (Hint. To get counter-examples, look at mappings from **N** to **N**.)

(7) (i) For any integer a define a mapping μ_a of **N** into itself by the rule $\mu_a: x \mapsto xa$. Show that $\mu_{ab} = \mu_a\mu_b$. (ii) For any integer a define a mapping α_a of **N** into itself by $\alpha_a: x \mapsto x+a$. Show that $\alpha_{a+b} = \alpha_a\alpha_b$.

(8) A mapping $f: S \to T$ is said to be *constant* if $xf = yf$ for all $x, y \in S$. Show that for any two distinct constant mappings of S into itself, $fg \neq gf$. What happens when S has only one element?

(9) In an infinite club indexed by the integers, $\{A_1, A_2, \ldots\}$, how much does A_n have to borrow from A_{n+1} in order that each member shall be £1 better off than before?

(10) Let S be the set of finite sequences of 0s and 1s and define a mapping f of S into itself by the rule: If $a = a_1a_2 \ldots a_n$ $(a_i = 0 \text{ or } 1)$, then $af = a_1'a_2' \ldots a_n'$, where $0' = 01$, $1' = 10$. Show that af has no block 000 or 111 and that in af^2 $(= aff)$ any block of length at least five contains 00 or 11.

1.4 Equivalence relations

By a *relation* on a set S we mean a correspondence of S with itself. E.g., 'being related' is a relation on the set of all humans (provided that we are equipped with an exhaustive genealogy). Let ω be a relation; we write $x \,\omega\, y$ to express the fact that x stands in the relation ω to y, i.e. that the pair (x, y) belongs to ω. Frequently relations are denoted by a symbol such as $\sim$, thus in place of $x \,\omega\, y$ we write $x \sim y$.

Many relations have one or more of the following three properties:

E. 1 *For every* $x \in S$, $x \sim x$ *(reflexive)*.

E. 2 *For all* $x, y \in S$, *if* $x \sim y$, *then* $y \sim x$ *(symmetric)*.

E. 3 *For all* $x, y, z \in S$, *if* $x \sim y$ *and* $y \sim z$, *then* $x \sim z$ *(transitive)*.

For example, the relation 'x is father of y' (on the set of all humans) has none of these properties. On the other hand, 'x has the same parents as y' has all three, 'x is ancestor of y' is transitive and 'x is brother of y' is symmetric on the set of all human males, but not on the set of all humans. This last point illustrates that we must always specify the set on which we are operating.

A relation on S which is reflexive, symmetric and transitive is called an *equivalence* on S. This is an important notion, which in some ways generalizes the notion of equality, for the relation of equality (on any set) trivially satisfies **E.** 1–3. An equivalence on S separates the elements of S into classes, grouping together objects which agree in some particular respect. E.g., 'x has the same parents as y' is an equivalence which groups siblings together. Similarly, the relation 'x has the same remainder after division by 2 as y' on **N** groups all the even numbers together and all the odd numbers.

Let us see how this can be done generally, for any equivalence on a set S. For any $x \in S$, we group together all the elements equivalent to x into an *equivalence class* or *block* S_x, i.e. we put

$$S_x = \{y \in S \mid x \sim y\}.$$

By the reflexivity, $x \in S_x$; we claim that any two blocks S_x, S_y either are disjoint or coincide. Suppose that S_x and S_y are not disjoint; we must prove that $S_x = S_y$ and we begin by showing that $x \sim y$. Since $S_x \cap S_y \neq \emptyset$, there exists $z \in S_x \cap S_y$; by definition this means that $x \sim z$ and $y \sim z$. By symmetry, $z \sim y$ and hence, by transitivity, $x \sim y$. Now let $u \in S_y$ then $y \sim u$, hence $x \sim u$ (by transitivity) and so $u \in S_x$; this proves that $S_y \subseteq S_x$. A similar argument shows that $S_x \subseteq S_y$ and so $S_x = S_y$, as claimed. Thus the different S_x provide a division of S into non-empty subsets, any two of which are disjoint. This is called a *partition* of S.

The example considered earlier, 'x and y leave the same remainder after division by 2' gives a partition of **N** into two blocks, the even numbers and the odd numbers. Similarly, in any given year, the relation 'x and y fall on the same day of the week' gives a partition of the days of the year into seven blocks, corresponding to the seven days of the week.

Any mapping $f: S \to T$ gives rise to an equivalence on S by the rule: $x \sim y$ if and only if $xf = yf$. The reader should verify that this is indeed an equivalence.

We note that conversely, every partition on a set S arises in this way from an equivalence. For suppose that S is partitioned into sets $A, B, \ldots$. Then each $x \in S$ belongs to just one set of the partition, say $x \in A$. We put $x \sim y$ if x and y lie in the same set. This is an equivalence on S with blocks $A, B, \ldots$.

Exercises

(1) Which of the following relations between positive integers are reflexive, which are symmetric and which are transitive? (i) $a \neq b$, (ii) $a < b$, (iii) a differs from b by less than 2, (iv) any positive integer dividing a also divides b.

(2) Which of the following are equivalence relations? (i) x is within sight of y (where the objects are points on the earth's surface), (ii) x is on the same latitude as y, (iii) x has the same number of digits as y (numbers in decimal notation).

(3) Let '$\sim$' be a reflexive relation. Show that '$\sim$' is symmetric and transitive iff $a \sim b, a \sim c \Rightarrow b \sim c$.

(4) What is wrong with the following 'proof' that every relation on S that is symmetric and transitive is reflexive? For any $a, b \in S$, $a \sim b$ implies $b \sim a$ (by symmetry) and hence, by transitivity, $a \sim a$. Give a counter-example to the assertion.

1.5 Ordered sets

Another basic notion we shall need is that of order; it may be defined abstractly as a relation with certain properties.

A relation ω on a set S is said to be an *ordering of S* or simply an *order on S* if it is reflexive, transitive and satisfies

E. 4 *$x \,\omega\, y$ and $y \,\omega\, x \Rightarrow x = y$ (antisymmetry).*

An obvious example is the natural order of the integers: $x \,\omega\, y$ iff $x \leqslant y$. For this reason an ordering relation is frequently denoted by '$\leqslant$'; we shall usually adopt this notation for an abstract ordering and write '$x < y$' to mean '$x \leqslant y$ but $x \neq y$'. We note that the ordering of the integers has a further property not shared by general orderings:

$$\text{given } x, y \in \mathbf{N}, \text{ either } x \leqslant y \text{ or } y \leqslant x. \qquad (1)$$

This is expressed by saying that the natural ordering of $\mathbf{N}$ is a *total* or *linear* ordering, in contrast to general orderings, which are also called *partial* orderings. Here are two examples of orderings that are *not* total:

(i) Let U be any set, then the relation '$S \subseteq T$' between subsets of U is an ordering; this is not total when U has more than one element.

(ii) On the set $\mathbf{N}$ of natural numbers define $x \mid y$ to mean: y is divisible by x. Again this is a partial ordering.

We leave the reader to check that these are orderings. If x, y are distinct elements of U, then neither of $\{x\}$, $\{y\}$ is a subset of the other and, in $\mathbf{N}$, neither of 2, 3 is divisible by the other.

Ordered sets are often represented by diagrams, as follows: Each element of the set is represented by a point in the diagram and, if $x < y$, we place x lower than y and join them by a line. E.g., if $S = \{a, b, c, d\}$ has an ordering in which $a < b < d$ and $a < c < d$ but no other relations hold, then we can represent S by the diagram shown here.

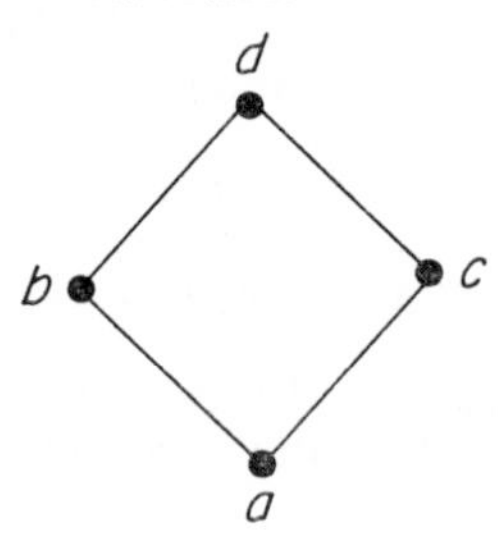

In an ordered set S we understand by a *greatest element* an element $g \in S$ such that $x \leqslant g$ for all $x \in S$, and by a *maximal element* an element $m \in S$ such that for all $x \in S$, $m \leqslant x$ implies $x = m$. In words, a greatest element surpasses all others, a maximal element is surpassed by none. For a totally ordered set (such as the integers) these two concepts are synonymous and this explains why they are not usually distinguished in everyday life. But it is important to note that in general ordered sets these notions are distinct: a greatest element is always maximal, but not conversely. In the set S illustrated above, d is the greatest element, while in the set shown here (obtained by removing d from the previous set) b and c are maximal. In an infinite

ordered set there may be no greatest element (e.g., **N**: there is no greatest integer), but when one does exist, it is unique. On the other hand, there may be more than one maximal element, as in the above example.

Least elements and minimal elements of an ordered set are defined in entirely analogous fashion: u is *least* if $u \leqslant x$ for all x and v is *minimal* if $x \leqslant v$ implies $x = v$, for all x.

Exercises

(1) Verify that the examples (i), (ii) in the text are partially ordered sets.

(2) Which of the following are orderings? Which of the orderings are total? (The universe is indicated after each example.) (i) x is a teacher of y (humans), (ii) x is north of y or $x = y$ (points on the earth's surface), (iii) x is a multiple of y (positive integers), (iv) x is a multiple of y (positive or negative integers), (v) $x' \leqslant y'$, where x' is the least non-negative remainder after division of x by 7.

(3) Let S, T be ordered sets and define a relation on the Cartesian product $S \times T$ by $(x, y) \leqslant (x', y')$ iff $x \leqslant x'$ and $y \leqslant y'$. Show that this is an ordering, but not total (even if S and T are each totally ordered) unless S or T consists of a single element.

(4) Let S, T be ordered sets and define a relation on $S \times T$ by $(x, y) \leqslant (x', y')$ iff $x < x'$, or $x = x'$ and $y \leqslant y'$. Show that this is an ordering of $S \times T$ and that it is total if S and T are both totally ordered (this is called the *lexicographic ordering* on $S \times T$).

(5) Let S be an ordered set (with ordering '$\leqslant$') and define '$x \,\omega\, y$' to mean '$y \leqslant x$'. Show that ω is again an ordering on S, total iff '$\leqslant$' is (this is called the ordering *opposite* to '$\leqslant$' and is usually denoted by '$\geqslant$'). Show that greatest, least, maximal, minimal elements for '$\leqslant$' become least, greatest, minimal, maximal elements respectively for ω.

(6) Let $<$ be a relation on S which is transitive and antireflexive (i.e. $x < x$ holds for no $x \in S$). Show that the relation '$x \leqslant y$' defined as '$x < y$ or $x = y$' is an ordering of S.

Further exercises on Chapter 1

(1) Verify that the following statement about a set S expresses the fact that S has exactly 2 elements:

$$(\exists x)(\exists y)(\forall z)[(x \neq y) \wedge ((z = x) \vee (z = y))].$$

Write down a formula expressing that S has exactly 3 elements.

(2) Let A, B be sets; the set of all mappings from A to B is called the *power set* and is usually denoted by B^A. If A and B are non-empty and finite, show that B^A has $|B|^{|A|}$ elements.

(3) Let S be a set and $2 = \{0, 1\}$ the 2-element set consisting of 0 and 1. For any subset T of S the *characteristic function* χ_T is defined as the mapping from S to 2 given by

$$x\chi_T = \begin{cases} 1 & \text{if } x \in T, \\ 0 & \text{if } x \notin T, \end{cases}$$

for all $x \in S$. Show that $\chi_T \in 2^S$ and that distinct subsets correspond to distinct members of 2^S; conversely show that every member of 2^S defines a unique subset of S in this way. Hence obtain a bijection between 2^S and the set of all subsets of S. Which subsets correspond to the constant functions?

(4) Show that the relation 'a is parallel to b' between lines in the plane is an equivalence relation and verify that the blocks of this equivalence are the directions in the plane.

(5) A reflexive and transitive relation on a set S is called a *preordering* on S. If μ is a preordering on S, show that '$(x \mu y) \wedge (y \mu x)$' is an equivalence on S. If the equivalence block containing x is denoted by $[x]$ and $[x] \leqslant [y]$ means: $x_1 \mu y_1$ for any $x_1 \in [x]$, $y_1 \in [y]$, show that this defines an ordering on the set of equivalence blocks.

(6) Let $f: S \to T$ be a mapping. If T is preordered by a relation $\leqslant$, show that the relation ω defined on S by the rule '$x \omega y$ iff $xf \leqslant yf$' is again a preordering. This is expressed by saying that f *reflects* preorderings. Which mappings reflect (i) total preorderings, (ii) orderings?

2

Integers and rational numbers

2.1 The integers

The integers are familiar to us from elementary arithmetic, but here we shall want to express that familiarity in precise terms. We do this by writing down a list of properties which the integers possess and on which we shall base all our deductions. Later, in Vol. 2, we shall see that all the properties listed here can actually be deduced from quite a brief list of axioms, but this is immaterial at present.

We denote the set of positive integers (also called natural numbers) 1, 2, 3, ... by **N** and write **Z** for the set of all integers, positive, negative and zero. Here **N** stands for *number* and **Z** for *Zahl*, the German for number; both abbreviations are generally used in mathematics (see Appendix 2).

The set **Z** admits three operations: addition, $x+y$, subtraction, $x-y$, and multiplication, $x.y$ or xy. Often it is convenient to express subtraction by adding the negative: $x-y = x+(-y)$. These operations are connected by the following laws:

Z. 1 *Associative law*: $(x+y)+z = x+(y+z)$, $\quad (xy)z = x(yz)$.

Z. 2 *Commutative law*: $x+y = y+x$, $\quad xy = yx$.

Z. 3 *Existence of neutral element*: $x+0 = x$, $\quad x1 = x$.

Z. 4 *Existence of* (*additive*) *inverse*: $x+(-x) = 0$.

The number 0 is said to be *neutral for addition* because adding it to any number x leaves x unchanged; likewise 1 is neutral for multiplication. Every number x has the additive inverse $-x$ (which undoes the effect of adding x), but apart from 1 and -1, no integer has a multiplicative inverse. However we shall find such inverses once we come to consider rational numbers in **2.4**.

In addition to the above laws, there is a further law, relating addition and multiplication:

Z. 5 *Distributive law*: $x(y+z) = xy+xz$.

The fact that these laws hold in **Z** is expressed by saying that **Z** is a *ring*; more precisely, it is a *commutative* ring (because the multiplication is commutative). However, these laws are not yet sufficient to determine **Z**; in Ch. **6** we shall give a general definition of a ring and we shall find that there are many different types.

We now look at some consequences of the above laws. It follows from the distributive law that $x0 = 0$ for all x. By the associative law, the sum of any number of terms is independent of the way in which brackets are placed, and by the commutative law the order of the terms is immaterial. A similar remark applies to multiplication; for the present we shall accept this without proof and return to this point in Ch. **3** to give a general proof.

Thus the sum of numbers $a_1, \ldots, a_n$ may be written $a_1 + \cdots + a_n$. Often one abbreviates this expression by writing down the general term a_ν with a capital sigma, $\sum$, to show that the sum is to be taken, with some indication of the range over which the terms are to be summed (unless this is clear from the context). So instead of $a_1 + \cdots + a_n$ we may write

$$\sum_{\nu=1}^{n} a_\nu \quad \text{or} \quad \textstyle\sum_1^n a_\nu \quad \textit{or} \quad \sum_\nu a_\nu \quad \text{or simply} \quad \sum a_\nu,$$

where in each case ν is a dummy variable (cf. **1.1**). When $n = 0$, the sum written here is empty and, by convention, this is taken to be 0. This notation is not only briefer; it can also help to make our formulae more perspicuous as well as more accurate. For instance, in the expression

$$1+2+\cdots+n,$$

the reader is expected to guess that he is dealing with an arithmetic progression; the expression

$$\textstyle\sum_1^n \nu$$

removes all doubt. Thus the formula for the sum of the first n natural numbers may be written $\sum_1^n \nu = \frac{1}{2}n(n+1)$. We observe that for $n = 0$ the right-hand side reduces to 0, so with our convention about empty sums, this formula still holds for $n = 0$.

For another example consider the distributive law. This has a generalized version which reads (cf. Ex. (3))

$$(a_1 + \cdots + a_r)(b_1 + \cdots + b_s) = a_1b_1 + a_1b_2 + \cdots + a_rb_s,$$

or in abbreviated form $\sum a_\mu . \sum b_\nu = \sum_{\mu,\nu} a_\mu b_\nu$. Here we have not indicated the precise range of summation, since it is immaterial, but only the indices of summation μ, ν.

A similar abbreviation exists for repeated products, using capital pi, $\prod$, in place of $\sum$. Thus instead of $a_1a_2 \ldots a_n$ we write

$$\prod_{\nu=1}^{n} a_\nu \quad \text{or} \quad \prod_1^n a_\nu \quad \text{or} \quad \prod_\nu a_\nu \quad \text{or simply} \quad \prod a_\nu.$$

For example, the factorial function may be defined as $n! = \prod_1^n \nu$. An empty product is taken to be 1; thus empty sums and products are neutral for addition and multiplication respectively.

It is an important property of the integers that the product of two non-zero integers is never zero:

Z. 6 *For any integers a, b, if $a \neq 0$ and $b \neq 0$, then $ab \neq 0$.*

This has the following useful consequence:

Cancellation law *For $a, b, c \in \mathbf{Z}$, if $ca = cb$ and $c \neq 0$, then $a = b$.*

This asserts that multiplication by a non-zero integer is an injective mapping of $\mathbf{Z}$ into itself. To prove it, suppose that $a \neq b$, then $a-b \neq 0$ and hence (by **Z.** 6) $c(a-b) \neq 0$, therefore $ca-cb = c(a-b) \neq 0$.

Besides the operations on $\mathbf{Z}$ we have an order relation, i.e. an ordering on $\mathbf{Z}$: $x \leqslant y$ or $y \geqslant x$. If $x \leqslant y$ but $x \neq y$, we write $x < y$ or also $y > x$. This relation satisfies the requirements for a total ordering (see **1.5**) and is related to the operations of $\mathbf{Z}$ by the following rules:

Z. 7 *If $x_1 \leqslant x_2$ and $y_1 \leqslant y_2$, then $x_1+y_1 \leqslant x_2+y_2$.*

Z. 8 *If $x \leqslant y$ and $z > 0$, then $zx \leqslant zy$.*

The presence of these rules means that $\mathbf{Z}$ is a *totally ordered ring*. Using the ordering we can describe the set $\mathbf{N}$ of positive integers as

$$\mathbf{N} = \{x \in \mathbf{Z} \mid x > 0\}. \tag{1}$$

Later we shall see how to reconstruct $\mathbf{Z}$ from $\mathbf{N}$; for the moment we note that, for every $x \in \mathbf{Z}$, either $x = 0$ or $x \in \mathbf{N}$ or $-x \in \mathbf{N}$ and that these three possibilities are mutually exclusive. In fact, this is true in any totally ordered ring, taking $\mathbf{N}$ to be defined by (1). For we know that just one of the following holds (because we have a total order): $x = 0$ or $x > 0$ or $x < 0$. Now $x+(-x) = 0$, hence, if $x < 0$, then $0 < -x$ by **Z.** 7. Thus either $x = 0$ or $x > 0$ or $-x > 0$, as asserted.

In order to fix $\mathbf{Z}$ completely, we use the following condition on the set $\mathbf{N}$ of positive integers:

I (*Principle of induction*): *Let S be a subset of* $\mathbf{N}$ *such that $1 \in S$ and $n+1 \in S$ whenever $n \in S$. Then $S =$* $\mathbf{N}$.

This principle forms the basis of the familiar method of proof by induction. Let $P(n)$ be an assertion about a positive integer n, e.g., $P(n)$ might be 'the sum of the first n positive integers is $n(n+1)/2$'. Suppose we wish to prove $P(n)$ for all n, i.e. $(\forall n)P(n)$. Then by **I** it will be enough to prove (i) $P(1)$ and (ii) $(\forall n)(P(n) \Rightarrow P(n+1))$. For this means that the set S of all n for which $P(n)$ holds contains 1 and contains $n+1$ whenever it contains n. Hence by **I**, $S = \mathbf{N}$, i.e. $P(n)$ holds for all $n \in \mathbf{N}$.

There are two alternative forms of **I** that are often useful.

I′ *Let S be a subset of* $\mathbf{N}$ *such that $1 \in S$ and $n \in S$ whenever $m \in S$ for all $m < n$; then $S =$* $\mathbf{N}$.

I″ (*Principle of the least element*): *Every non-empty set of positive integers has a least element.*

To prove **I**, **I′** and **I″** equivalent we shall establish the implications **I** $\Rightarrow$ **I′** $\Rightarrow$ **I″** $\Rightarrow$ **I**.

I $\Rightarrow$ **I′**. Let S be such that $1 \in S$ and $n \in S$ whenever $m \in S$ for all $m < n$. Define $T = \{x \in \mathbf{N} \mid y \in S \text{ for all } y \leqslant x\}$, thus $x \in T$ precisely when all the numbers from 1 to x lie in S. Clearly $T \subseteq S$, so it will be enough to show that $T = \mathbf{N}$. Since $1 \in S$, we have $1 \in T$ and, if $n \in T$, then $y \in S$ for all $y \leqslant n$, hence $n+1 \in S$ and so $y \in S$ for all $y \leqslant n+1$; but this means that $n+1 \in T$. Applying **I**, we see that $T = \mathbf{N}$.

I′ $\Rightarrow$ **I″**. Let S be a set of positive integers without a least element; we shall show that S is empty. Denoting the complement of S by S', we must show that $S' = \mathbf{N}$. Now, since S has no least element, $1 \notin S$, so $1 \in S'$; moreover, if $m \in S'$ for all $m < n$, then $n \in S'$, for otherwise n would be the least element in S. Thus by **I′**, $S' = \mathbf{N}$ and S must be empty.

I″ $\Rightarrow$ **I**. Let S be a subset of $\mathbf{N}$ such that $1 \in S$ and $n+1 \in S$ whenever $n \in S$, then the complement S' of S in $\mathbf{N}$ has no least element. For $1 \notin S'$ and if $n \in S'$, then $n-1 \in S'$, hence, by **I″**, $S' = \varnothing$ and so $S = \mathbf{N}$ as we wished to show.

We end this section with a practical remark on proofs by induction. Generally a theorem is easier to prove, the stronger the hypothesis and the weaker the conclusion. But in an induction proof the conclusion at the nth step becomes the hypothesis at the $(n+1)$th step and the theorem may actually become easier to prove if the conclusion is strengthened; for an instance of this see Lemma 4, **4.3**.

Exercises

(1) Prove that for any integers a, b the equation $a+x = b$ has a unique solution.

(2) Prove that $(a+b)(c+d) = ac+bc+ad+bd$.

(3) Prove $a(\sum b_\nu) = \sum ab_\nu$ by induction. Deduce the general distributive law: $(\sum a_\mu)(\sum b_\nu) = \sum a_\mu b_\nu$.

(4) Prove that for any integer a, $a0 = 0$.

(5) Prove the rule of signs: $(-a)b = -ab$, $(-a)(-b) = ab$.

(6) Prove that for any integers a, b, if $ab = 1$, then $a = b = \pm 1$.

(7) If **Z** is totally ordered in any way, subject only to the conditions **Z**.7–8 in the text, show that (i) $a > 0$ implies $-a < 0$ and (ii) $1 > 0$. Deduce that the usual ordering of **Z** is the only one satisfying **Z**.7–8.

(8) For any integer a define the *absolute value* $|a|$ as a if $a \geqslant 0$ and $-a$ otherwise. Verify the following rules: (i) $|a| \geqslant 0$, with equality iff $a = 0$, (ii) $|a+b| \leqslant |a|+|b|$, (iii) $|ab| = |a| \, . \, |b|$, (iv) $||a|-|b|| \leqslant |a-b|$. Under what conditions on a and b does equality hold in (ii) or (iv)?

(9) Show that the sum of the first n positive integers is $n(n+1)/2$.

(10) Find the sum of the first n odd integers.

(11) Find the sum of the first n cubes.

(12) Show that the product of any n successive integers is divisible by $n!$. Define $\binom{n}{k} = n(n-1)\ldots(n-k+1)/k!$ and show that $\binom{n}{k} = \binom{n-1}{k-1}+\binom{n-1}{k}$; deduce (by induction) that $(a+b)^n = \sum_{k=0}^{n}\binom{n}{k}a^k b^{n-k}$ (the binomial theorem).

2.2 Divisibility and factorization in Z

Given $a, b \in \mathbf{Z}$, we write $b \mid a$ (read: b divides a) to indicate that a is divisible by b, i.e. $a = bc$ for some $c \in \mathbf{Z}$. Since any multiple of 0 is 0, $0 \mid a$ is true only when $a = 0$. For this reason one usually takes $b \neq 0$ in $b \mid a$, although a is allowed to be 0, in fact $b \mid 0$ holds for all $b \in \mathbf{Z}$. The negation of $b \mid a$ is written $b \nmid a$; thus $b \nmid a$ means 'a is not divisible by b'. Divisibility on $\mathbf{Z}$ satisfies the following easily verified rules:

D. 1 *$c \mid b$ and $b \mid a$ imply $c \mid a$.*

D. 2 *$a \mid a$ for all $a \in \mathbf{Z}$.*

D. 3 *If $a \mid b$ and $b \mid a$, then $a = \pm b$.*

D. 1–3 show that divisibility defines a (partial) ordering on the set of positive integers.

D. 4 *$b \mid a_1$ and $b \mid a_2$ imply $b \mid (a_1 - a_2)$.*

D. 5 *$b \mid a$ implies $b \mid ac$ for any $c \in \mathbf{Z}$.*

We leave the verification to the reader, but, as an example, let us check **D.** 3. If $a \mid b$, $b \mid a$, then $a = bc$, $b = ad$, hence $a = bc = adc$. It follows that either $a = b = 0$, or $a \neq 0$ and $dc = 1$, whence $c = d = \pm 1$, because 1 and -1 are the only integers which have inverses. Two integers a, b such that $a \mid b$ and $b \mid a$ are said to be *associated*. Thus every non-zero integer is associated to exactly one positive integer.

A *prime number* or *prime* is an integer p greater than 1 whose only positive factors are p and 1. For example, 2, 3, 19 are prime numbers, but not 1, 9 or 15. The primes may be thought of as the constituents from which every natural number can be constructed by multiplication, just as every natural number can be obtained from 1 by repeated addition. For, as we shall see in Th. 3 below, every positive integer can be written as a product of prime numbers in essentially only one way.

Two integers a, b are said to be *coprime* and a is said to be *prime to* b if there is no integer other than ± 1 dividing both a and b; e.g. 12 and 25 are coprime, though neither is a prime number. In fact a prime number p may be characterized by the property that p is greater than 1 and prime to all positive integers less than p. We also note that $a, 0$ are coprime precisely when $a = \pm 1$; in particular, two coprime numbers cannot both be 0.

The basic tool for studying divisibility in $\mathbf{Z}$ is a lemma going back to Euclid, which depends on the *division algorithm*:

Given $a, b \in \mathbf{Z}$, if $b > 0$, then there exist $q, r \in \mathbf{Z}$ such that

$$a = bq+r, \qquad 0 \leqslant r < b. \tag{1}$$

It is easy to establish (1) using rational numbers: let q be the largest integer not exceeding a/b, then $0 \leqslant (a/b)-q < 1$, hence the number r given by $r = a-bq$ satisfies the inequality in (1). But (1) can be proved without invoking rational numbers; and this is more appropriate here, since they have not been officially defined (however, it is of interest to note that (1) is not used in defining rational numbers, see Ch. 6).

To establish (1) using the principle of the least element, let S be the set of all non-negative integers of the form $a-bn$. This set S is not empty (take $n = -a^2$), hence it has a least element $r = a-bq$, say. By definition $r \geqslant 0$ and we must show that $r < b$. Suppose that $r \geqslant b$, then $r-b = a-b(q+1)$ is an element of S smaller than r, a contradiction. Thus $r < b$ and (1) holds.

LEMMA 1 *Given two integers a and b, there exist integers u and v such that*

$$au+bv = 1 \tag{2}$$

if and only if a and b are coprime.

Proof. Any two integers a, b satisfying (2) are necessarily coprime, for any factor common to a and b must also divide $au+bv = 1$ and so must be ± 1.

Conversely, if a and b are coprime, they are not both 0. Therefore the set $S = \{am+bn \mid m, n \in \mathbf{Z}\}$ contains positive integers, e.g. a^2+b^2. Let d be the least positive integer in S, say

$$d = ax+by; \tag{3}$$

the result will follow if we can prove $d = 1$. We do this by showing that $d \mid a$, $d \mid b$, for then d is a common factor of a, b and so must be 1. Let us divide a by d, using the division algorithm:

$$a = dq+r \qquad 0 \leqslant r < d.$$

Then $r = a-dq = a-(ax+by)q = a(1-xq)+b(-yq) \in S$ and $0 \leqslant r < d$. By the definition of d (as the least positive element in S) this means that $r = 0$ and so $a = dq$, i.e. $d \mid a$. Similarly $d \mid b$ and the result follows. ∎

To illustrate Lemma 1, 5 and 18 are coprime and we have $18 \cdot 2 - 5 \cdot 7 = 1$. This can be found from the proof of the lemma: In order to find the least element of S we take any element d of S and apply the division algorithm, as in the lemma. If d is not the least element of S, this produces a smaller element d' of S, and after a finite number of steps we reach the desired least element. We shall return to this point in Ch. 6, when we come to discuss the Euclidean algorithm in detail.

LEMMA 2 *Let p be a prime number and* $a_1, \ldots, a_n \in \mathbf{Z}$ *such that* $p \mid a_1 a_2 \ldots a_n$, *then* $p \mid a_i$ *for some* $i = 1, \ldots, n$.

Proof. We shall prove the contrapositive form of this assertion, i.e. $p \nmid a_i$ for $i = 1, \ldots, n$ implies $p \nmid a_1 a_2 \ldots a_n$.

For $n = 1$ there is nothing to prove, so we may begin with the case $n = 2$; thus we are given $p \nmid a$, $p \nmid b$ and we must show $p \nmid ab$. Since p is prime, a and p are coprime, so by Lemma 1, $ax+py = 1$ for some $x, y \in \mathbf{Z}$; similarly, $bu+pv = 1$, hence

$$1 = (ax+py)(bu+pv) = abxu+p(ybu+axv+ypv)$$

and this shows ab and p to be coprime, whence $p \nmid ab$.

To establish the general case we use induction on n. Thus we assume that $n > 2$ and that the result has been proved for values less than n. Given that $p \nmid a_i$ $(i = 1, \ldots, n)$, we have, using the induction hypothesis, $p \nmid a_2 \ldots a_n$ and $p \nmid a_1$ is given, hence (by the case $n = 2$), $p \nmid a_1 a_2 \ldots a_n$. ■

We are now in a position to prove the unique factorization of integers into primes. This is sometimes known as the *fundamental theorem of arithmetic*:

THEOREM 3 *Every positive integer can be written as a product of prime numbers*:

$$a = p_1 p_2 \ldots p_r \tag{4}$$

and, if a has a second such factorization,

$$a = q_1 q_2 \ldots q_s, \tag{5}$$

then $s = r$ *and when the* q_i *are suitably renumbered, then* $p_i = q_i$.

Proof. The number 1 can be represented as an empty product, by the convention about such products. If there are positive integers *not* expressible as a product of primes, let c be the least. Then c is not 1 or a prime, hence $c = c_1 c_2$, where $1 < c_i < c$ $(i = 1, 2)$. By the choice of c, each of c_1, c_2 is a product of primes, hence so is c. This contradiction shows that no such c can exist, i.e. every positive integer can be written as a product of primes.

Now let a be expressed as a product of primes in two ways, say as (4) and (5). Then $p_1 \mid a$, hence $p_1 \mid q_1 q_2 \ldots q_s$ and by Lemma 2, $p_1 \mid q_i$ for some i. Renumber the qs so that $p_1 \mid q_1$, say $q_1 = p_1 u$, then $u = 1$ because q_1 is a prime, i.e. $p_1 = q_1$. On dividing by p_1 we obtain from (4) and (5) $p_2 \ldots p_r = q_2 \ldots q_s$; by induction on r we find that $r-1 = s-1$ and for a suitable renumbering of the qs, $p_i = q_i$ for $i = 2, \ldots, r$. But this also holds for $i = 1$ and so $p_i = q_i$ for $i = 1, \ldots, r$ and $r = s$, as claimed. ■

If $p_1, p_2, \ldots$ are all the primes in some order, then every positive integer can be written uniquely in the form

$$a = p_1^{\alpha_1} p_2^{\alpha_2} \ldots p_n^{\alpha_n} \qquad (\alpha_i \geqslant 0), \tag{6}$$

provided that we disregard factors p_i^0, which are 1 by convention. In this way every positive integer corresponds to a finite sequence of non-negative integers $(\alpha_1, \alpha_2, \ldots, \alpha_n)$ and conversely, every such sequence corresponds to a unique positive integer. Let us complete each of these sequences by zeros, so that we have infinite sequences, each with only finitely many non-zero elements. Then (assuming the number of primes to be infinite) we have a bijection between **N** and the set of all sequences of non-negative integers (each with only finitely many non-zero terms). This at first sight somewhat surprising result will be taken up in more detail in Vol. 2.

So far the question whether there are infinitely many prime numbers has been left open. The answer does not follow from Th. 3, for even if there were only n primes, the right-hand side of (6) would still have infinitely many different values. The issue is decided in the next result, which goes back to Euclid:

THEOREM 4 (Euclid's theorem) *The number of prime numbers is infinite.*

Proof. Suppose there were only finitely many primes, $p_1, p_2, \ldots, p_n$, say, and consider $c = p_1p_2 \ldots p_n+1$. Then $c > 1$ and by Th. 3, c is divisible by at least one of the ps, say p_1 (by suitable renumbering): $c = p_1d$. Then $p_1(d-p_2 \ldots p_n) = p_1d-p_1p_2 \ldots p_n = 1$ and this is a contradiction, for no prime divides 1. ■

The first few primes are 2, 3, 5, 7, 11, 13, 17, 19, 23, 29, ... They become progressively more sparse and are rather irregularly distributed, though the 'average' distribution is very regular (cf. *Ex.* (11), p. 38).

Exercises

(1) If $au+bv = 1$, show that u, v are unique up to multiples of b and a respectively. For a given coprime pair a, b, how many pairs u, v exist satisfying $au+bv = 1$ and $|u| < |b|$, $|v| < |a|$?

(2) Show that any integer $n > 1$ is either a prime or has as factor a prime $\leqslant \sqrt{n}$.

(3) Show that $\binom{2n}{n}$ is divisible by every prime p such that $n \leqslant p \leqslant 2n$.

(4) For any fraction a/b, denote by $[a/b]$ the largest integer not exceeding a/b. If $b > 1$, show that the series $\sum_1^\infty [a/b^n]$ has at most $(\log a)/(\log b)$ non-zero terms.

(5) Given a prime p, show that for any positive integer c the largest power of p dividing $c!$ is p^α, where $\alpha = \sum_1^\infty [c/p^n]$.

(6) Show that for any integer c and any integer $n > 1$, the equation $x^n = c$ has no rational solution which is not integral.

(7)* If p_n is the nth prime (in order of magnitude), show that $p_n < 2^{2^n}$. (Use the method of proof of Euclid's theorem and induction on n.)

(8) Show that every product of numbers of the form $4n+1$ is again of this form. Deduce that there are infinitely many prime numbers of the form $4n-1$. (Observe that every odd prime is of the form $4n+1$ or $4n-1$.)

(9) Show that there are infinitely many prime numbers of the form $6n-1$. (Ex. (8) and Ex. (9) are special cases of Dirichlet's theorem on arithmetic progressions, which states: If a, b are any coprime integers, then there are infinitely many primes of the form $an+b$. For a proof see, e.g., Serre, *Cours d'arithmétique*.)

2.3 Congruences

In calculations with integers involving division it often happens that we are interested in the remainder, but not the quotient. E.g., to find the day of the week on which a given date falls, we can ignore multiples of 7 at any stage, because they represent complete weeks. Similarly, to see if a number is even or odd, we can ignore multiples of 2. The number whose multiples are being ignored is called the *modulus*. If the modulus is m (a positive integer), we say that *a is congruent to b modulo m* and write

$$a \equiv b \pmod{m}, \tag{1}$$

if $a-b$ is divisible by m. The relation (1) between a and b is called a *congruence*. To express the negation of (1), i.e. $m \nmid a-b$, we write $a \not\equiv b \pmod{m}$. This notation (1) was introduced by Gauss (1801); although just a different notation for the statement $m \mid a-b$, it has advantages in many situations in that it suggests an analogy to ordinary equality. This is expressed more precisely in Th. 1 below and, in a much more general context, in Ch. **10**.

THEOREM 1 *Let m be any positive integer. Then the relation of congruence* mod m *is an equivalence on* $\mathbf{Z}$. *Moreover,*

(i) *if* $a \equiv a', b \equiv b' \pmod{m}$ *then* $a+b \equiv a'+b' \pmod{m}$,

(ii) *if* $a \equiv a', b \equiv b' \pmod{m}$ *then* $ab \equiv a'b' \pmod{m}$,

(iii) *if* $ca \equiv cb \pmod{m}$ *and c is prime to m, then* $a \equiv b \pmod{m}$,

(iv) *if* $a \equiv b \pmod{km}$ *and* $k \neq 0$, *then* $a \equiv b \pmod{m}$.

All these assertions follow quite easily from the definition of congruence, except possibly (iii), so we prove the latter and leave the rest to the reader.

Assume that $ca \equiv cb \pmod{m}$. By hypothesis, c and m are coprime, hence $cu+mv = 1$ for some $u, v \in \mathbf{Z}$, i.e.

$$cu \equiv 1 \pmod{m}. \tag{2}$$

By (ii), $acu \equiv a$, $bcu \equiv b \pmod{m}$ and multiplying the given congruence by u we find $acu \equiv bcu \pmod{m}$, hence $a \equiv acu \equiv bcu \equiv b \pmod{m}$. ■

It is instructive to write this proof out in full: in particular this will give a practical demonstration of the advantage of using the congruence notation. We also note that the proviso in (iii) (that c and m be coprime) cannot be omitted; thus $3 \equiv 15 \pmod{12}$, but $1 \not\equiv 5 \pmod{12}$.

Let m again be a fixed positive integer. From the division algorithm we know that for each $a \in \mathbf{Z}$ there is an equation

$$a = mq+r, \qquad \text{where } 0 \leqslant r < m. \tag{3}$$

This means that each $a \in \mathbf{Z}$ is congruent (mod m) to one of $0, 1, 2, \ldots, m-1$. These numbers are the least residues mod m and it is clear that no two of them are congruent mod m. Thus the equivalence on $\mathbf{Z}$ defined by (1) partitions $\mathbf{Z}$ into m blocks, called the *residue classes of* $\mathbf{Z}$ mod m. The set of these residue classes is denoted by $\mathbf{Z}/m$; thus $\mathbf{Z}/m$ is a set with m elements. If we denote by (x) the residue class mod m containing x, then the different elements of $\mathbf{Z}/m$ are

$$(0), (1), \ldots, (m-1). \tag{4}$$

Generally, $(x) = (y)$ iff $x \equiv y \pmod{m}$ and for any $a \in \mathbf{Z}$, to find (a) in the list (4) we divide a by m as in (3), then $(a) = (r)$ and the latter occurs in (4).

The elements of $\mathbf{Z}/m$ may be added and multiplied in a rather natural fashion. Let $\alpha, \beta \in \mathbf{Z}/m$, say $\alpha = (a)$, $\beta = (b)$, then we can form $(a+b)$ and this is independent of the choice of a and b within their residue classes. For if a', b' are other numbers in the residue classes α, β then $(a) = (a')$, $(b) = (b')$, hence by Th. 1(i), $(a+b) = (a'+b')$. Thus $(a+b)$ depends only on the classes α, β and we may define the sum $\alpha+\beta$ as the residue class $(a+b)$. Similarly, by Th. 1 (ii), $(ab) = (a'b')$, therefore it makes sense to define the product $\alpha\beta$ as the residue class (ab). In this way an addition and multiplication are defined on $\mathbf{Z}/m$ and it is not hard to verify that **Z**. 1–5 (of **2.1**) hold for these operations. We shall prove **Z**. 5 as an example: Thus we must show that

$$\alpha(\beta+\gamma) = \alpha\beta+\alpha\gamma. \tag{5}$$

Write $\alpha = (a)$, $\beta = (b)$, $\gamma = (c)$, then $\alpha(\beta+\gamma) = (a(b+c))$ while $\alpha\beta+\alpha\gamma = (ab+ac)$. By **Z**. 5 for integers, $a(b+c) = ab+ac$, and (5) is an immediate consequence.

The set $\mathbf{Z}/m$ with the addition and multiplication just defined is called the *ring of integers* mod m.

In the course of proving Th. 1 we saw that the congruence $cx \equiv 1 \pmod{m}$ has a solution whenever c is prime to m (cf. (2)). This is a useful fact, worth stating separately; indeed the condition given is necessary as well as sufficient:

PROPOSITION 2 *The congruence*

$$ax \equiv 1 \pmod{m} \tag{6}$$

has a solution if and only if a is prime to m.

For (6) is equivalent to $ax+my = 1$ for some $x, y \in \mathbf{Z}$ and by Lemma 1, **2.2**, this has a solution precisely when a and m are coprime. ■

This result can also be expressed in terms of the ring of integers mod m. If $\alpha \in \mathbf{Z}/m$ and we have two representatives for α, say $\alpha = (a) = (a')$, then a' is prime to m iff a is; in that case we say that α is prime to m. Now Prop. 2

states that an element α of $\mathbf{Z}/m$ is invertible (i.e. has a multiplicative inverse) precisely when it is prime to m.

For a prime number p, all numbers not divisible by p are prime to p and, in that case, Prop. 2 takes on the following form:

COROLLARY *Let p be a prime number, then*

$$ax \equiv 1 \pmod{p}$$

has a solution for any $a \not\equiv 0 \pmod{p}$. ■

Thus when p is a prime, every non-zero element of $\mathbf{Z}/p$ is invertible.

There is another more precise and more striking version of this last result. Let α be any residue class mod p, then multiplication by α defines a mapping of $\mathbf{Z}/p$ into itself. If $\alpha \neq 0$, this mapping is injective, for then α is prime to p, and by Th. 1 (iii), $\alpha\beta = \alpha\beta'$ implies $\beta = \beta'$. Hence if $\gamma_1, \ldots, \gamma_{p-1}$ are the non-zero residue classes mod p, then $\alpha\gamma_1, \ldots, \alpha\gamma_{p-1}$ are the same classes, though possibly in a different order (this result, easily seen directly, is an instance of Lemma 3, **1.3**). It follows that the two products are equal:

$$\prod_1^{p-1} \gamma_\nu = \prod_1^{p-1} \alpha\gamma_\nu = \alpha^{p-1} \prod_1^{p-1} \gamma_\nu.$$

Since the product $\prod \gamma_\nu$ is non-zero, we can divide by it and find that $\alpha^{p-1} = (1)$. Thus we obtain

THEOREM 3 (Fermat's theorem) *Let p be a prime number and $a \not\equiv 0 \pmod{p}$, then*

$$a^{p-1} \equiv 1 \pmod{p}. \quad ■$$

If we multiply by a we get a congruence which holds for $a \equiv 0 \pmod{p}$ as well:

COROLLARY *Let p be a prime number, then*

$$a^p \equiv a \pmod{p} \qquad \textit{for all integers } a. \quad ■$$

We note that by Th. 3, the congruence $ax \equiv 1 \pmod{p}$ has, for a prime to p, the solution $x \equiv a^{p-2}$, hence $ax \equiv b \pmod{p}$ has the solution $x \equiv a^{p-2}b \pmod{p}$.

Let us return to Prop. 2. This can be made the basis of a study of linear congruences (in analogy with linear equations):

THEOREM 4 *Let a, m be coprime, then for any b the congruence*

$$ax \equiv b \pmod{m} \tag{7}$$

has a solution which is unique mod m.

The last assertion means: Any two solutions x_1, x_2 of (7) satisfy $x_1 \equiv x_2 \pmod{m}$; thus there is just one congruence class satisfying (7). This type of answer is to be expected, since with x_1 any element congruent to $x_1 \pmod{m}$ is also a solution of (7).

To prove the result, let x_0 be a solution of $ax \equiv 1 \pmod{m}$ (which exists by Prop. 2), then $ax_0 \equiv 1 \pmod{m}$, hence $abx_0 \equiv b \pmod{m}$ and so (7) has the solution $x \equiv bx_0 \pmod{m}$. If x_1, x_2 are two solutions, then $a(x_1 - x_2) \equiv ax_1 - ax_2 \equiv b - b \equiv 0 \pmod{m}$ and since a is prime to m, we can cancel it (Th. 1(iii)) and obtain $x_1 - x_2 \equiv 0 \pmod{m}$, as asserted. ■

The case of several congruences and unknowns can be treated similarly (see the exercises to Ch. 5). But we may also have several congruences in one unknown, to different moduli, say $a_i x \equiv b_i \pmod{m_i}$. If a_i, m_i are coprime for each i, we can find a unique solution $x_i \pmod{m_i}$ for each i, by the methods of Th. 4. In the case where the moduli are coprime in pairs, we can moreover find a common solution to all the congruences. This principle, which was used by the Chinese mathematicians of the first century A.D., is usually called the *Chinese remainder theorem*. We shall state the result for two congruences and leave the general case to the reader (Ex. (4)).

THEOREM 5 *Let r and s be coprime, then the congruences*

$$x \equiv u \pmod{r}, \qquad x \equiv v \pmod{s} \tag{8}$$

have a common solution, unique mod rs.

Proof. The general solution of the first congruence is $x = u + ra$; this satisfies the second congruence iff $ra \equiv v - u \pmod{s}$ and, since s is prime to r, we can solve for a by Th. 4.

Thus the congruences (8) have a common solution x_1. If x_2 is another common solution, then $x_2 - x_1$ is divisible by r and s and, since they are coprime, by their product, i.e. $x_2 \equiv x_1 \pmod{rs}$ as asserted. ■

Let us restate Th. 5 in terms of the rings $\mathbf{Z}/r$, $\mathbf{Z}/s$. If m is any positive integer, we shall write $(x)_m$ for the residue class of x mod m. We can define a mapping $\mathbf{Z}/rs \to \mathbf{Z}/r$ by the rule

$$(x)_{rs} \mapsto (x)_r.$$

This is a mapping of residue classes, because if $(x)_{rs} = (x')_{rs}$, then $x \equiv x' \pmod{rs}$ and hence, by Th. 1(iv), $x \equiv x' \pmod{r}$. In the same way we can define a mapping $\mathbf{Z}/rs \to \mathbf{Z}/s$; now we combine these two into a mapping f of $\mathbf{Z}/rs$ into the Cartesian product $\mathbf{Z}/r \times \mathbf{Z}/s$:

$$f\colon (x)_{rs} \mapsto ((x)_r, (x)_s). \tag{9}$$

Th. 5 may now be expressed by saying that this mapping is a bijection between $\mathbf{Z}/rs$ and $\mathbf{Z}/r \times \mathbf{Z}/s$; it is surjective because the congruences (8) always have a common solution, for any u, v and it is injective because this solution is unique mod rs.

Observe that on the set $\mathbf{Z}/r \times \mathbf{Z}/s$ we again have an addition and multiplication if we carry out the operations componentwise: $(\alpha, \beta) + (\alpha', \beta') = (\alpha + \alpha', \beta + \beta')$, $(\alpha, \beta)(\alpha', \beta') = (\alpha\alpha', \beta\beta')$. The set $\mathbf{Z}/r \times \mathbf{Z}/s$ with these operations is called the *direct product* of $\mathbf{Z}/r$ and $\mathbf{Z}/s$. Now it may be verified that the mapping (9) preserves the operations, i.e. $(x+y)f = xf+yf$, $(xy)f = xf.yf$. A bijection with these properties is called an *isomorphism*; thus f is an isomorphism between $\mathbf{Z}/rs$ and $\mathbf{Z}/r \times \mathbf{Z}/s$, we also say that $\mathbf{Z}/rs$ is *isomorphic* to $\mathbf{Z}/r \times \mathbf{Z}/s$ and write

$$\mathbf{Z}/rs \cong \mathbf{Z}/r \times \mathbf{Z}/s.$$

It is important to bear in mind that r, s must be coprime for this relation to hold. Of course, $\mathbf{Z}/r \times \mathbf{Z}/s$ can be defined as a direct product without any restriction on r and s and the mapping f can also be defined, but it will not be an isomorphism unless r and s are coprime. The reader is advised to follow this construction through in a particular case, say $r = 2$, $s = 3$, and also to look where it fails for $r = 2$, $s = 4$.

More generally it can be shown in the same way that for integers $m_1, \ldots, m_k$ that are pairwise coprime (i.e. any pair m_i, m_j with $i \neq j$ are coprime), if $m = m_1 m_2 \ldots m_k$, we have an isomorphism

$$\mathbf{Z}/m \cong \mathbf{Z}/m_1 \times \cdots \times \mathbf{Z}/m_k. \tag{10}$$

This is a useful result because it reduces the structure of $\mathbf{Z}/m$ to that of the $\mathbf{Z}/m_i$. From Th. 3 we know that every positive integer may be written as a product of prime numbers; grouping powers of the same prime together we may thus write m as

$$m = p_1^{\alpha_1} \ldots p_k^{\alpha_k}, \tag{11}$$

where the p_i are distinct primes. If we put $m_i = p_i^{\alpha_i}$, we see that the m_i are pairwise coprime; thus (10) expresses $\mathbf{Z}/m$ as a direct product of the $\mathbf{Z}/m_i$ where each m_i is a prime power.

As an illustration we shall calculate the number of invertible elements in $\mathbf{Z}/m$. This number is denoted by $\varphi(m)$ and is called *Euler's function*. If we recall that the residue class $(a)_m$ is invertible in $\mathbf{Z}/m$ precisely when a is prime to m, we see that $\varphi(m)$ is just the number of positive integers less than and prime to m; e.g. $\varphi(2) = 1$, $\varphi(3) = 2$, $\varphi(4) = 2$, $\varphi(5) = 4$. Now an element in a direct product $\mathbf{Z}/r \times \mathbf{Z}/s$ is invertible iff each component is: $(x, y)(x', y') = (1, 1)$ iff $xx' = 1$, $yy' = 1$. Therefore

$$\varphi(rs) = \varphi(r)\varphi(s) \quad \text{whenever } r \text{ and } s \text{ are coprime.}$$

Any function φ on $\mathbf{Z}$ with this property is said to be *multiplicative*. It follows that when m is given by (11), then

$$\varphi(m) = \varphi(p_1^{\alpha_1}) \ldots \varphi(p_k^{\alpha_k}), \tag{12}$$

and now it only remains to find $\varphi(m)$ when m is a prime power. Consider p^α: the numbers *not* prime to p^α are just the multiples of p, and there are $p^{\alpha-1}$ not exceeding p^α. The rest are prime to p^α, so

$$\varphi(p^\alpha) = p^\alpha - p^{\alpha-1} = p^\alpha(1-p^{-1}).$$

Combining this result with (12), we obtain the following formula for the Euler function:

$$\varphi(m) = m \prod \left(1 - \frac{1}{p}\right),$$

where the product is taken over all distinct primes dividing m.

Exercises

(1) Give a proof of Th. 1.

(2) Solve the following congruences: (i) $4x \equiv 3 \pmod 7$, (ii) $3x+2 \equiv 0 \pmod 4$, (iii) $2x-1 \equiv 0 \pmod{15}$, (iv) $3x+6 \equiv 0 \pmod{12}$.

(3) Solve the following systems of congruences:
(i) $2x \equiv 3 \pmod 5$, $3x \equiv 2 \pmod 4$,
(ii) $x \equiv 1 \pmod 2$, $x \equiv 2 \pmod 3$, $x \equiv 3 \pmod 5$.

(4) By induction on k, prove that if $m_1, \ldots, m_k$ are coprime in pairs and $x \equiv x_i \pmod{m_i}$ for $i = 1, \ldots, k$, then there is a common solution, unique mod $m_1 m_2 \ldots m_k$ (Chinese remainder theorem).

(5) If a is prime to m, show that $a^{\varphi(m)} \equiv 1 \pmod m$, where $\varphi(m)$ is Euler's function. (This generalization of Fermat's theorem is known as Euler's theorem.)

(6) Verify **Z**.1–5 for $\mathbf{Z}/m$, and likewise for $\mathbf{Z}/r \times \mathbf{Z}/s$.

(7) For which integers m does $\mathbf{Z}/m$ satisfy **Z**.6 of **2.1**? Can any $\mathbf{Z}/m$ be ordered so as to satisfy **Z**.7–8?

(8) If p is a prime number, show that $\binom{p}{1}, \binom{p}{2}, \ldots, \binom{p}{p-1}$ are all divisible by p. Deduce that $(a+b)^p \equiv a^p + b^p \pmod p$; generalize this result to n summands and deduce another proof of Fermat's theorem.

(9) Show that any integer, written as $a_n a_{n-1} \ldots a_0$ in decimal notation, where $0 \leqslant a_i \leqslant 9$, is congruent to $a_0 + a_1 + \ldots + a_n \pmod 9$. Hence obtain a quick method of finding the least residue (mod 9) of any given integer.

(10) A well-known rule for checking calculations ('casting out nines') consists in performing the calculation mod 9, using the method of Ex. (9) to find the remainders. Show that an accidental transposition of two digits cannot be detected by this method.

(11) Devise a similar rule for 11 ('casting out elevens'), based on the fact that $a_n a_{n-1} \ldots a_1 a_0 \equiv a_0 - a_1 + \ldots + (-1)^n a_n \pmod{11}$ and show that any single error due to a transposition of neighbouring digits will be detected by this method.

(12) Show how the methods of Ex. (9) and Ex. (11) can be used to test for divisibility by 9 and 11. Describe a corresponding test for 21 and hence obtain a test for divisibility by 7.

(13) Find all solutions of $x^2 - 1 \equiv 0 \pmod{16}$.

(14) Show that the congruence $x^2 + y^2 \equiv 3 \pmod 4$ has no solutions.

(15) Show that the congruence $x^2+y^2+z^2 \equiv 7 \pmod 8$ has no solutions. Deduce that neither of the congruences $x^2+2y^2 \equiv 7 \pmod 8$, $x^2+2y^2 \equiv 5 \pmod 8$ has a solution.

(16)* Show that $\sum_{d|m} \varphi(d) = m$, where the sum is taken over all divisors of m. (Hint. Prove first that the left-hand side is a multiplicative function of m.)

2.4 The rational numbers and some finite fields

We know from our schooldays how to handle fractions; here we shall look at the collection of all fractions as a set with certain operations defined on it: the field of rational numbers. Later, in **6.2**, we shall examine the construction by which we obtain fractions from the integers.

The rational numbers share with the integers the laws **Z.** 1–5 listed in **2.1.** Furthermore, every non-zero element x has a multiplicative inverse x^{-1} and the four basic operations of arithmetic: addition, subtraction, multiplication and division can always be carried out, with the exception of division by zero. (Since $0x = 0$ for all x, the equation $0x = a$ has no solution for $a \neq 0$.) These facts are expressed by saying that the rational numbers form a *field*; we shall follow general usage by denoting the field of rational numbers by **Q** (for 'quotient field of **Z**').

There are other fields besides the rational numbers; e.g., the real numbers admit the four operations of arithmetic and satisfy the laws **Z.** 1–5. Thus the real numbers also form a field, generally denoted by **R**. Again the details of constructing the real numbers will be postponed to Vol. 2; we may think of a real number as being specified by its decimal expansion. This includes all rational numbers, and others besides, such as $\sqrt{2}$ which is irrational (cf. **1.1**). In the same way the complex numbers form a field, denoted by **C**, which contains **R** as subfield.

But quite different fields are possible: if p is a prime number, the ring $\mathbf{Z}/p$ of integers mod p admits division by any non-zero element and so is a field (Prop. 2, **2.3**, Cor.); in this way we obtain a finite field of p elements for each prime number p. Looking at the problem from a different point of view we may try to construct fields on few elements abstractly. The case of a 1-element set is rather trivial, because then $0 = 1$. If we have only two elements, with 0 and 1 and try to make a field from these two elements, with 0 and 1 playing the roles of zero (neutral for addition) and one (neutral for multiplication) respectively, we get the following addition and multiplication tables:

+	0	1
0	0	1
1	1	0

×	0	1
0	0	0
1	0	1

The only entry which does not follow from the special properties of 0 and 1 is $1+1$. This must be 0 or 1; if $1+1 = 1$, then by subtraction, $1 = 0$, which is false, hence $1+1 = 0$. The field we have found is $\mathbf{Z}/2$, which is therefore the only field of two elements.

Next take the case of three elements: 0, 1, a. Again part of the addition and multiplication tables is fixed in advance. To complete them we note that in the addition table no row or column can have a repetition, for if $x+y = x+y'$ say (corresponding to a repetition in the row for x) then by subtraction $y = y'$. The multiplication table has a similar property if we omit the row and column of zeros (because the non-zero elements admit division). Hence e.g. $1+a = 0$, $1+1 = a$ and, if we write 2 ($= 1+1$) for a, we obtain the tables

+	0	1	2
0	0	1	2
1	1	2	0
2	2	0	1

×	0	1	2
0	0	0	0
1	0	1	2
2	0	2	1

This system is realized by the ring $\mathbf{Z}/3$ of integers mod 3, and the discussion shows that there are no other fields of three elements. Later, in Vol. 2, we shall find that for any positive integer n there exists a field of n elements iff n is a prime power; this field, unique up to isomorphism, is denoted by $\mathbf{F}_n$. When n is a prime p, we have of course $\mathbf{F}_p = \mathbf{Z}/p$ and we shall also use the notation $\mathbf{F}_p$ in what follows.

We observe that in $\mathbf{Z}/m$,

$$1+1+\ldots+1 = 0 \qquad (m \text{ summands}) \tag{1}$$

thus a multiple of 1 is 0, whereas in $\mathbf{Q}$, $1+1+\ldots+1 \neq 0$ for any (positive) number of summands.

We shall discuss fields in a more general context in Ch. **6**; but from our examples we already notice one feature of general fields: Different fields show very different behaviour depending on the least positive multiple of 1 which is 0. We shall find: either no such multiple exists, the field is then said to have *characteristic* 0 (e.g., $\mathbf{Q}$) or when some multiple of 1 is 0, the least such multiple is a prime number p; then the field is said to have *characteristic* p, e.g., $\mathbf{F}_p$.

Exercises

(1) Show that every rational number can be expressed in just one way as m/n, where $n > 0$ and m, n are coprime.

(2) Given two rational numbers $\alpha < \beta$, show that there exist m and r such that $\alpha < m/2^r < \beta$. Can r be found so that the inequality holds (with this r) both for an odd value and an even value of m?

(3) Show that there is no non-empty proper subset of **Q** admitting the four operations of arithmetic.

(4) Make a multiplication table for **Z**/4, **Z**/5, **Z**/6.

(5) Prove that if F is any field with 4 elements, then every element of F satisfies $2x = 0$. Use this fact to construct addition and multiplication tables for F.

(6) Fix a prime number p and let A_p be the set of rational numbers whose numerator is divisible by p. Show that A_p is closed under addition and multiplication and that any set closed under addition and multiplication and containing A_p properly also contains **Z**. Prove the same for the set A_∞ of non-negative rational numbers.

(7) Show that in $\mathbf{F}_p$, where p is an odd prime, exactly half the non-zero elements are squares.

(8) Show that in $\mathbf{F}_p$, where p is any prime, every element can be written as a sum of two squares.

Further exercises on Chapter 2

(1) If r and s are coprime integers and n is divisible by r and s, show that n is divisible by rs.

(2) Show that the highest common factor (greatest common divisor) of $\binom{n}{1}$, $\binom{n}{2}$, $\ldots$, $\binom{n}{n-1}$ is p if n is a positive power of a prime p and 1 otherwise.

(3) Let a, b be coprime positive integers. Show that the set $\{am+bn \mid m, n \in \mathbf{N}\}$ contains all integers $\geqslant (a+1)(b+1)$.

(4) Show that the product of n terms of an arithmetic progression is divisible by $n!$ provided that the common difference in the progression is prime to $n!$.

(5) If q, m, n are any positive integers, show that $q-1 \mid q^n-1$; deduce that for $m \mid n$, $q^m-1 \mid q^n-1$.

(6) Given integers $q > 1$, m, $n > 0$, if $q^m-1 \mid q^n-1$, show that $m \mid n$. (Hint. Write $n = ma+b$, where $0 \leqslant b < m$.)

(7) Show that if q^n-1 is prime then $q = 2$ and n is prime. (The numbers $M_n = 2^n-1$ are the *Mersenne numbers*; the first few that are prime correspond to $n = 2, 3, 5, 7, 13, 17, 19$.)

(8) If $q > 1$, $n > 0$, show that $q+1 \mid q^n+1$ iff n is odd, and $2 \mid q^n+1$ iff q is odd. Deduce that q^n+1 can be prime only when q is even and $n = 2^m$. (The numbers $F_n = 2^{2^n}+1$ are the *Fermat numbers*; the first four Fermat numbers are prime and Fermat conjectured that all are prime, but Euler showed that F_5 is composite.)

(9) If F_n is again the nth Fermat number, verify that for any $k \geqslant 0$, $F_{n+k}-1 = (F_n-1)^{2^k}$. Deduce that any two Fs are coprime, and hence obtain another proof of Euclid's theorem. (Polya)

(10) Fill in the details in the following proof of the fundamental theorem of arithmetic (due to Zermelo). Let n be the least number which does not have a unique factorization into primes. Show that the least factor (>1) of n is a prime p; if $n = pn'$, deduce that n' has a unique factorization into primes and hence obtain one factorization of n into primes. If there is another one, let q be a prime occurring in it and write $n = qm$. Now show that $n-pm$ has two distinct factorizations and is less than n, and obtain a contradiction.

(11)* For any real x, the number of primes not exceeding x is denoted by $\pi(x)$, e.g. $\pi(1) = 0$, $\pi(\pi) = 2$, $\pi(5{\cdot}5) = 3$. Using Ex. (5), **2.2**, prove that $\binom{2n}{n}$ divides $\prod p^{\nu_p}$, where ν_p is the largest integer such that $p^{\nu_p} \leqslant 2n$ and p runs over all primes $\leqslant 2n$. Deduce the inequality

$$n^{\pi(2n)-\pi(n)} \leqslant \binom{2n}{n} \leqslant (2n)^{\pi(2n)}.$$

From this inequality and the estimate

$$\frac{2^{2n}}{2n} \leqslant \binom{2n}{n} \leqslant 2^{2n}, \text{ obtained by expanding } (1+1)^{2n} = 2^{2n},$$

deduce the existence of positive constants c, c' such that

$$c\frac{n}{\log n} \leqslant \pi(n) \leqslant c'\frac{n}{\log n}.$$

(This inequality was found by Čebyšev, 1850; Gauss conjectured in 1801 that $\pi(n) \,.\, \log n/n \to 1$ as $n \to \infty$ (the prime number theorem); this was proved in 1896 by analytic methods.)

(12) Find a multiple of 7 which has remainder 1, 2, 3 when divided by 2, 3, 4 respectively.

(13) If $r > 0$, p is prime and $c \equiv 1 \pmod{p^r}$, show that $c^p \equiv 1 \pmod{p^{r+1}}$.

(14) Let p be an odd prime and $a \not\equiv 0 \pmod p$; show that the number of roots of $x^2 \equiv a \pmod{p^n}$ is the same for all $n > 0$.

(15) Show that the number of fractions r/s with denominator (when expressed in lowest terms) not exceeding n is $\sum_1^n \varphi(\nu)$.

(16)* Let m be an integer and p a prime. Prove that for $r \geqslant 0$,

$$\binom{mp^r}{p^r} \equiv m \pmod p.$$

Deduce that if $p \nmid m$, the number of subsets with p^r elements of a set of mp^r elements is not divisible by p. Deduce also that if in some field k,

$$(\alpha+\beta)^n = \alpha^n+\beta^n \quad \text{for some } n > 1 \text{ and all } \alpha, \beta \in k,$$

then n is a prime power, $n = p^r$, and k has characteristic p.

3

Groups

3.1 Monoids

The number systems described in Ch. 2 had two basic operations: addition and multiplication, and sometimes their inverses: subtraction and division. We now want to study the general properties of an associative but not necessarily commutative operation. The method of doing this is not to look at a particular system such as the integers or the rational numbers, but to assume that we have an arbitrary set with an operation satisfying the associative law and see what consequences can be derived in this way. It will be useful to begin with a formal definition.

By a *monoid* we understand a set S with a mapping $\mu\colon S^2 \to S$ such that if $\mu(x, y)$ is the result of applying μ to the pair $x, y \in S$, then

M. 1 $\mu(x, \mu(y, z)) = \mu(\mu(x, y), z)$ *for all* $x, y, z \in S$.

M. 2 *There exists* $e \in S$ *such that* $\mu(e, x) = \mu(x, e) = x$ *for all* $x \in S$.

Note that by **M.** 2, a monoid is never empty. A mapping such as μ which acts on pairs of elements of S is called a *binary operation*, with values in S, and an element e satisfying **M.** 2 is said to be a *neutral element* for μ. There cannot be more than one neutral, for if we also have $\mu(e', x) = \mu(x, e') = x$, then $e = \mu(e, e') = e'$.

Clearly both addition and multiplication are instances of such a binary operation, and in fact **Z** is a monoid under addition (with 0 as neutral) as well as multiplication (with 1 as neutral). This example illustrates that in discussing particular instances of monoids one must name not only the set but also the operation, unless this is clear from the context.

It will simplify the notation as well as help our intuition if we adopt the multiplicative terminology in general monoids. Thus we shall speak of the *multiplication* of elements, write xy instead of $\mu(x, y)$ and call xy the *product* of x and y. The neutral element e is called the *unit-element* and is usually denoted by 1. Of course in adopting this notation we must be careful not to assume anything that has not been proved; for example xy will in general be different from yx. If it happens that $xy = yx$ for particular elements x and y, we say that x and y *commute*; if $xy = yx$ for all x, y, the monoid is said to be *commutative*.

Sometimes one may wish to use the additive terminology, e.g., in discussing the addition of the integers. Then the element $\mu(x, y)$ is called *sum* and is written as $x+y$ and the neutral is called *zero* and is written 0.

It is a consequence of the associative law that any product can be written without bracketing. To prove this fact we shall show that a product of n factors, $x_1x_2 \ldots x_n$ has the same value for any way of bracketing the factors (so long as we take care to keep the factors in the right order). We shall use induction on n; for $n = 1$ there is nothing to prove, so we assume that $n > 1$ and that for $r < n$ the value of a product $y_1y_2 \ldots y_r$ is independent of bracketing. If we bracket $x_1x_2 \ldots x_n$ in two different ways, we thus have to show that

$$(x_1 \ldots x_i)(x_{i+1} \ldots x_n) = (x_1 \ldots x_j)(x_{j+1} \ldots x_n), \tag{1}$$

where we have only indicated the outermost pairs of brackets, because the terms within them have less than n factors and so are already unambiguous, by the induction hypothesis. For $i = j$ there is nothing to prove; otherwise we may take $i < j$ and write $x_1 \ldots x_i = a, x_{i+1} \ldots x_j = b, x_{j+1} \ldots x_n = c$, then (1) may be rewritten as

$$a(bc) = (ab)c,$$

and this is just an instance of the associative law.

As a result we can write any product in a monoid without brackets, taking care however to preserve the order of the factors. A product of equal factors is usually abbreviated as in the case of numbers. Thus we put x^2 for xx, and generally write x^n for the product of n factors equal to x. It is easily verified that with this definition we have

$$x^rx^s = x^{r+s}, \tag{2}$$

$$(x^r)^n = x^{rn} \tag{3}$$

and

$$(xy)^n = x^ny^n, \quad \text{provided that } xy = yx. \tag{4}$$

Consider two monoids S and T and let $f\colon S \to T$ be a mapping between them; f is said to be a *homomorphism* if

$$(xy)f = xf \,.\, yf \tag{5}$$

and

$$1_Sf = 1_T, \tag{6}$$

where 1_S, 1_T are the neutrals in S and T respectively; here we have indicated the monoids in which the neutrals lie, although this reference will often be omitted, when the meaning is clear from the context.

A bijective homomorphism is called an *isomorphism* and we say that S and T are *isomorphic*, in symbols, $S \cong T$, if there is an isomorphism from S to T. Special names are used when $T = S$; thus a homomorphism of S into itself is called an *endomorphism* of S and an isomorphism of S with itself is an *automorphism*. These four names are used quite generally, in discussing algebraic systems, to denote these four types of mapping preserving the structure of the systems. Thus in Ch. **6** we shall define the analogous notion for rings and it will then be clear how the definition would run in other cases.

Examples of monoids. (i) We have already seen that $\mathbf{Z}$ can be regarded as a monoid (in two ways). The same applies to the other number systems considered in Ch. **2**: $\mathbf{Q}$, $\mathbf{R}$, $\mathbf{C}$. The natural numbers $\mathbf{N}$ form a monoid under multiplication, but not addition. However, the set $\mathbf{Z}^+$ of non-negative integers is a monoid under addition.

(ii) The set $\mathbf{R}^+$ of non-negative real numbers is a monoid under the operation $\max\{x, y\}$, with neutral 0, where $\max\{x, y\}$ stands for the greater of the numbers x, y.

(iii) Let $X = \{x_1, x_2, \ldots, x_n\}$ be a finite set, the *alphabet*, and denote by X^* the set of 'words' formed from X, i.e. the set of all finite sequences of elements of X. To multiply two such words we write them one after the other. For example, if $X = \{x, y\}$, then $X^* = \{1, x, y, x^2, xy, yx, y^2, x^3, \ldots\}$, where 1 stands for the empty word (which acts as neutral). X^* is called the *free* monoid on X. An even simpler case is that of $X = \{x\}$, then $X^* = \{1, x, x^2, \ldots\}$; we note that $\{x\}^* \cong \mathbf{Z}^+$, an isomorphism being $n \mapsto x^n$.

An element a of a monoid S is said to be *invertible* if there exists $a' \in S$ such that

$$aa' = a'a = 1.$$

There cannot be more than one such element a', for if a'' is another with the same properties, then $a'' = a''1 = a''aa' = 1a' = a'$. We shall call a' the *inverse* of a and generally denote it by a^{-1}. The following rules are easily verified:

(i) *If a is invertible, then so is a^{-1}, and*

$$(a^{-1})^{-1} = a. \tag{7}$$

(ii) *If a, b are invertible, then so is ab and*

$$(ab)^{-1} = b^{-1}a^{-1}. \tag{8}$$

It is important to observe the reversal of the factors which occurs in (8). To give an illustration from everyday life, in the morning we *first* put on socks, *then* shoes, but at night we have to start by taking off our shoes before we can take off our socks.

Exercises

(1) Which of the following are monoids under addition? (i) All integers, (ii) all even integers, (iii) all odd integers, (iv) all positive even integers, (v) all integers or halves of integers.

(2) Verify that for any set S, the set of all mappings of S into itself is a monoid under composition, with the identity mapping as neutral element.

(3) Let $M = \{x_0, x_1, \ldots, x_6\}$ have a multiplication defined by the rule: $x_i x_j = x_{i+j-k}$, where k is the largest multiple of 3 contained in $i+j-4$. Verify that M is a monoid.

(4) If M is any monoid and $a \in M$, show that the multiplication defined on M by $x * y = xay$ is associative and that M is a monoid under this multiplication iff a is invertible. What is the neutral in this case?

(5) Show that the operation of taking the least common multiple of two numbers defines a monoid structure on **N**. What about the operation of taking the highest common factor (= greatest common divisor) of two numbers?

(6) An element z of a monoid M is called a *zero element* if $xz = zx = z$ for all $x \in M$. Show that a monoid cannot have more than one zero.

(7) Verify the rules (7) and (8) in the text.

(8) Show that for any invertible element x in a monoid, and any integer n, $(x^{-1})^n = (x^n)^{-1}$.

(9) Let f be a surjective mapping of monoids satisfying (5). Show that f is a homomorphism and also give an example of a mapping satisfying (5) but not (6).

(10) Let M be a monoid and $x \in M$. Show that there is exactly one homomorphism of $\mathbf{Z}^+$ (as additive monoid) into M such that $1 \mapsto x$.

(11) Let M be a monoid and $a \in M$ such that $a^3 = 1$. Show that the mapping $x \mapsto axa^2$ is an automorphism of M.

(12) Let S be a set with a binary operation xy defined on it satisfying **M.2** (existence of neutral) but not **M.1** (associativity). Show that the neutral e is necessarily unique, as in the associative case, but that the inverse of x, defined as solution of $xx' = x'x = e$, need not be unique.

(13) Show that an element a in a monoid M is invertible iff there exists $x \in M$ such that $axa = 1$.

3.2 Groups; the axioms

By a *group* one understands a monoid in which every element is invertible.

The commutative law is not assumed; when it holds, i.e. if $xy = yx$ for all x, y, the group is said to be *commutative* or *abelian* (after N. H. Abel).

Thus a group G is defined by the following four conditions:

G. 1 *G has a binary operation xy defined on it.*

G. 2 *The operation is associative*: $(xy)z = x(yz)$ *for all* $x, y, z \in G$.

G. 3 *G has a neutral element* 1: $1x = x1 = x$ *for all* $x \in G$.

G. 4 *Every element* $x \in G$ *has an inverse* x^{-1}: $xx^{-1} = x^{-1}x = 1$.

It is a remarkable fact that the whole of group theory, which includes many deep results and is far from being fully explored yet, rests ultimately on this simple set of axioms. Of course we shall only be able to touch on the most basic properties in this chapter and in Ch. 9, but it is hoped that even this brief account will convey something of the inherent simplicity and beauty of the theory.

We begin by observing that the axioms **G.** 1–4 are to some extent redundant: less is needed to verify that we have a group. This remark is of practical use as it may help to shorten our verifications. Before proving it we note another useful consequence of the group axioms:

G. 5 *Given* $a, b \in G$, *the equations* $bx = a$, $yb = a$ *each have a unique solution, namely* $x = b^{-1}a$, $y = ab^{-1}$.

For if $bx = a$, then $b^{-1}a = b^{-1}bx = 1x = x$, so there is at most one solution and, in fact, $b(b^{-1}a) = bb^{-1}a = 1a = a$. Thus $bx = a$ has exactly one solution; by symmetry the same holds for $yb = a$.

THEOREM 1 *Let G be a set with a binary operation which is associative. Then the following conditions on G are equivalent*:

(a) *G is a group*,

(b) *G is not empty and for all* $a, b \in G$, *the equations* $bx = a$, $yb = a$ *each have a solution*,

(c) *there exists* $e \in G$ *such that* $xe = x$ *for all* $x \in G$ *and if we fix such e, then for each* $x \in G$ *there exists* $x' \in G$ *such that* $xx' = e$.

Note that we do not assume that the solutions in (b) are unique.

We shall prove the theorem according to the scheme (a) ⇒ (b) ⇒ (c) ⇒ (a). The remarks preceding the theorem show that (a) ⇒ (b). Now assume (b), take $a \in G$ and find $e \in G$ to satisfy $ae = a$. For any $b \in G$ there exists $x \in G$ such that $xa = b$ (by (b)) and hence $b = xa = xae = be$; moreover, $aa' = e$ always has a solution a', therefore (c) follows.

Next assume (c); we must show that G is a group. Given $x \in G$, we can find $x' \in G$ with $xx' = e$ and then $x'' \in G$ with $x'x'' = e$. Hence $x'x = x'xe = x'xx'x'' = x'ex'' = x'x'' = e$; thus $xx' = e = x'x$. Moreover, $x = xe = xx'x = ex$, hence e is neutral and x' is an inverse of x, i.e. G is indeed a group. ■

As a first application we have

PROPOSITION 2 *A finite monoid S with the properties*

$$xz = yz \Rightarrow x = y \quad \text{for all } x, y, z \in S \text{ (right cancellation)}, \tag{1}$$

$$zx = zy \Rightarrow x = y \quad \text{for all } x, y, z \in S \text{ (left cancellation)} \tag{2}$$

is a group.

For by (2) the mapping $\lambda_z : x \mapsto zx$ of S into itself is injective, hence bijective (by Lemma 3, **1.3**, because S is finite). Thus $bx = a$ has a solution for any $a, b \in S$. By symmetry (1) shows that $yb = a$ has a solution for all $a, b \in S$, and S is not empty, hence it satisfies (b) of Th. 1 and is therefore a group. ■

The notions of homomorphism, isomorphism, endomorphism and automorphism defined for monoids carry over to groups as a special case. We note: to verify that a mapping $f: G \to H$ between groups is a homomorphism we need only check that $(xy)f = xf \,.\, yf$, for if this holds then $(1_G f)^2 = 1_G f$, where 1_G is the neutral of G; by division it follows that $1_G f$ is the neutral of H. Since homomorphisms preserve neutrals it will cause no confusion to denote the neutral in any group by the same symbol 1. We also note that for any group homomorphism f, we have $(xf)(x^{-1}f) = (xx^{-1})f = 1f = 1$, and similarly $(x^{-1}f)(xf) = 1$, hence

$$(xf)^{-1} = x^{-1}f. \tag{3}$$

Let G be a group, then by a *subgroup* of G one understands a subset H of G which is a group relative to the operations in G. Thus H is a subgroup of G iff (i) $1 \in H$, (ii) $x, y \in H \Rightarrow xy \in H$, (iii) $x \in H \Rightarrow x^{-1} \in H$. Alternatively, H is a subgroup of G iff (iv) $H \neq \varnothing$ and $x, y \in H \Rightarrow xy^{-1} \in H$. Clearly (iv) holds for any subgroup of G; conversely, when it holds, take $a \in H$, then $1 = aa^{-1} \in H$, hence for any $x \in H$, $x^{-1} = 1x^{-1} \in H$, and whenever $x, y \in H$ then $y^{-1} \in H$ by what has just been proved, and so by (iv), $xy = x(y^{-1})^{-1} \in H$, therefore H is indeed a subgroup of G.

In discussing groups the following notation is often useful. Let A, B be any subsets of a group G and write

$$AB = \{xy \mid x \in A, y \in B\}, \qquad A^{-1} = \{x^{-1} \mid x \in A\}.$$

If A reduces to a single element: $A = \{a\}$, we write aB instead of $\{a\}B$, and similarly for B. To give an example, the condition for a non-empty subset H of G to be a subgroup may be restated as $HH \subseteq H$ and $H^{-1} \subseteq H$, or equivalently, $HH^{-1} \subseteq H$. Some care must be taken to write HH and not H^2 (which is sometimes taken to mean $\{x^2 \mid x \in H\}$).

PROPOSITION 3 *Let G be a group. If H and K are any subgroups of G, then so is their intersection $H \cap K$.*

To prove this result we need only verify that $H \cap K$ satisfies the conditions (i)–(iii) for a subgroup, a task which may be left to the reader. ∎

In precisely the same way it may be shown that the intersection of any set of subgroups of G (even infinitely many) is again a subgroup. Thus if X is any subset of G we can take the collection of all subgroups H of G such that $X \subseteq H$ and form their intersection, T, say. Then T will be a subgroup of G and it may be characterized as the least subgroup containing X. A more constructive way of describing T is as follows. Let T' be the set of all products in the elements of $X \cup X^{-1}$, i.e. the products formed from elements of X and their inverses. It is not hard to verify that T' is a subgroup and $T' \supseteq X$. Moreover, any subgroup containing X also contains T'. Therefore $T \supseteq T'$, but from the definition of T as the intersection of all subgroups containing X we have $T \subseteq T'$, hence $T = T'$. This subgroup T is called the *subgroup generated by* X and is written $gp\{X\}$, while X is called a *generating set* of T. In particular, a generating set of G is a subset X such that $G = gp\{X\}$.

Examples of groups. (i) **Z**, **Q**, **R**, **C** are all groups under addition, with 0 as neutral and $-x$ as inverse of x; in fact they are abelian groups, each a subgroup of the next.

(ii) Let F be any field, such as **Q**, **R**, **C** or $\mathbf{F}_p$, and denote the set of non-zero elements in F by F^*. Then F^* is a group under multiplication, with 1 as neutral and $x^{-1} = 1/x$ as inverse. Again these groups are abelian. Note that in the chain $\mathbf{C}^* \supseteq \mathbf{R}^* \supseteq \mathbf{Q}^*$ each term is a subgroup of the preceding group. In $\mathbf{Q}^*$ we have a subgroup of two elements, namely $\{1, -1\}$, and this (like every group) contains the *trivial subgroup* $\{1\}$ consisting of the unit-element alone. We shall usually denote the trivial subgroup by **1** or, for additive groups, by **0**.

(iii) Let S be any set; by a *permutation* of S we understand a bijection of S with itself. The set of all permutations of S is denoted by $\Sigma(S)$. Any permutation has an inverse, again a permutation, and the product of two permutations (obtained by applying them in succession) is also a permutation. In this way $\Sigma(S)$ forms a group, with the identity mapping on S as the unit-element. It is only necessary to check the associative law, and this follows from the way products were defined, as the equations

$$x[(fg)h] = [x(fg)]h = ((xf)g)h, \qquad x[f(gh)] = (xf)(gh) = ((xf)g)h$$

show, where $x \in S$, $f, g, h \in \Sigma(S)$.

The group $\Sigma(S)$ is called the *symmetric group* on S. We note that as soon as S has more than 2 elements, $\Sigma(S)$ is non-abelian. For, given distinct elements $x, y, z \in S$, let α be the permutation interchanging x and y and leaving all other elements fixed, and let β be the permutation interchanging y and z and leaving the rest fixed, then $\alpha\beta$ maps x, y, z to z, x, y respectively, while $\beta\alpha$ maps them to y, z, x respectively.

(iv) The degree of symmetry of a geometrical figure is expressed by the different ways of moving the figure so as to present the same appearance, e.g., a square has 'more' symmetry than an oblong rectangle. To make this precise we observe that the different movements a figure can undergo without changing its appearance form a group (the symmetry group of the figure), if we 'multiply' two such movements by performing them in succession. E.g., the group of symmetries of an equilateral triangle ABC is the symmetric group on A, B, C, while the group of an isosceles triangle with base BC (and not equilateral) is the symmetric group on B, C.

(v) Any finite group may be described by its multiplication table: write the group elements along the top row and the left-hand column of a square table and put the product ab in the intersection of the row for a and the column for b. From the property **G.** 5 of groups we see that each row and each column must be a permutation of the group elements. This provides a useful check in the construction of multiplication tables. E.g., to construct all groups of 3 elements, we write the elements as 1, a, b. By **G.** 5, ab cannot be a or b, hence $ab = 1$ and similarly $ba = 1$. Now there is only one way of completing the table, which looks as follows:

	1	a	b
1	1	a	b
a	a	b	1
b	b	1	a

It is easily checked that we do indeed have a group; this still has to be checked, all the preceding argument shows is that there is at most one group with 3 elements. Thus we find that (up to isomorphism) there is exactly one group with 3 elements.

(vi) The set of all rotations about a fixed point in 3-space forms a group, where the product of two rotations is obtained by carrying them out in succession. This group may also be regarded as the symmetry group of the unit-sphere in 3-space. Unlike the symmetry groups considered earlier, the rotation group is infinite; the rotations about a fixed axis through the given point form a subgroup.

Exercises

(1) Show that the complex numbers ± 1, $\pm i\,(i = \sqrt{-1})$ form a group under multiplication. More generally, show that for any $n > 1$, the complex nth roots of 1 form a group under multiplication.

(2) Show that the mappings $f_{ab}: x \mapsto ax+b(a, b \in \mathbf{R}, a \neq 0)$ form a group under composition, the 1-dimensional affine group $\mathbf{Af}_1(\mathbf{R})$. Write down formulae for the product and inverse.

(3) Let P be the set of all rotations of a horizontal circular disc about a vertical axis through its centre. Which of the following subsets of P are subgroups? (i) All rotations, (ii) all rotations through less than $\pi/2$, (iii) all rotations through more than $\pi/2$, (iv) all rotations through a multiple of $\pi/2$, (v) all rotations through a multiple of a radian.

(4) Let $A = \{a, b\}$ be a set with a multiplication defined by the table

	a	b
a	a	b
b	b	b

Show that A is a monoid, but not a group. Which rules holding in groups are violated by A? Do the same for the set $\{1, a, b\}$ with multiplication table

	1	a	b
1	1	a	b
a	a	a	b
b	b	a	b

(5) Make a group table for the symmetric group on 3 symbols.

(6) If a group G is generated by elements a, b and $ba = ab^k$, $a^n = 1$, show that every element of G can be written in the form $a^r b^s$ $(0 \leqslant r < n)$; show also that when $k \neq 1$, then $b^m = 1$ for some $m > 1$.

(7) Find all groups of four elements and show that they are abelian.

(8) Show that $x \mapsto \exp x$ is an isomorphism between the additive group of $\mathbf{R}$ and the multiplicative group of the positive elements of $\mathbf{R}$.

(9) Show that in any group G, the mapping $x \mapsto x^2$ is an endomorphism iff G is abelian. What are the conditions on a finite group G for $x \mapsto x^2$ to be an automorphism?

(10) Let G be any group and $a \in G$. Show that the mapping $x \mapsto a^{-1}xa$ is an automorphism. What is its inverse?

(11) Let G be a group and $a, b, c \in G$. If c commutes with a and b, show that c commutes with ab^{-1}. Deduce that the set $\mathbf{C}_a(G) = \{x \in G \mid xa = ax\}$ is a subgroup of G ($\mathbf{C}_a(G)$ is called the *centralizer* of a in G). Deduce that, for any subset X of G, the set $\mathbf{C}_X(G) = \bigcap\{\mathbf{C}_x(G) \mid x \in G\}$ is a subgroup. In particular, $\mathbf{C}(G) = \mathbf{C}_G(G)$ is a subgroup contained in each centralizer ($\mathbf{C}(G)$ is called the *centre* of G).

3.3 Group actions and coset decompositions

Let S be a set and G any group. By a *group action of G* or a *G-action* on S we understand a mapping $\mu\colon S \times G \to S$—i.e. a binary operation associating with any $p \in S$, $g \in G$ an element $\mu(p, g) \in S$—such that

A. 1 $\mu(p, gh) = \mu(\mu(p, g), h)$ *for all* $p \in S$, $g, h \in G$,

A. 2 $\mu(p, 1) = p$ *for all* $p \in S$.

We express this fact also by saying that G *acts* on S and call S a *G-set*. Generally we write pg instead of $\mu(p, g)$ and rely on the context to indicate whether the group multiplication or the group action is intended; with a little practice and a judicious choice of letters to denote the different types of elements this causes no confusion.

When we have a G-action on S, each $g \in G$ defines a mapping $\varphi_g\colon p \mapsto pg$ of S into itself. In terms of these mappings the rules **A.** 1–2 are expressed by the equations:

$$\varphi_{gh} = \varphi_g \varphi_h, \qquad \varphi_1 = 1. \tag{1}$$

Thus the mapping

$$g \mapsto \varphi_g \tag{2}$$

is a homomorphism (of monoids) from G to the monoid of all mappings of S into itself. By (1), $\varphi_g \varphi_{g^{-1}} = \varphi_{g^{-1}} \varphi_g = 1$, which shows that each φ_g is actually a permutation of S. Thus (2) is a homomorphism of G into $\Sigma(S)$, the symmetric group on S. Conversely, given any set S and any homomorphism

$g \mapsto \varphi_g$ of G into $\Sigma(S)$, we can define a G-action on S by putting $pg = p\varphi_g$ $(p \in S)$; the conditions **A**. 1–2 follow from the equations (1). The action is said to be *faithful* if the mapping (2) is injective.

Let S be a set, then any group action on S can be used to define an equivalence on S. Let us put

$$x \sim y \text{ iff } y = xg \text{ for some } g \in G. \tag{3}$$

The group laws just ensure that this is an equivalence relation. It is reflexive because $x1 = x$ for all $x \in S$, symmetric because $y = xg$ implies $x = yg^{-1}$ and transitive because $y = xg$, $z = yh$ imply $z = (xg)h = x(gh)$. The equivalence classes are called the *orbits* of the action and the orbit containing x is written xG. If S consists of a single orbit, G is said to act *transitively*.

Examples of group actions. (i) Let G be the group of rotations in 3-space about a fixed point O. This is a G-action on 3-space; the orbits are the spheres with O as centre.

(ii) If S is any set, the symmetric group $\Sigma(S)$ acts on S in a natural way. We can pass from any element of S to any other by a suitable permutation, thus Σ acts transitively.

(iii) If S is any set and G any group, we can define a G-action by putting $xg = x$ for all $x \in S$, $g \in G$. This is called the *trivial* G-action; the orbits are the 1-element subsets of S.

(iv) Let G be any group. We can let G act on itself by right multiplication, i.e. for each $a \in G$ we define a mapping of G into itself by the rule

$$\rho_a: x \mapsto xa.$$

This is also called the *regular representation* of G. The associative law in G shows that this is indeed a group action: $\rho_{ab} = \rho_a\rho_b$. By **G**. 5, G acts transitively on itself.

If we try to let G act on itself by left multiplication we obtain

$$\lambda_a: x \mapsto ax,$$

but this need not define a group action: the effect of λ_{ab} on x is abx and this is obtained by applying first λ_b and then λ_a, thus

$$\lambda_{ab} = \lambda_b\lambda_a, \tag{4}$$

and this need not equal $\lambda_a\lambda_b$ as required by (1). There are two ways out of this dilemma: either we can distinguish *right* actions, satisfying (1) and *left* actions, satisfying (4), or we observe that left multiplication *can* be used to define a group action satisfying (1), provided that we combine it with the inverse:

$$\mu_a: x \mapsto a^{-1}x.$$

Here $x\mu_a\mu_b = b^{-1}(a^{-1}x) = (ab)^{-1}x = x\mu_{ab}$, hence $\mu_{ab} = \mu_a\mu_b$ as required. For this reason we shall not need to introduce actions satisfying (4). Again G acts transitively under the action so defined.

(v) There is another way in which a group G can act on itself. Consider the mapping

$$\alpha_a: x \mapsto a^{-1}xa.$$

We find that $x\alpha_a\alpha_b = b^{-1}(a^{-1}xa)b = (ab)^{-1}x(ab) = x\alpha_{ab}$, thus $\alpha_{ab} = \alpha_a\alpha_b$ and we have a group action, called *conjugation*. This action is never transitive, unless G is trivial, for each α_a leaves 1 fixed. It is the trivial action on G precisely when G is abelian.

Most of these examples will be used later; for the moment let us consider more closely the group action by right multiplication. Given a group G and a subgroup H, we consider the natural H-action on G obtained by right multiplication. When H is a proper subgroup, this is no longer transitive; the orbits are of the form xH $(x \in G)$, they are called the *right cosets* † of H in G; note that $H = 1H$ is itself a right coset. From the definition of cosets as the classes of an equivalence relation we obtain a partition of the group G in the form

$$G = \bigcup xH, \tag{5}$$

called the *right coset decomposition*. A subset of G containing just one element from each right coset xH is called a *left transversal* of H in G, and right transversals are defined correspondingly.

Suppose that G is finite; the number of its elements is called the *order* of G and will be denoted by $|G|$. The number of right cosets of H in G is then also finite; it is called the *index* of H in G and is written $(G:H)$. Further H itself is then finite, of order $|H|$, and each right coset xH has $|H|$ elements, because xH is bijective with H, via the mapping $h \mapsto xh$ $(h \in H)$. The decomposition (5) therefore yields the important formula

$$|G| = (G:H)|H|. \tag{6}$$

In particular this proves

THEOREM 1 (Lagrange's theorem) *Let G be a finite group, then the order of any subgroup divides the order of G.* ■

This is a useful result which greatly facilitates finding subgroups of finite groups. E.g., it shows that a group of prime order p can have no subgroups apart from itself and the trivial subgroup.

Consider the action of H on G by left multiplication; the orbits in this case are the *left cosets* Hx, and in this way we obtain a decomposition of G into left cosets:

$$G = \bigcup Hx.$$

The argument leading to (6) can again be applied, and it shows that in a finite group the number of left cosets of H is the same as the number of right

† Sometimes they are called *left* cosets and the Hx are called right cosets. The above usage is based on the analogy with ideals (cf. Ch. **10**).

cosets. But this can be proved more generally and more simply by observing that in any group G the mapping $x \mapsto x^{-1}$ is a permutation of G, which for any subgroup H of G induces a bijection from the set of right cosets to the set of left cosets, namely $xH \mapsto Hx^{-1}$.

Let us return to an arbitrary G-set S. With each $p \in S$ we associate the set of group elements leaving p fixed:

$$G_p = \{g \in G \mid pg = p\}.$$

Clearly G_p is a subgroup: if g, $h \in G_p$ then $p(gh) = (pg)h = ph = p$ and $pg^{-1} = p$, moreover $p1 = p$, so that G_p is indeed a subgroup of G. It is called the *stabilizer* of p under the action of G.

There is a useful formula for the size of orbits in terms of the stabilizer. Let us fix an element p of S and consider the mapping from G to S given by $x \mapsto px$. The image is, by definition, the orbit pG of p. In general this mapping is not injective; $px = py$ iff $pxy^{-1} = p$, i.e. $xy^{-1} \in G_p$, where G_p is the stabilizer of p:

$$px = py \Leftrightarrow G_p x = G_p y.$$

Thus there is a bijection between the set of left cosets of G_p and the orbit containing p, and we obtain the following formula for the size of an orbit:

THEOREM 2 *If a group G acts on a set S and $p \in S$ is a point lying in a finite orbit pG, then the number of elements in pG equals the index in G of the stabilizer of p:*

$$|pG| = (G : G_p). \blacksquare$$

To illustrate this result we consider conjugation in a group. Two elements x, y in a group G are said to be *conjugate* if there exists $c \in G$ such that

$$y = c^{-1}xc.$$

It is easily verified that this defines an equivalence relation on G; the equivalence classes are called the *conjugacy classes* of G. Now recall that the mappings

$$\alpha_c : x \mapsto c^{-1}xc$$

define a group action of G on itself; the orbits under this action are just the conjugacy classes. The stabilizer of $x \in G$ is the set of elements commuting with x, i.e. the centralizer of x in G:

$$\mathbf{C}_x(G) = \{y \in G \mid xy = yx\}.$$

Now the orbit formula (Th. 2) shows that the number of elements in the conjugacy class containing x is the index in G of the centralizer of x. By Lagrange's theorem, the index of any subgroup of G divides the order of G, hence we see that the number of elements in each conjugacy class of a finite group G is a factor of the order of G.

Exercises

(1) In any G-action, if x is a fixed point of $g \in G$, show that xh is a fixed point of $h^{-1}gh$. Deduce that conjugate subgroups have the same index (two subgroups H and K are said to be *conjugate* if $K = g^{-1}Hg$, for some $g \in G$).

(2) Let $\mathbf{R}_\infty$ be the set consisting of the real numbers and a symbol ∞. Show that the operations $x \mapsto 1-x$, $x \mapsto x^{-1}$ (where $0^{-1} = \infty$, $\infty^{-1} = 0$, $1-\infty = \infty$) define an action of the symmetric group on three symbols on $\mathbf{R}_\infty$. Show that there are two orbits of 3 points, while all other orbits have 6 points. Describe the action when $\mathbf{R}$ is replaced by $\mathbf{C}$, the complex numbers.

(3) If a group G acts on sets S and T, we can define a G-action on the Cartesian product $S \times T$ by the rule $(x, y)g = (xg, yg)$. Show that the stabilizer of (x, y) is $A \cap B$, where A is the stabilizer of x and B the stabilizer of y.

(4) Let G be a group and H, K subgroups of index r, s respectively in G. Show that $H \cap K$ has index at most rs.

(5) Let G be a group and H, K conjugate subgroups of index r. Show that $H \cap K$ has index at most $r(r-1)$. (Hint. Find a transitive G-action on an r-element set and consider the action of G on ordered pairs.)

(6) Let G be a group and for $x, a, b \in G$ define $x^a = a^{-1}xa$, $x^{a+b} = x^a x^b$. Show that the addition so defined is associative but not in general commutative; show also that $(x^a)^{b+c} = x^{ab+ac}$, $(x^{a+b})^c = x^{ac+bc}$.

(7) Show that in any group, ab is conjugate to ba.

(8) Let $C_1, \ldots, C_k$ be the conjugacy classes of a group. Show that each product C_iC_j is a union of conjugacy classes.

(9) Determine all finite groups with only two conjugacy classes,

(10) If H, K are subgroups of G, show that $|HK| \,.\, |H \cap K| = |H| \,.\, |K|$. (Caution. HK need not be a subgroup, cf. **9.1**.)

(11) Let G act on S. If $a, b \in G$ are such that $ab = ba$, and X is the subset of S of points fixed by a, show that b maps X into itself.

(12) If G acts on S in two different ways: μ, μ', and these two actions commute: $\mu_a\mu'_b = \mu'_b\mu_a$ for all $a, b \in G$, show that we obtain another G-action on S by putting $\nu_a = \mu_a\mu'_a$.

Verify that the actions ρ_a, μ_a defined in the text commute and identify the action $\rho_a\mu_a$.

3.4 Cyclic groups

A group is said to be *cyclic* if it can be generated by a single element; thus all its elements can be expressed as powers of the generator. E.g., $\mathbf{Z}$ is a cyclic group under addition, with generator 1: every integer can be written as a multiple of 1 (of course **Z** can equally well be generated by -1). Similarly **Z**/m for any $m > 1$ is a cyclic group under addition, again with 1 as generator. We shall now show that every cyclic group is isomorphic to just one of these types. That no two of the groups **Z**, **Z**/m ($m > 1$) are isomorphic is clear, because they all have different orders. Let C be a cyclic group with generator c, then two cases are possible:

(i) all powers of c are distinct, thus the elements in C are

$$\ldots, c^{-2}, c^{-1}, 1, c, c^2, c^3, \ldots$$

The mapping $n \mapsto c^n$ is an isomorphism from **Z** to C, so that $C \cong \mathbf{Z}$. We also say that C is *infinite cyclic.*

(ii) Not all powers of c are distinct. Let $c^r = c^s$ for $r > s$, then $c^{r-s} = 1$. There is a least positive integer, m say, such that $c^m = 1$; we claim that the distinct elements in C are $1, c, c^2, \ldots, c^{m-1}$ and that $C \cong \mathbf{Z}/m$. This will follow if we show that

$$c^r = c^s \Leftrightarrow r \equiv s \pmod{m}. \tag{1}$$

Let $r \equiv s \pmod{m}$, say $r = s+km$, then $c^r = c^{s+km} = c^s(c^m)^k = c^s$, because $c^m = 1$. Conversely, if $c^r = c^s$, let us divide $r-s$ by m: $r-s = mq+t$, where $0 \leqslant t < m$, then $1 = c^{r-s} = c^{mq+t} = (c^m)^q c^t = c^t$, but m was the least positive integer satisfying $c^m = 1$, hence $t = 0$, i.e. $r \equiv s \pmod{m}$. This proves (1) and it shows that $C \cong \mathbf{Z}/m$. We also express this fact by saying that C is *cyclic of order m*, and denote it by $\mathbf{C}_m$.

Let us determine the subgroups of cyclic groups. Turning first to **Z**, we claim that every subgroup is of the form **0** or $n\mathbf{Z}$ ($n > 1$). For let H be a subgroup of **Z** and suppose that H is non-trivial, then H contains a non-zero integer k and either k or $-k$ is positive, hence H contains positive integers. Let n be the least positive integer in H, then all multiples of n lie in H, i.e. $n\mathbf{Z} \subseteq H$; we want to establish equality here. Let $h \in H$ and write $h = nq+r$, $0 \leqslant r < n$. Then $r = h-nq \in H$, but n was the least positive integer in H, so $r = 0$ and $h = nq \in n\mathbf{Z}$ as asserted. Observe that $n\mathbf{Z}$ is again infinite cyclic, with generator n; in fact $n\mathbf{Z}$ consists precisely of the n-fold multiples of elements of **Z**.

Now let C be a finite cyclic group, of order m say, with generator c. If $k \mid m$, then c^k generates a subgroup $C(k)$ of C, of order $h = m/k$, its elements being

$$1, c^k, c^{2k}, \ldots, c^{(h-1)k}. \tag{2}$$

Conversely, if H is a subgroup of C, let k be the least positive integer such that $c^k \in H$, then as in the infinite cyclic case we find that $c^r \in H$ iff $k \mid r$. In particular, since $c^m = 1 \in H$, we have $k \mid m$ and H consists of the elements (2), i.e. $H = C(k)$. We note that these elements are precisely all the kth powers of elements of C. Again each subgroup is cyclic; these results can be summed up as

THEOREM 1 *Any subgroup of a cyclic group is cyclic. The subgroups of the infinite cyclic group* $\mathbf{Z}$ *have the form* $n\mathbf{Z}$*, where* n *is a non-negative integer, while the subgroups of the finite cyclic group of order* m *have order* h *dividing* m*; for each factor* h *of* m*, if* $m = hk$*, there is just one subgroup of order* h*, consisting of all* kth *powers of the elements of* C. ■

Cyclic groups are the simplest type of group, but they are important generally. If G is any group and $a \in G$, then the subgroup A generated by a consists precisely of all the powers of a:

$$A = gp\{a\} = \{a^n \mid n \in \mathbf{Z}\}.$$

Thus A is cyclic, and is therefore determined up to isomorphism by its order. This order is also called the *order* of the element a in G; it may be described as the least positive integer m such that $a^m = 1$, or ∞ if no such integer exists. In a finite group the order of each element is finite and divides the order of the group, by Lagrange's theorem.

As an example, consider the multiplicative group $\mathbf{U}(m)$ of residue classes prime to m, i.e. the subgroup of elements of $\mathbf{Z}/m$ which have a multiplicative inverse. The order of this group is the Euler function $\varphi(m)$, by definition of the latter. Hence we obtain Euler's theorem (Ex. 5, **2.3**):

$$a^{\varphi(m)} \equiv 1 \pmod{m} \quad \text{for all } a \text{ prime to } m.$$

The exact order d of a is called the *order of a* (mod m); it is a factor of $\varphi(m)$. If there is a number c of exact order $\varphi(m)$, it is called a *primitive root* (mod m), it means that $\mathbf{U}(m)$ is cyclic, with c as generator. E.g., 2 is a primitive root (mod 5). Later (in Vol. 2) we shall see that for any prime number p, $\mathbf{U}(p)$ is cyclic, i.e. every prime has primitive roots. More precisely, for an odd prime p, p^r has primitive roots for all $r \geqslant 1$.

Exercises

(1) Show that $\mathbf{C}_m$ has $\varphi(m)$ elements which are generators, where φ is the Euler function.

(2) Find all cyclic groups which have exactly two elements that are generators.

(3) Let G be any group and let a, b be two commuting elements in G, of orders r, s respectively, where r and s are coprime. Show that ab has order rs. If c is an element of order rs in G, where r, s are coprime, show that there exist elements a, b of orders r and s respectively, both powers of c, such that $c = ab$. If two elements of coprime order in a group commute, show that they are powers of the same element.

(4) Show that an abelian group of finite square free order (i.e. of order not divisible by a square) is cyclic.

(5) In any group G, show that the equation $x^k = a$ has a unique solution for any k prime to the order of a.

(6) Show that in any group ab and ba have the same order.

(7) Let G be any group and $n > 2$. Show that the number of elements in G of order n is even.

(8) Show that every group of even order has an element of order 2.

(9) Show that a group has no proper subgroups iff it is cyclic of prime order.

(10) Give a direct proof that a finite cyclic group contains at most one subgroup of any given order.

(11) Show that every finitely generated subgroup of the additive group of $\mathbf{Q}$ is cyclic.

3.5 Permutation groups

The action of a group G on a set S may be studied from two points of view. We may be primarily interested in the set S; in that case we have a set with a G-action, already considered in **3.3**. Or, we may wish to consider the group G as the primary object. This means that we have a group of permutations where the nature of the set permuted is now immaterial. In particular, if S is finite, there is no loss of generality in taking the objects of S to be $1, 2, \ldots, n$. A permutation of S is said to be of *degree n*. The group of all such permutations, i.e. the symmetric group on $\{1, 2, \ldots, n\}$ is written Sym_n and called the *symmetric group of degree n*. To find its order we observe that Sym_n acts transitively on $\{1, 2, \ldots, n\}$ and the stabilizer of n is precisely Sym_{n-1}. Hence by the orbit formula,

$$(\mathrm{Sym}_n : \mathrm{Sym}_{n-1}) = n.$$

By induction we find that Sym_n has order $n(n-1)\ldots 2\cdot 1 = n!$. Of course this order can also be obtained in a more elementary fashion as the number of permutations of n symbols.

For a closer study of symmetric groups we need a concise way of writing permutations. There are several notations, useful in different contexts. In the first place, to describe any permutation σ we may write down $1, 2, \ldots, n$ in any order, and under i put $i\sigma$. E.g.,

$$\begin{pmatrix} 1 & 2 & 3 \\ 2 & 3 & 1 \end{pmatrix} = \begin{pmatrix} 3 & 1 & 2 \\ 1 & 2 & 3 \end{pmatrix} = \begin{pmatrix} 3 & 2 & 1 \\ 1 & 3 & 2 \end{pmatrix} \text{ and } \begin{pmatrix} 1 & 2 & 3 \\ 2 & 3 & 1 \end{pmatrix}^{-1} = \begin{pmatrix} 1 & 2 & 3 \\ 3 & 1 & 2 \end{pmatrix}.$$

This notation makes it easy to multiply permutations:

$$\begin{pmatrix} 1 & 2 & 3 \\ 1 & 3 & 2 \end{pmatrix}\begin{pmatrix} 1 & 2 & 3 \\ 2 & 1 & 3 \end{pmatrix} = \begin{pmatrix} 1 & 2 & 3 \\ 2 & 3 & 1 \end{pmatrix}, \quad \begin{pmatrix} 1 & 2 & 3 \\ 2 & 1 & 3 \end{pmatrix}\begin{pmatrix} 1 & 2 & 3 \\ 1 & 3 & 2 \end{pmatrix} = \begin{pmatrix} 1 & 2 & 3 \\ 3 & 1 & 2 \end{pmatrix}.$$

However, it is cumbersome to write and it gives no insight into the structure of the permutation. To overcome these defects one uses the cycle notation:

a *circular permutation* or *cycle* $(abc\dots f)$ is a permutation in which each symbol is replaced by the next, and the last one replaced by the first; thus $a \mapsto b$, $b \mapsto c, \dots, f \mapsto a$. Every permutation σ of $1, 2, \dots, n$ can be written as a product of cycles without a common symbol: the first cycle is $1, 1\sigma, 1\sigma^2, \dots$, where we stop just before reaching 1 again (we must reach 1 eventually, because σ has finite order). If any of $1, 2, \dots, n$, say r, is not included in this cycle, we form another cycle starting from r, and so on, until all n symbols are exhausted. E.g.,

$$\begin{pmatrix} 1 & 2 & 3 & 4 & 5 & 6 \\ 4 & 1 & 6 & 2 & 5 & 3 \end{pmatrix} = (1\ 4\ 2)(3\ 6)(5).$$

A cycle consisting of a single symbol is the identity mapping and may therefore be omitted. Thus the above permutation may also be written as $(1\ 4\ 2)(3\ 6)$. In this way every permutation σ of $1, 2, \dots, n$ can be written as a product of cycles without common symbols:

$$\sigma = (a_1, \dots, a_r)(a_{r+1}, \dots, a_s)\dots(a_{t+1}, \dots, a_n). \tag{1}$$

This representation is unique, except for the order in which the cycles are written, and the symbols in each cycle can be permuted cyclically without affecting σ.

Various properties, such as the order of a permutation, may be read off from its cycle structure: suppose the permutation σ, when written in cycle form (1), has α_1 cycles of length 1, α_2 of length 2 and generally α_i of length i. We say briefly: σ has the *cycle structure* $1^{\alpha_1}2^{\alpha_2}\dots r^{\alpha_r}$. Any cycle of length k has order k; hence $\sigma^m = 1$ whenever m is a common multiple of those numbers i among $1, 2, \dots, r$ for which $\alpha_i > 0$. This condition is also necessary, because there can be no cancellation between different cycles. Thus the order of a permutation is the least common multiple of the lengths of its cycles (in the representation (1)).

The cycle structure also determines the conjugacy class within Sym_n:

THEOREM 1 *Two permutations are conjugate in* Sym_n *if and only if they have the same cycle structure.*

For let $\sigma = (a_1, \dots, a_r)(a_{r+1}, \dots, a_s)\dots(a_{t+1}, \dots, a_n)$ and $\tau = (b_1, \dots, b_r)(b_{r+1}, \dots, b_s)\dots(b_{t+1}, \dots, b_n)$, then

$$\alpha = \begin{pmatrix} a_1 & a_2 & \dots & a_n \\ b_1 & b_2 & \dots & b_n \end{pmatrix}$$

is such that $\alpha^{-1}\sigma\alpha = \tau$. For $\alpha^{-1}\sigma\alpha$ maps $b_1 \mapsto a_1 \mapsto a_2 \mapsto b_2$, and similarly for other symbols. The converse is clear. ■

A third method of writing permutations is as a product of transpositions. By a *transposition* one understands a cycle of length 2. Every cycle is a product of transpositions:

$$(1\ 2\dots n) = (1\ 2)(1\ 3)\dots(1\ n). \tag{2}$$

Since every permutation can be expressed as a product of cycles, it follows that every permutation can be expressed as a product of transpositions. Of course this expression is far from unique, but we can make some economies by restricting the type of transposition used:

THEOREM 2 Sym_n *is generated by* $(1\ 2), (1\ 3), \ldots, (1\ n)$.

This means that every permutation is a product of the transpositions shown (and their inverses, which, however, are not needed in this case). Since we have already seen how to express every permutation in terms of transpositions, it only remains to express the general transposition in terms of those of the form $(1\ i)$. If $i, j \neq 1$, this is done by the formula

$$(i\ j) = (1\ i)(1\ j)(1\ i);$$

together with the relation $(i\ 1) = (1\ i)$ this establishes the result. ■

Although there are many ways of expressing a given permutation σ as a product of transpositions, the parity is always the same, i.e. all such expressions for σ have an odd number of factors, or all have an even number of factors. To prove this fact let us write σ in cycle notation (1) and let us call σ *even* or *odd* according to whether the number of cycles of even length in (1) is even or odd. Correspondingly we define the sign of σ, $\mathrm{sgn}\,\sigma$, to be $+1$ or -1. Note that a cycle of even length is odd, and one of odd length even.

Consider how the sign is affected if we multiply by a transposition, $(1\ i)$ say. Two cases are possible:

(a) 1 and i occur in the same cycle in (1), say in $(1\ 2 \ldots i \ldots r)$. Then

$$(1\ 2 \ldots i \ldots r)(1\ i) = (1\ 2 \ldots i-1)(i\ i+1 \ldots r) \tag{3}$$

and this changes the sign of σ: if r is even, both cycles on the right are even or both are odd; if r is odd, just one cycle on the right is even; so in each case the number of even cycles is changed by one.

(b) 1 and i occur in different cycles; by suitable renumbering we may assume that they occur in $(1\ 2 \ldots i-1)$ and $(i\ i+1 \ldots r)$ and then

$$(1\ 2 \ldots i-1)(i\ i+1 \ldots r)(1\ i) = (1\ 2 \ldots i \ldots r). \tag{4}$$

This is the inverse of (a) and the same argument shows that the sign is changed. Thus each multiplication by a transposition changes the sign. By induction we see that any odd number of transpositions change the sign and any even number leave it unchanged. Moreover, the product of two permutations is even if both have the same sign and odd if they have different signs. These results may be summed up as

THEOREM 3 *With every permutation* σ *we can associate a sign* $+1$ *or* -1, *written* $\mathrm{sgn}\,\sigma$, *in such a way that*

$$\mathrm{sgn}\,\sigma\tau = \mathrm{sgn}\,\sigma \,.\, \mathrm{sgn}\,\tau. \ ■$$

In terms of $\mathbf{C}_2$, the cyclic group of order 2, we can express this result by saying that there is a homomorphism of Sym_n onto $\mathbf{C}_2$. The even permutations in

Sym_n form a subgroup, called the *alternating group* of degree n, and denoted by Alt_n. It is of index 2 in Sym_n, with the odd permutations forming the other coset. For $n = 1, 2$, Alt_n is the trivial group, while Alt_3 is cyclic of order 3. Generally Alt_n has order $n!/2$, since it is of index 2 in Sym_n.

From the definition we see that Alt_n contains all cycles of length 3. In fact it is generated by these 3-cycles; more precisely, we have

THEOREM 4 *Alt_n is generated by* $(1\ 2\ 3), (1\ 2\ 4), \ldots, (1\ 2\ n)$.

Proof. The formulae $(i\ j)(k\ l) = (i\ j\ k)(k\ i\ l)$, $(i\ j)(i\ k) = (i\ j\ k)$ show that Alt_n is generated by all 3-cycles. Now we use the formulae $(i\ j\ k) = (1\ 2\ i)^{-1}(2\ j\ k)(1\ 2\ i)$, $(2\ j\ k) = (1\ 2\ j)^{-1}(1\ 2\ k)(1\ 2\ j)$, $(1\ j\ k) = (1\ 2\ k)(1\ 2\ j)(1\ 2\ k)^{-1}$ to express any 3-cycle in terms of the cycles $(1\ 2\ i)$. ■

The notion of a *permutation group*—i.e. a group of permutations—is useful because it provides a method of representing group elements, from which the whole structure of the group can in principle be recovered. This makes it important to know what groups occur as permutation groups. In fact all groups do; this is the content of

THEOREM 5 (Cayley's theorem) *Every group is isomorphic to a group of permutations on a set.*

Proof. Let G be the given group; as our set we take G itself. Each $a \in G$ acts on G by right multiplication $\rho_a\colon x \mapsto xa$. We have seen that this is a G-action, thus we have a homomorphism

$$G \to \Sigma(G) \tag{5}$$

given by $a \mapsto \rho_a$, the regular representation of G, and it only remains to show that this is injective. Suppose that $\rho_a = \rho_b$, then $a = 1\rho_a = 1\rho_b = b$, hence (5) is injective and so G is isomorphic to a subgroup of $\Sigma(G)$, i.e. to a group of permutations of G. ■

The result shows in particular that every *finite* group, of order n say, is isomorphic to a subgroup of Sym_n. Of course this applies only to finite groups, because Sym_n is itself finite.

Exercises

(1) Show that an n-cycle $(1\ 2 \ldots n)$ can be expressed as a product of $n-1$ transpositions but no fewer.

(2) Find $\alpha^{-1}\beta^{-1}\alpha\beta$ in the following cases: (i) $\alpha = (1\ 2\ 3)$, $\beta = (1\ 4\ 5)$, (ii) $\alpha = (1\ 2\ 3\ 4)$, $\beta = (1\ 3\ 5)$, (iii) $\alpha = (1\ 2\ 3)$, $\beta = (4\ 5\ 6)$.

(3) Show that the elements 1, $(1\ 2)(3\ 4)$, $(1\ 3)(2\ 4)$, $(1\ 4)(2\ 3)$ form a subgroup of Alt_4. (This permutation group, and sometimes any abstract group isomorphic to it, is called the *four group*.)

(4) Show that Sym_n is generated by $(1\ 2), (2\ 3), \ldots, (n-1\ n)$.

(5) Show that Sym_n is generated by $(1\ 2 \ldots n)$ and $(1\ 2)$.

(6) Show that Alt_n is generated by $(1\ 2\ 3)$ and $(1\ 2 \ldots n)$ or by $(1\ 2\ 3)$ and $(2\ 3 \ldots n)$ according to whether n is odd or even.

(7) Show that every finite group can be represented as a group of even permutations.

(8) Let G be a subgroup of Sym_n not contained in Alt_n. Show that exactly half the permutations in G are even.

(9) Show that any group of order $4n+2$ has a subgroup of index 2. (Hint. Use Ex. (8).)

(10) A permutation is said to be *regular* if (in cycle notation) all its cycles have the same length. Show that a permutation is regular iff it is a power of a cycle.

(11) A permutation group is said to be *regular* if it consists of regular permutations. Show that a permutation group is regular iff each permutation moves all points or no point. Use Cayley's theorem to show that every finite group can be represented as a regular permutation group.

(12) Show that every abelian transitive permutation group is regular.

3.6 Symmetry

Groups form an important tool in the study of geometrical symmetry. Many mosaics and tapestries show symmetries of varying kinds and the natural way to classify them is by means of the groups they admit. Below we give some examples. For a fuller discussion we refer to the books by Speiser, Weyl and Coxeter (see Appendix 1).) These examples will not be needed elsewhere in the book.

We shall need a formula for the number of orbits in a G-set:

THEOREM 1 *Let G be a finite group acting on a finite G-set S. For each $g \in G$ let c_g be the number of points fixed by g, then the number of orbits is*

$$t = \frac{1}{|G|} \sum_{g \in G} c_g. \tag{1}$$

Thus t is the 'average' number of points fixed by a permutation. To prove (1) we count the number of pairs $(x, g) \in S \times G$ such that $xg = x$ in two ways: on the one hand, for each $g \in G$, the number of pairs occurring is c_g; on the other hand, for each orbit, of k points say, each point x is fixed by the elements of its stabilizer, which by the orbit formula has $|G|/k$ elements. Thus each orbit contributes $|G|$ pairs in all and so $\sum c_g = |G| \cdot t$, where t is the number of orbits. Now (1) follows on dividing by $|G|$. ■

As a first example consider all finite rotation groups about a fixed point O in 3-space. Any rotation $\alpha \neq 1$ about O has an axis through O, whose points it leaves fixed. In fact the rotation is completely determined by the way it permutes the points of the unit-sphere S^2 about O as centre. The points where the axis of α meets S^2 (and which are left fixed by α) are called the *poles* of α.

Let G be a finite group of rotations and write $|G| = n$. A pole P is said to have multiplicity k or be *k-tuple* if it is a pole of precisely k rotations (including the identity) of G. Thus the stabilizer of P has order k and the G-orbit of P has n/k members. Clearly these are all poles: if P is a pole of α, then $P\beta$ is a pole of $\beta^{-1}\alpha\beta$. Let the poles fall into t orbits, and let the ith orbit have r_i poles; if $n = r_i m_i$, these poles must be m_i-tuple. Each non-identity rotation fixes two points, while the identity fixes all. Thus, letting G act on the set of poles of elements $\neq 1$ of G, we have, by Th. 1,

$$n \,.\, t = 2(n-1) + \sum r_i.$$

On dividing by n and remembering that $r_i = n/m_i$, we find (after a little rearrangement)

$$2\left(1-\frac{1}{n}\right) = \sum\left(1-\frac{1}{m_i}\right). \tag{2}$$

Excluding the trivial group, we have $n > 1$, hence the left-hand side of (2) is at least 1 but less than 2. Now $m_i \geqslant 2$, so on the right each term is at least $\frac{1}{2}$ but less than 1, so there are at least 2 terms but not more than 3 on the right of (2). We take these two cases separately:

(i) There are two orbits. Then (2) becomes

$$\frac{2}{n} = \frac{1}{m_1} + \frac{1}{m_2}, \qquad \text{i.e. } 2 = r_1 + r_2.$$

Since each r_i is a positive integer, $r_1 = r_2 = 1$ and $m_1 = m_2 = n$. There are just two poles; if α is a rotation through the least angle θ say, then any rotation in G is a power of α. For if β is a rotation through an angle λ, then $r\theta \leqslant \lambda < (r+1)\theta$ for some $r \in \mathbf{Z}$, hence $\beta\alpha^{-r}$ is a rotation through an angle $\lambda - r\theta$, which is non-negative but less than θ. Hence $\lambda - r\theta = 0$ and $\beta = \alpha^r$; this shows every rotation to be a power of α, so G is cyclic of order n: $G \cong \mathbf{C}_n$. This may be regarded as the group of a pyramid with a regular n-gon as base.

(ii) There are 3 orbits. We now have

$$\frac{1}{m_1} + \frac{1}{m_2} + \frac{1}{m_3} = 1 + \frac{2}{n}.$$

Each m_i is at least 2, but not all are greater, because $1/3 + 1/3 + 1/3 = 1 < 1 + 2/n$. Hence we may take $m_1 = 2$ and then

$$\frac{1}{m_2} + \frac{1}{m_3} = \frac{1}{2} + \frac{2}{n}.$$

Not both of m_2, m_3 are greater than 3, so we may take m_2 to be 2 or 3. We consider these cases in turn.

(ii, a) $m_1 = m_2 = 2$, $m_3 = n/2$. Put $m = n/2$, then one orbit consists of two m-tuple poles, P, Q say, and the other two orbits consist of m poles each. The stabilizer of P is a subgroup H of index 2, and any rotation not in H interchanges P and Q. Clearly H is a rotation group with only two poles; by case (i) it is a cyclic group of order m, consisting of the rotations of a regular m-gon in its plane, while the remaining elements of G rotate the m-gon about an axis in its plane. G is called the *dihedral group* of order $2m$; it consists of all the symmetries (in 3-space) of a regular plane m-gon.

In the remaining cases $m_1 = 2$, $m_2 = 3$, $m_3 \geqslant 3$ and we find

$$\frac{1}{m_3} = \frac{1}{6} + \frac{2}{n}.$$

This holds when $m_3 = 3$, $n = 12$; $m_3 = 4$, $n = 24$; $m_3 = 5$, $n = 60$ and for no other values.

(ii, b) $n = 12$, the multiplicities m_i are (2, 3, 3). There is an orbit of 4 points A, B, C, D. If we join these 4 points, we obtain a regular tetrahedron; its 4 faces are equal and equilateral triangles. A second orbit of 4 points is obtained by intersecting the unit-sphere S^2 with the perpendiculars in the midpoints of the faces, and an orbit of 6 points by joining the midpoints of opposite edges and intersecting these joins with S^2. The group is called the *tetrahedral group*. If we represent it as a permutation group on A, B, C, D we see that it contains all 3-cycles, and hence it contains the alternating group on these 4 letters, Alt_4. Since it has the same order as Alt_4, viz. 12, it coincides with Alt_4 as abstract group.

(ii, c) $n = 24$, multiplicities (2, 3, 4). There is an orbit of 6 points; they form the vertices of a regular octahedron. Its faces are 8 equal and equilateral triangles and the perpendiculars in their midpoints meet S^2 in 8 points of a second orbit. The joins of midpoints of opposite sides meet S^2 in 12 points of a third orbit. This group is called the *octahedral group*.

(ii, d) $n = 60$, multiplicities (2, 3, 5). The smallest orbit has 12 points, which form the vertices of a regular icosahedron, whose faces are 20 equal equilateral triangles. Again we get an orbit of 20 points from the faces and an orbit of 30 points from the edges. The group is called the *icosahedral group*. We shall show that it can be regarded as a permutation group of degree 5. Consider the orbit of 30 poles; they fall into 15 pairs of opposite poles. Each pole belongs to just one rotation apart from the identity and the pole pairs are permuted transitively by G. Hence the stabilizer H of such a pair (PQ) say, has order 4. If α lies in H but does not fix P, it interchanges P and Q, hence α^2 fixes P, Q as well as the poles of α and so must be 1. Thus $H = \{1, \alpha, \beta, \gamma\}$ consists of 3 elements of order 2 besides the unit element. Since α, β, γ all have even order, their poles are all double and form a set of 3 pairs among the 15. In this way the 30 double poles fall into 5 sets of 6, which are permuted transitively by G.

It may be shown that G is Alt_5, by identifying G as the alternating group on the 5 sets of 6 poles, by a closer analysis of the icosahedron (Ex. (7)).

The finite rotation groups are closely connected with the five regular solids, which were already determined by Euclid. Some of them appeared naturally in the above discussion; the rest are obtained by dualizing, i.e. interpreting faces as vertices and vice versa. The complete list is as follows, with the number of faces, edges and vertices listed after each name: (i) *tetrahedron* (4, 6, 4), (ii) *hexahedron* or *cube*, (6, 12, 8), (iii) *octahedron* (8, 12, 6), (iv) *dodecahedron* (12, 30, 20), (v) *icosahedron* (20, 30, 12). From the numbers given we see that the icosahedron and dodecahedron have the same group, as do the octahedron and cube.

A second example is taken from topology. In any topological space X fix a point x and consider *loops* on x, i.e. continuous images of a finite interval beginning and ending at x. Two loops are multiplied by traversing them in succession. An equivalence between loops is defined by declaring α equivalent to β if α can be deformed continuously into β. It is now possible to consider the multiplication of equivalence classes and to show that we have a group; the neutral element consists of all loops that can be contracted to x. The resulting group $\pi_1(X)$ is called the *first homotopy group* of X; when X is connected, the isomorphism type of $\pi_1(X)$ does not depend on the choice of the base point x. This group is an important topological invariant of the space, e.g., for the plane or a 2-sphere it is trivial, but for the circumference $x^2+y^2 = 1$ in the plane (or also for the infinite cylinder in 3-space given by the same equation) it is the infinite cyclic group.

Exercises

(1) Let G be a group acting transitively on a set of n elements. Show that the average number of points moved by an element of G is $n-1$. Deduce that G is a regular permutation group iff $|G| = n$.

(2) Let $\mathbf{D}_n$ be the dihedral group of order $2n$. Show that the centre of $\mathbf{D}_n$ is $\mathbf{1}$ or $\mathbf{C}_2$ according to whether n is odd or even.

(3) Let a, b be two elements of order 2 in a group. If ab has order n, show that the subgroup generated by a and b is isomorphic to $\mathbf{D}_n$.

(4) In the dihedral group $\mathbf{D}_n$ of order $2n$, considered as acting on a regular n-gon, let α be a rotation through $2\pi/n$, and β a rotation through π about an axis in the plane of the n-gon. Prove that $\alpha^n = \beta^2 = 1$, $\beta\alpha\beta = \alpha^{-1}$ and deduce that every element of $\mathbf{D}_n$ can be written uniquely as $\alpha^r\beta^s$ where $0 \leqslant r < n$, $0 \leqslant s < 2$.

(5)* Determine the conjugacy classes of the dihedral group. Similarly for the tetrahedral, octahedral and icosahedral groups.

(6) Show that the octahedral group permutes the four diagonals of the cube and deduce that it is isomorphic to Sym_4.

(7)* In the description of the icosahedral group given, show that there is a rotation permuting 3 of the 5 sets of 6 poles cyclically. Deduce that the group of the icosahedron is Alt_5.

(8) Consider the action of Sym_7 on the columns of the following array (where each column is regarded as an unordered set):

$$\begin{matrix} 1 & 2 & 3 & 4 & 5 & 6 & 7 \\ 2 & 3 & 4 & 5 & 6 & 7 & 1 \\ 4 & 5 & 6 & 7 & 1 & 2 & 3 \end{matrix}$$

Let T be the subgroup of Sym_7 which maps each column into another column. Show that T acts transitively on the columns. If T' denotes the stabilizer of the column $\{1, 2, 4\}$, show that T' is isomorphic to the symmetric group on 3, 5, 6, 7. Deduce that $|T| = 168$.

Further exercises on Chapter 3

(1) Verify that the inverse of a bijective homomorphism (between monoids or groups) is again a homomorphism and show that the relation of isomorphism is an equivalence relation (i.e. reflexive, symmetric and transitive).

(2) Let G be a group, M a monoid and $f: G \to M$ a homomorphism of monoids. Show that the image of G under f is a group.

(3) Let G be a group and S a G-set. Given $p \in S$, let H be its stabilizer and denote by $c_1, \ldots, c_r, \ldots$ a right transversal of H in G. Show that the set $pc_1, \ldots, pc_r, \ldots$ represents the points of the orbit containing p without repetition.

(4) How can one recognize from the multiplication table of a group whether it is abelian?

(5) If the group table is arranged so that the column corresponding to the row of a is headed by a^{-1}, then the main diagonal of the table (from the NW to the SE corner) consists of 1s. Show that in this form any rectangle with $1, a, b$ at 3 corners has ab (or ba) at the fourth corner.

(6) If H, K are subgroups of a group, show that $H \cup K$ is not a subgroup unless one contains the other.

(7) In any group G, if $a^2 = (ab)^2 = 1$, then $a^{-1}ba = b^{-1}$.

(8) For any field F, the affine transformations $x \mapsto ax+b$, where $a, b \in F$, $a \neq 0$, form a group $\mathbf{Af}_1(F)$. Find the order of $\mathbf{Af}_1(F)$ when F is finite.

(9) Show that the mapping $x \mapsto x^{-1}$ in a group is an automorphism iff the group is abelian.

(10) Show that a group satisfying $(xy)^2 = x^2y^2$ is abelian. In any group express $a^{-1}b^{-1}ab$ as a product of 3 squares.

(11) Let G be a group and H a subgroup. Find a G-set S on which G acts transitively, with H as the stabilizer of a point.

(12) If a finite group acts transitively on a set S, show that some $g \in G$ moves all points.

(13) Let G be a group and H, K subgroups. If $Hx = Ky$ for some $x, y \in G$, show that $H = K$. What can we conclude from $Hx = yK$?

(14) Let G be a group and H, K two subgroups (not necessarily distinct). Define a relation on G by putting $x \sim y$ whenever $HxK = HyK$; verify that this is an equivalence relation on G. Show that $|HxK| = |H| . |K|/|H^x \cap K|$, where $H^x = x^{-1}Hx$. (The sets HxK are called *double cosets* of G with respect to H and K, and the expression of G as a disjoint union $G = \bigcup Hx_iK$ is called the *double coset decomposition.*)

(15) Let c be an element of order n in a group; show that c^k has order n/d, where d is the highest common factor of n and k.

(16) Show that any group with exactly one maximal proper subgroup is cyclic of prime power order.

(17) Let G be a group, H a subgroup and c an element of order n in G. If r is the least positive integer such that $c^r \in H$, show that $r \mid n$.

(18)* Let G be a group and H a subgroup of index n. If $c \in G$, show that $c^k \in H$ for some k such that $0 < k \leqslant n$, but the least positive k need not be a factor of n. (Hint. Consider the action of G on the left cosets Hx and look for the orbit of H under the action of c.)

(19) Find all subgroups of Sym_4.

(20) Show that in Sym_n a cycle of length n commutes only with its powers; does this still hold for a cycle of length $n-1$? Show that Sym_n has trivial centre for $n \geqslant 3$, and Alt_n has trivial centre for $n \geqslant 4$.

(21) For any $\sigma \in \mathrm{Sym}_n$ an *inversion* is defined as a pair (i, j) in the range $\{1, \ldots, n\}$ such that $i < j$ and $i\sigma > j\sigma$. By considering the action of σ on the product $\prod_{i<j} (x_i - x_j)$ show that $\mathrm{sgn}\, \sigma = (-1)^{\nu(\sigma)}$, where $\nu(\sigma)$ is the number of inversions of σ.

(22) Describe the group of the square as a subgroup of Sym_4. Find its conjugates in Sym_4 and show that they all have the four group in common.

(23) Let G be a group of odd order. Show that for each element a of G, the equation $x^2 = a$ has a unique solution. Denoting this solution by $a^{\frac{1}{2}}$, define a new operation '$\circ$' on the set G by the formula

$$x \circ y = (x^{\frac{1}{2}}y^{\frac{1}{2}})^2.$$

Prove that G forms a group under this operation and that the two group operations on G coincide iff the original group is abelian.

4

Vector spaces and linear mappings

Many mathematical problems have the property that for any two solutions u and v, $u+v$ is also a solution. Such problems are called *linear* and are usually very much easier to solve than more general problems. Furthermore, many problems arising in applications are in fact linear, at least in the first approximation, so that this method is of great practical use. For example, the principle of superposition in physics is just an expression of the fact that the differential equations satisfied by heat, light, electricity and other phenomena are linear.

Since these problems have a common mathematical content, it is natural to study this in its abstract form, guided at first by geometrical analogy. We recall that in geometry, vectors are used to represent 'directed lengths'; more generally, any physical quantity such as velocity, which is described by a magnitude and direction, can be represented by a line segment OP of the appropriate length and direction, from the origin O of coordinates. If the endpoint P has coordinates (a_1, a_2, a_3), the vector $a = OP$ is completely specified by these coordinates and we may write $a = (a_1, a_2, a_3)$. Any triple of real numbers a_1, a_2, a_3 defines a vector in this way and distinct triples correspond to distinct vectors, e.g. $(2, 1, 3) \neq (2, 3, 1)$.

If $a = (a_1, a_2, a_3)$ and $b = (b_1, b_2, b_3)$ are two vectors, then the *sum* of a and b is formed by adding the components: $a+b = (a_1+b_1, a_2+b_2, a_3+b_3)$. This corresponds to the well-known parallelogram rule illustrated in Fig. 1.

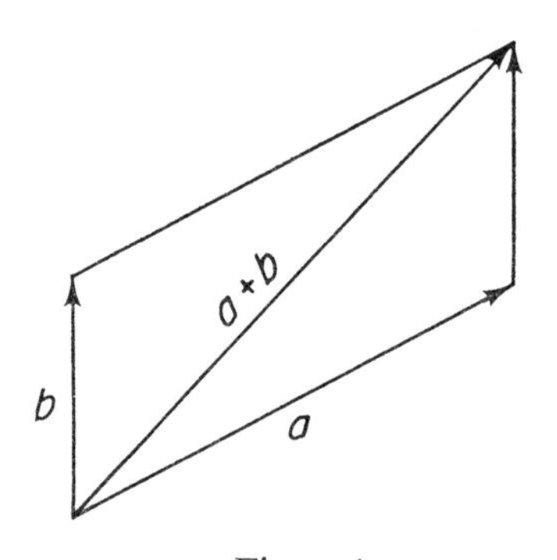

Figure 1

Similarly, on multiplying a by a real number α (a 'scalar') we obtain another vector $\alpha a = (\alpha a_1, \alpha a_2, \alpha a_3)$ in the same straight line as a. Any linear combination of a and b such as $\alpha a+\beta b$, where α and β are scalars, represents a vector in the same plane as a and b and, provided that a, b define a plane (i.e. if they do not lie in the same straight line), every vector in this plane can be written as a linear combination of a and b; more briefly, a and b *span* the plane.

In one's first encounter with vectors it is of course helpful to keep the geometrical picture in mind. But it is important to realize that this particular interpretation, though it aids one's intuition, is not an essential part of the theory. This becomes clear if, e.g., we make a statistical study of the measurements of 17 different characteristics of an organism. These are most naturally plotted in 17-dimensional space, in which the vectors are sequences with 17 components. Here the geometrical analogy, though still highly suggestive, is no longer relevant. All that matters are the relations between vectors and the operations by which they are combined. This is brought out most clearly by an axiomatic development of vectors, a course which has other advantages too. In the first place, it allows us to treat vectors of any finite number of dimensions with the same ease. Secondly, we deal directly with the vectors, as far as possible, without referring to a basis. Thirdly, we can allow the scalars to lie in any field, not necessarily the real numbers.

4.1 Vectors and linear dependence

We begin by defining a vector space over a field F of scalars. Here F may be any field, although the reader is advised to keep in mind a concrete case such as $F = \mathbf{R}$ at first. Likewise he should keep in mind the case of geometrical vectors as an illustration, although, when it comes to proofs, we can of course only use the axioms or results proved earlier.

A *vector space* over a field F is an abelian group V, written additively, with a mapping of $F\times V$ into V (to represent multiplication by scalars), subject to the axioms **V**. 1–4 below. We shall write the sum of the vectors x and y as $x+y$ and the image of (α, x), for $\alpha \in F$, $x \in V$, as αx; then the axioms read:

V.1 $\quad \alpha(x+y) = \alpha x+\alpha y.$

V.2 $\quad (\alpha+\beta)x = \alpha x+\beta x. \qquad (\alpha, \beta \in F,\ x, y \in V)$

V.3 $\quad (\alpha\beta)x = \alpha(\beta x).$

V.4 $\quad 1 \,.\, x = x.$

The elements of V are called *vectors*; by contrast the field elements are called *scalars* and F itself is called the *ground field*. We shall mostly use the latin alphabet for vectors and greek characters for scalars.

Of the above axioms, **V**. 1 states that each $\alpha \in F$ defines an endomorphism of V, qua abelian group, while **V**. 2–4 just show that the field operations correspond to the way scalars act on V. Note that in **V**. 2 the addition on the left is in F and on the right in V; similarly on the left of **V. 3** we have the field multiplication, while on the right we have the action of F on V.

We note the following consequences of the definition:

(i) $0x = \alpha 0 = 0$. Here the first zero is a scalar, while the other two are vectors. This rule is analogous to the corresponding rule for integers or indeed field elements. To prove the rule, we write $0x = (0+0)x = 0x+0x$ by **V. 2**, hence $0x = 0$; the other part follows similarly, using **V. 1**.

(ii) $\alpha x = 0$ implies $\alpha = 0$ or $x = 0$. For if $\alpha \neq 0$, then $x = 1 . x = (\alpha^{-1}\alpha)x = \alpha^{-1}(\alpha x) = 0$.

(iii) If m is any positive integer, then F contains an element $m . 1$, and $(m . 1)x = x+x \ldots +x$ (m summands). This is easily proved by induction; the details are left to the reader. As a result there is no ambiguity in writing mx for $(m . 1)x$; note however, that for a field of prime characteristic p, where $p . 1 = 0$, we have $px = 0$ for all x.

(iv) If the negative of x (i.e. the additive inverse of x in V) is written $-x$, then $(-1)x = -x$. For we have $x+(-1)x = (1-1)x = 0x = 0$.

Examples of vector spaces. (i) The geometrical vectors in the plane or in space form a vector space over the field $\mathbf{R}$ of real numbers.

(ii) For any field F and any integer $n \geqslant 1$, denote by F^n the set of all n-tuples $(\xi_1, \ldots, \xi_n)$ where $\xi_i \in F$ and $(\xi_1, \ldots, \xi_n)$ and $(\eta_1, \ldots, \eta_n)$ are distinct unless $\xi_1 = \eta_1, \ldots, \xi_n = \eta_n$. This set F^n forms a vector space under the operations

$$(\xi_1, \ldots, \xi_n)+(\eta_1, \ldots, \eta_n) = (\xi_1+\eta_1, \ldots, \xi_n+\eta_n), \tag{1}$$

$$\alpha(\xi_1, \ldots, \xi_n) = (\alpha\xi_1, \ldots, \alpha\xi_n) \qquad \alpha \in F. \tag{2}$$

This is an important example, which in many ways is typical (cf. Th. 1, **4.3**). In particular, for $n = 1$ this shows that each field may be regarded as a vector space over itself.

(iii) For any field F consider the set of infinite sequences $(\xi_1, \xi_2, \ldots)$. Addition and multiplication by scalars can be defined as in (1) and (2) and it is easily seen that we then have a vector space, F^N say. Similarly the subset $F^{(N)}$ consisting of all sequences with only finitely many non-zero coordinates is a vector space.

(iv) The set of all real functions is a vector space if we define $f+g$ as the function whose value at x is $f(x)+g(x)$, and αf as the function whose value at x is $\alpha f(x)$.

(v) The complex numbers $\mathbf{C}$ form a vector space over $\mathbf{R}$; both $\mathbf{C}$ and $\mathbf{R}$ are vector spaces over the rational numbers $\mathbf{Q}$.

(vi) The set consisting of a single element 0 is a vector space, called the *trivial* space. The abelian group structure is clear and we define $\alpha 0 = 0$ for all $\alpha \in F$.

Let V be a vector space over F. A *subspace* of V is a subgroup of V admitting multiplication by scalars. Thus U is a subspace of V if U is a subset of V such that (i) $0 \in U$, (ii) $x, y \in U \Rightarrow x+y \in U$, and $\alpha x \in U$ for all $\alpha \in F$ (for on taking $\alpha = -1$ we see that $-x \in U$, so the conditions for a subgroup are satisfied). E.g., V is always a subspace of itself; a subspace of V other than V is called a *proper* subspace. Of course the trivial space is always a subspace of V. Other examples of subspaces: In Example (v) above, $\mathbf{R}$ is a subspace of $\mathbf{C}$, in Example (ii) F^n has a subspace obtained by restricting the last coordinate to be 0, and in Example (iii) $F^{(N)}$ is a subspace of F^N.

The intersection of two (or indeed, any number of) subspaces of a vector space V is again a subspace: if U_1, U_2 are subspaces, then clearly $0 \in U_1 \cap U_2$, and if $x, y \in U_i$ $(i = 1, 2)$, then $x+y$ and αx belong to U_i and hence to $U_1 \cap U_2$. A similar argument works for more than two subspaces.

On the other hand, the union of two subspaces is not generally a subspace, as a concrete example will quickly show: in Example (i) above, the union of two distinct lines clearly does not form a subspace. However, for any subspaces U_1 and U_2 there is a least subspace containing U_1 and U_2, called their *sum* and written U_1+U_2. It consists of all vectors u_1+u_2 $(u_i \in U_i)$; for clearly any subspace containing U_1 and U_2 must contain all vectors $u_1 \in U_1$ and $u_2 \in U_2$, and hence all vectors u_1+u_2. To show that U_1+U_2 is actually the least subspace containing U_1 and U_2 it only remains to show that the set U_1+U_2 is a subspace, a task that may be left to the reader (cf. **3.2** for a similar discussion on subgroups). More generally, the sum of any finite set of subspaces $U_1, \ldots, U_k$ is $U_1+\cdots+U_k = \{\sum x_i \mid x_i \in U_i\}$.

Let V be a vector space and take vectors $x_1, \ldots, x_n$ in V. Then there is a least subspace containing $x_1, \ldots, x_n$, namely the intersection of all subspaces containing $x_1, \ldots, x_n$. This space is denoted by $\langle x_1, \ldots, x_n\rangle$; here is a more constructive way of forming it: By definition, $\langle x_1, \ldots, x_n\rangle$ contains $x_1, \ldots, x_n$; being a subspace, it must contain $\alpha_1 x_1, \ldots, \alpha_n x_n$, for any $\alpha_1, \ldots, \alpha_n \in F$, and hence also the linear combination $\alpha_1 x_1 + \ldots + \alpha_n x_n$. But the set of all these linear combinations is already a subspace and therefore coincides with $\langle x_1, \ldots, x_n\rangle$. Thus

$$\langle x_1, \ldots, x_n\rangle = \{\alpha_1 x_1 + \cdots + \alpha_n x_n \mid \alpha_1, \ldots, \alpha_n \in F\}.$$

E.g., $\langle x_1\rangle$ consists of all αx_1 $(\alpha \in F)$ and $\langle\,\rangle$, the least subspace containing the empty set, is the zero space $\mathbf{0}$.

The space $\langle x_1, \ldots, x_n\rangle$ is called the subspace *spanned* by $x_1, \ldots, x_n$. If $\langle x_1, \ldots, x_n\rangle = V$, we say that $x_1, \ldots, x_n$ *span* V, or that they form a *spanning set* for V. If $x \in \langle x_1, \ldots, x_n\rangle$, so that x is a linear combination of $x_1, \ldots, x_n$:

$$x = \alpha_1 x_1 + \cdots + \alpha_n x_n \quad \text{for some } \alpha_i \in F, \tag{3}$$

we also say that x is *linearly dependent* on $x_1, \ldots, x_n$. E.g., to say that a vector in 3-space is linearly dependent on a pair of vectors a, b means that it lies in the subspace spanned by a and b; this subspace may be a plane, or a line (if a and b are parallel) or a point (if $a = b = 0$). Thus (3, 5, 7) is linearly dependent on the pair (1, 1, 1) and (1, 2, 3), as the equation

$$(3, 5, 7) = (1, 1, 1)+2(1, 2, 3)$$

shows; but (3, 5, 6) is not, as some trials will convince the reader. Soon we shall meet general methods for testing when a vector is linearly dependent on a given set.

Since we shall frequently have to write such expressions as (3) for x, we shall abbreviate it by using the $\sum$ notation already encountered in Ch. **2**. Thus instead of (3) we write $x = \sum_1^n \alpha_i x_i$ or simply $\sum \alpha_i x_i$ or also

$$\sum_{i=1}^{n} \alpha_i x_i,$$

when the precise range is not clear from the context.

A family of vectors $x_1, \ldots, x_n$ is said to be *linearly dependent*, if one of them is linearly dependent on the rest; otherwise the family is *linearly independent*. E.g., any family including the zero vector is linearly dependent, for if $x_1 = 0$, say, then $0 \in \langle x_2, \ldots, x_n \rangle$. Note that this holds even if the family consists only of the single vector 0. Similarly any family in which a vector occurs more than once is linearly dependent. On the other hand, the empty set is linearly independent.

We shall often use the following criterion for the linear independence of a family of vectors:

THEOREM 1 *A family of vectors $x_1, \ldots, x_n$ is linearly independent if and only if there is no non-trivial relation between them, i.e. if for any $\alpha_1, \ldots, \alpha_n \in F$,*

$$\sum \alpha_i x_i = 0 \textit{ implies } \alpha_1 = \cdots = \alpha_n = 0.$$

For if one of the xs, say x_1, is linearly dependent on the rest: $x_1 = \lambda_2 x_2 + \cdots + \lambda_n x_n$, then the relation

$$x_1 - \lambda_2 x_2 - \cdots - \lambda_n x_n = 0$$

is non-trivial, because the coefficient of x_1 is non-zero. Conversely, suppose that we have a non-trivial relation $\sum \alpha_i x_i = 0$, say $\alpha_1 \neq 0$, then the equation

$$x_1 = -\alpha_1^{-1}(\alpha_2 x_2 + \cdots + \alpha_n x_n)$$

shows the linear dependence of x_1 on $x_2, \ldots, x_n$. ■

Exercises

(1) Which of the following families of vectors are linearly dependent? (i) (1, 2), (4, −2), (ii) (1, 2), (4, −2), (3, 5), (iii) (−2, 1), (4, −2), (iv) (3, 1, 4), (2, 5, −1), (4, −3, 7), (v) (1, 0, 0, 0), (1, 1, 0, 0), (1, 1, 1, 0), (1, 1, 1, 1), (vi) (3, 1, 2, 3), (2, 1, 3, 2), (3, 1, ,2 3), (1, 3, 2, 1).

(2) Test the following sets for linear dependence and, when they are linearly dependent, express each vector when possible as a linear combination of the rest: (i) $(1, 2, -3, 1)$, $(2, 0, 4, 1)$, $(5, -4, 14, -3)$, (ii) $(2, 1, 2, 1)$, $(6, 3, 6, 3)$, $(5, 1, 4, 3)$, (iii) $(1, 1, 0)$, $(1, 0, 1)$, $(0, 1, 1)$.

(3) If x, y, z are linearly independent vectors in a space, show that $x+y$, $y+z$, $z+x$ are linearly independent, provided that the field of coefficients has characteristic $\neq 2$ (so that we can divide by 2). Are $x-y$, $y-z$, $z-x$ ever linearly independent?

(4) Check the assertions made in Examples (i)–(vi) in the text.

(5) Prove the general distributive law for vector spaces: $(\sum \alpha_i)(\sum x_j) = \sum_{ij} \alpha_i x_j$.

(6) Which of the following sets are subspaces of $\mathbf{R}^n$: (i) all vectors $a = (\alpha_1, \ldots, \alpha_n)$ satisfying $\alpha_1 = 2\alpha_2$, (ii) all vectors a satisfying $\sum \alpha_i = 0$, (iii) all vectors a satisfying $\sum \alpha_i = 1$, (iv) all vectors a satisfying $\sum \alpha_i^2 = 0$. Do any of these answers change if $\mathbf{R}$ is replaced by $\mathbf{C}$?

(7) Let F be a field and I a non-empty set and define F^I as the set of all mappings from I to F. Given two elements $f, g \in F^I$, define their sum $f+g$ as the mapping $i \mapsto f_i + g_i$ (where f_i, g_i are the images of i under f, g) and define αf as $i \mapsto \alpha f_i$. Show that F^I is a vector space over F and verify that Examples (ii), (iii) and (iv) are special cases of this construction.

(8) If F is a finite field with q elements, show that F^n has q^n elements. Show that every non-zero space over an infinite field is infinite.

4.2 Linear mappings

For a closer study of vector spaces we need to define their 'homomorphisms', i.e. mappings preserving the vector space structure. For historical reasons they are usually called 'linear mappings'.

Let U, V be vector spaces over the same field F. A mapping $\theta: U \to V$ is said to be *linear* if

$$(x+y)\theta = x\theta + y\theta \qquad x, y \in U, \tag{1}$$

$$(\lambda x)\theta = \lambda(x\theta) \qquad \lambda \in F. \tag{2}$$

We note that (1) just expresses the fact that θ is a homomorphism of U into V, regarded as additive groups.

Any linear mapping satisfies

$$(\alpha_1 x_1 + \ldots + \alpha_n x_n)\theta = \alpha_1(x_1\theta) + \ldots + \alpha_n(x_n\theta) \qquad (x_i \in U, \alpha_i \in F), \tag{3}$$

a remark that will often be used. To prove it we note that for $n = 1$ it reduces to (2), and so we may use induction on n. Let $n > 1$, then $(\sum \alpha_i x_i)\theta = (\alpha_1 x_1)\theta + (\sum_2^n \alpha_i x_i)\theta$ by (1); applying (2) to the first term and induction on n to the sum, we find $(\sum \alpha_i x_i)\theta = \alpha_1(x_1\theta) + \sum_2^n \alpha_i(x_i\theta)$, i.e. (3). In particular, θ satisfies

$$(\lambda x + \mu y)\theta = \lambda(x\theta) + \mu(y\theta) \qquad (x, y \in U, \lambda, \mu \in F). \tag{4}$$

We observe that linear mappings can also be characterized by (4), for, when this holds, we obtain (1) by taking $\lambda = \mu = 1$ and (2) by taking $y = 0$.

Examples of linear mappings. (i) For any $\alpha \in F$, $x \mapsto \alpha x$ is a linear mapping. Note in particular the cases $\alpha = 0, 1$, which show the zero mapping and the identity mapping to be linear.

(ii) $(\xi_1, \xi_2) \mapsto (\xi_1, \xi_2, \xi_1+\xi_2)$, as a mapping of $F^2 \to F^3$.

(iii) $(\xi_1, \xi_2, \xi_3) \mapsto (\xi_1, \xi_2)$, as a mapping of $F^3 \to F^2$. Geometrically this represents a projection on the plane $\xi_3 = 0$.

An important example of a linear mapping is obtained as follows. Let V be a vector space and take any $x_1, \ldots, x_n \in V$. Then the mapping $\theta: F^n \to V$ defined by the rule

$$(\xi_1, \ldots, \xi_n) \mapsto \xi_1 x_1 + \ldots + \xi_n x_n \tag{5}$$

is easily seen to be linear; the verification is left to the reader. Let us denote by e_i the element of F^n whose ith component is 1 while the others are 0. Then it is clear from (5) that

$$e_i\theta = x_i \qquad (i = 1, \ldots, n). \tag{6}$$

We assert that θ is the only linear mapping from F^n to V with the property (6): for if (6) holds, then by (3),

$$(\xi_1, \ldots, \xi_n)\theta = (\sum \xi_i e_i)\theta = \sum \xi_i(e_i\theta) = \sum \xi_i x_i$$

and so θ satisfies (5), as claimed. In other words, every mapping of the family $(e_1, \ldots, e_n)$ into V (such as (6)) can be extended to a unique linear mapping of F^n into V. This property of the space F^n is an instance of the *universal mapping property*, which we shall meet in many forms later.

Exercises

(1) Let A be the space of all real-valued functions on $\mathbf{R}$. Which of the following mappings of A into itself are linear? (i) $f \mapsto f_1$ where $f_1(x) = f(x+1)$, (ii) $f \mapsto f+1$, (iii) $f \mapsto f^2$, (iv) $f \mapsto f^{(2)}(x)$, where $f^{(2)}(x) = f(f(x))$.

(2) Show that a mapping $\theta: U \mapsto V$ is linear iff

$$(x+\lambda y)\theta = x\theta + \lambda(y\theta) \quad \text{for all } x, y \in U, \lambda \in F.$$

(3) Show that every homomorphism of vector spaces over $\mathbf{Q}$ (regarded as abelian groups) is linear. Do the same for spaces over $\mathbf{F}_p$.

(4) A mapping $\varphi: U \to V$ between vector spaces over F is said to be *affine* if $x\varphi = x\theta + c$, where θ is linear and c is a fixed vector of V. Show that a mapping $x \mapsto x'$ is affine iff $(\sum \alpha_i x_i)' = \sum \alpha_i x'_i$ for any family of elements $x_i \in U$, and $\alpha_i \in F$ such that $\sum \alpha_i = 1$.

4.3 Bases and dimension

In any vector space V, we define a *basis* to be a linearly independent spanning set of V. The importance of a basis $v_1, \ldots, v_n$ is that every vector of V can be written as a linear combination of $v_1, \ldots, v_n$, with uniquely determined scalar coefficients. For, given $x \in V$, since the v_i span V, we have

$$x = \xi_1 v_1 + \ldots + \xi_n v_n, \tag{1}$$

for some $\xi_i \in F$ and, if we also had $x = \sum \xi_i' v_i$, then $\sum (\xi_i - \xi_i') v_i = x - x = 0$, hence by the linear independence, $\xi_i' = \xi_i$. Thus the coefficients ξ_i in (1) are uniquely determined by x. Conversely, if every $x \in V$ has a unique expression (1), then the v_i span V and, by uniqueness, no linear combination of the vs is 0, except the trivial one in which all the coefficients are 0; thus the v_i are linearly independent. This shows that the existence and uniqueness of the expression (1) for each $x \in V$ characterizes a basis of V.

For example, the vectors $e_1, \ldots, e_n$ defined in **4.2** form a basis of F^n, since every vector $(\xi_1, \ldots, \xi_n)$ can be uniquely written as $\sum \xi_i e_i$. This basis will be called the *standard basis* of F^n.

Given a vector space V, let us take any vectors $x_1, \ldots, x_n \in V$ and consider again the linear mapping $\theta: F^n \to V$ defined in (5), **4.2**:

$$\theta: (\xi_1, \ldots, \xi_n) \mapsto \sum \xi_i x_i. \tag{2}$$

We observe that θ is surjective iff $x_1, \ldots, x_n$ span V, while it is injective iff $x_1, \ldots, x_n$ are linearly independent, hence θ is bijective precisely when $x_1, \ldots, x_n$ form a basis of V. Thus when V has a basis of n elements, we can use the mapping (2) to establish a linear bijection between F^n and V. A linear bijection between vector spaces has the property that its inverse is again linear (as the reader will verify without difficulty). A linear bijection is also called an *isomorphism*; if an isomorphism exists between spaces U and V, we write $U \cong V$ and say that the spaces are *isomorphic*. The importance of this notion is that any abstract property of vector spaces is preserved under isomorphism. The observations made earlier can now be summed up as

THEOREM 1 *Any vector space over a field F, with a basis of n elements, is isomorphic to F^n.* ■

As in the case of groups, the notion of isomorphism between vector spaces is an equivalence relation. Thus if two spaces each have a basis of n elements, for the same n, then both are isomorphic to F^n and hence to each other; this proves the

COROLLARY *Any two vector spaces with bases of the same number of elements are isomorphic.* ■

Th. 1 shows the importance of the spaces F^n, but it leaves open two questions:

(i) *When does a space have a (finite) basis?*

(ii) *Can F^m be isomorphic to F^n for $m \neq n$?*

Certainly not every space has a finite basis; as an example we cite the spaces of sequences F^N and $F^{(N)}$ defined earlier. The reader should have no difficulty in showing that neither has a finite basis (this will in any case follow easily from Th. 5). Of course we have not defined what is meant by an infinite basis, but it is not hard to give appropriate definitions and show that $F^{(N)}$ has a basis (cf. Ex. (8)). As a matter of fact, we shall see in Vol. 2 that every vector space has a basis. The answer to question 2 is 'no' (as one would hope), see Th. 5.

We begin with an obvious remark which is sometimes useful.

PROPOSITION 2 *If* $X = \{x_1, \ldots, x_n\}$ *is a maximal linearly independent set in a vector space* V, *in the sense that no set of vectors containing* X *as a proper subset is linearly independent, then* X *is a basis of* V.

Proof. We must show that X spans V, i.e. that every $x \in V$ is linearly dependent on $x_1, \ldots, x_n$. This is clear if x is one of the x_i; otherwise $\{x, x_1, \ldots, x_n\}$ contains X as proper subset and so is linearly dependent. Let

$$\alpha x + \sum \alpha_i x_i = 0$$

be a non-trivial relation. If $\alpha = 0$, then by the linear independence of $x_1, \ldots, x_n$ all the α_i must also vanish and the relation is trivial. Hence $\alpha \neq 0$ and on dividing by α, we can express x as a linear combination of $x_1, \ldots, x_n$, as claimed. ■

A first (and important) method of getting a basis is given in

PROPOSITION 3 *Let* V *be a vector space with a finite spanning set* $X = \{x_1, \ldots, x_n\}$, *and assume that* $x_1, \ldots, x_k$ *are linearly independent, for some* k *in the range* $0 \leqslant k \leqslant n$. *Then there is a subset of the* xs *including* $x_1, \ldots, x_k$ *which forms a basis of* V.

We note the extreme cases: $k = n$ means that X is a linearly independent spanning set, then the result holds by definition. When $k = 0$, we start from the empty set, which is always linearly independent; in this case the result shows that any vector space with a finite spanning set has a basis.

To prove the proposition we write down the xs in order, $x_1, x_2, \ldots, x_k, \ldots, x_n$ and in turn cross out each one that is linearly dependent on the preceding ones. Thus x_1 gets crossed out precisely if it is 0, x_2 gets crossed out precisely if it is a multiple of x_1 and so on. Of course, none of $x_1, \ldots, x_k$ can be crossed out, because they are linearly independent, but, as we saw, k may be 0. Let B be the set of xs that remain; this will include $x_1, \ldots, x_k$. Further, every x_i is linearly dependent on B, hence B like X spans V. Moreover, B is linearly independent, for if not, then there would be a non-trivial relation between the vectors of B and hence some x_i in B would be linearly dependent on earlier xs in B, which is a contradiction. Thus B is a basis of the required form. ■

Suppose that X is a finite spanning set of V which is *minimal* in the sense that no proper subset of X spans V. If we apply the above construction we find that X itself is a basis of V, hence we obtain the

COROLLARY *Any minimal finite spanning set of a vector space is a basis.* ∎

Here is a typical application of Prop. 3. We are given a linearly independent set $X = \{x_1, \ldots, x_r\}$ and a finite spanning set Y, and we wish to construct a basis from X and Y that includes all of X. We simply apply Prop. 3 to the union $X \cup Y$, with $x_1, \ldots, x_r$ written down first. It is clear from the construction that the number of elements in the basis is at least r, but there is no indication how it is related to the number of elements in Y. We next show that the number of basis elements cannot exceed the number of elements in Y. In order to make the induction go through we have to prove a little more:

LEMMA 4 (Steinitz exchange lemma) *Let $x_1, \ldots, x_r$ be a linearly independent set of elements of a vector space V and let Y be a spanning set consisting of s elements. Then $r \leqslant s$ and we can find a spanning set of the form $\{x_1, \ldots, x_r, y_{r+1}, \ldots, y_s\}$, where $y_i \in Y$ for $i = r+1, \ldots, s$.*

Proof. We use induction on r. When $r = 0$, there are no xs and there is then nothing to prove, so let $r > 0$. Clearly $\{x_1, \ldots, x_{r-1}\}$ is linearly independent, hence by the induction hypothesis,

$$r-1 \leqslant s, \tag{3}$$

and there are vectors $y_r, \ldots, y_s \in Y$ such that $\{x_1, \ldots, x_{r-1}, y_r, \ldots, y_s\}$ spans V. We can therefore express x_r as a linear combination of these vectors, say

$$x_r = \sum_1^{r-1} \alpha_i x_i + \sum_r^s \beta_j y_j. \tag{4}$$

If $s = r-1$, the second sum in (4) is empty and hence is 0; likewise this sum is 0 when all the βs vanish; in either case we get a linear dependence between the xs, which is a contradiction. Hence $r-1 < s$, i.e. $r \leqslant s$, and some $\beta_j \neq 0$. By renumbering the ys, if necessary, we may assume that $\beta_r \neq 0$, and then by (4),

$$y_r = -\beta_r^{-1}\left(\sum_1^{r-1} \alpha_i x_i - x_r + \sum_{r+1}^{s} \beta_j y_j\right),$$

i.e. y_r is linearly dependent on $x_1, \ldots, x_r, y_{r+1}, \ldots, y_s$ and by hypothesis every vector is linearly dependent on $x_1, \ldots, x_{r-1}, y_r, \ldots, y_s$, hence the set $\{x_1, \ldots, x_r, y_{r+1}, \ldots, y_s\}$ spans V, as claimed. ∎

We now give a method of constructing bases which is in essence an elaboration of Prop. 3.

THEOREM 5 (Main theorem for vector spaces) *Let V be a vector space with a finite spanning set. Then V has a finite basis, and any two bases of V have the same number of elements. Moreover,*

(i) *any linearly independent subset of V is contained in a basis of V; it is a basis if and only if it cannot be enlarged* (*i.e. is a maximal linearly independent set*),

(ii) *any spanning set contains a basis if and only if it cannot be diminished* (*i.e. is a minimal spanning set*).

Proof. Prop. 3 tells us how to pick a basis from the finite spanning set, hence V has at least one (finite) basis, B say. By Lemma 4, no basis can have more elements than B because it is linearly independent, whereas B spans V. By symmetry it cannot have less elements than B has and this proves the first part. Now let n be the common value of the number of elements in any basis. Then we know from Lemma 4 that any spanning set has at least n elements and any linearly independent set has at most n elements. In particular, this assures us that any linearly independent set is finite.

Let X be any linearly independent set. Together with a basis B this spans V, hence by Prop. 3 we can find a basis containing X. It equals X precisely when we cannot adjoin any vectors to X without destroying the linear independence, i.e. when X is maximal linearly independent (note that this also follows from Prop. 2).

Let Y be a spanning set, and let us pick vectors $y_1, y_2, \ldots$ from it, as in the proof of Prop. 3, in such a way that none is linearly dependent on the preceding ones. In this way we get a linearly independent set; the process stops only when every vector of Y is linearly dependent on the ones we have selected, i.e. (since Y was a spanning set) when we have a basis of V. ∎

The common number in each basis of V is called the *dimension* of V and is written dim V. For example, the space F^n has a basis consisting of n elements and so is n-dimensional, as one would expect. Th. 5 also shows that any space with a finite spanning set is finite-dimensional. Any space without a finite spanning set is called *infinite-dimensional.*

Any isomorphism between vector spaces U and V transforms a basis of U into a basis of V. It follows that two isomorphic spaces have the same dimension. Hence F^m and F^n are not isomorphic unless $m = n$. This answers the second question raised earlier; combining this result with Th. 1, Cor. we obtain

COROLLARY 1 *If U, V are any finite-dimensional vector spaces, then $U \cong V$ if and only if* $\dim U = \dim V$. ∎

From Th. 5 we also obtain the following characterization of n-dimensional vector spaces:

COROLLARY 2 *Let V be a vector space and n a positive integer, then*

$$\dim V \leqslant n$$

if and only if every set of $n+1$ vectors in V is linearly dependent.

For if every set of $n+1$ vectors is linearly dependent and B is a linearly independent set with as many elements as possible, (necessarily $\leqslant n$), then B is a maximal linearly independent set and hence is a basis; it follows that $\dim V \leqslant n$. Conversely, when $\dim V \leqslant n$, no linearly independent set can have more than n elements. ∎

This corollary is the quickest way of determining the dimension of a vector space. Of course to be able to use it we need a simple method of finding whether a given set of vectors is linearly dependent. We shall meet such methods in Ch. **5**.

Exercises

(1) Complete the following sets to bases of $\mathbf{R}^4$: (i) (2, 1, 4, 3), (2 ,1, 2, 0), (ii) (0, 1, 2, 3), (1, 2, 3, 4), (0, 0, 0, 1), (iii) (0, 2, 1, 0), (0, 1, 2, 0).

(2) Find two vectors which together with each of the following 3 pairs of vectors form a basis of $\mathbf{R}^4$: (i) (1, 2, 3, 4), (4, 3, 2, 1), (ii) (2, 3, 0, 0,) (1, 0, 1, 1), (iii) (1, −1, 1, −1), (0, 3, 2, 1).

(3) Prove the transitivity of linear dependence, i.e. if X, Y, Z are three sets of vectors such that each vector of X is linearly dependent on Y and each vector of Y is linearly dependent on Z, then each vector of X is linearly dependent on Z.

(4) If a vector u is linearly dependent on $X \cup \{v\}$, but not on X, show that v is linearly dependent on $X \cup \{u\}$. Show how to obtain a proof of the Steinitz exchange lemma from this fact.

(5) If U is a subspace of an n-dimensional space, show that $\dim U \leqslant n$. In what circumstances is equality possible? (See **4·4**.)

(6) Give a proof that $F^m \not\cong F^n$ for $m \neq n$, when F is a finite field, using Ex. (8), **4·1**.

(7) An infinite set of vectors is defined to be linearly dependent if it has a finite subset which is linearly dependent and if it is linearly independent otherwise. It is said to span the space V if every vector of V is linearly dependent on a finite subset and a basis of V is (as before) a linearly independent spanning set. Show that Prop. 2 still holds for infinite sets.

(8) Show that in $F^{(N)}$ the vectors e_i $(i = 1, 2, \ldots)$ with 1 in the ith place and 0 elsewhere form a basis.

4.4 Direct sums

Let V be a vector space and $U_1, \ldots, U_k$ any subspaces of V. We say that V is the *direct sum* of the spaces $U_1, \ldots, U_k$ and write

$$V = U_1 \oplus \cdots \oplus U_k,$$

if each $x \in V$ can be uniquely expressed in the form

$$x = x_1 + \cdots + x_k \qquad \text{where } x_i \in U_i. \tag{1}$$

If V is a direct sum of $U_1, \ldots, U_k$ we clearly have

$$V = U_1 + \cdots + U_k, \tag{2}$$

for in accordance with the definition in **4.1** this just means that x can be written in the form (1). Given (2), the sum is direct iff

$$\text{for any } x_i, y_i \in U_i, \quad \sum x_i = \sum y_i \Rightarrow x_i = y_i \quad \text{for } i = 1, 2, \ldots, k. \tag{3}$$

We observe that this holds iff

$$\text{for all } x_i \in U_i, \quad \sum x_i = 0 \Rightarrow x_i = 0 \quad \text{for } i = 1, 2, \ldots, k. \tag{4}$$

For on the one hand, (4) is the special case $y_i = 0$ of (3); on the other hand if (4) holds and $\sum x_i = \sum y_i$, then $\sum (x_i - y_i) = 0$ and $x_i - y_i \in U_i$, hence by (4), $x_i - y_i = 0$, i.e. $x_i = y_i$.

Assuming V to be finite-dimensional, let $V = U_1 \oplus \cdots \oplus U_k$ and take a basis B_i for U_i. Then each x_i in U_i is linearly dependent on B_i, hence on writing $x \in V$ as in (1) we see that x is linearly dependent on $B = B_1 \cup \ldots \cup B_k$, i.e. B spans V. Clearly no proper subset of B will span V, for if we omit a vector from B_1, say, then the rest of B_1 will fail to span U_1 and this cannot be made up by the other vectors in B, because of the uniqueness in (1). Thus B is a basis of V and, since the B_i are clearly disjoint, we obtain the following formula by comparing the number of elements in B and in each B_i:

$$\text{If } V = U_1 \oplus \cdots \oplus U_k \text{ then } \dim V = \dim U_1 + \cdots + \dim U_k. \tag{5}$$

For $k = 2$ there is a criterion for direct sums which is often useful.

PROPOSITION 1 *A vector space V is the direct sum of two subspaces U' and U'':*

$$V = U' \oplus U'' \tag{6}$$

if and only if (i) $U' + U'' = V$ *and* (ii) $U' \cap U'' = \mathbf{0}$.

For if (6) holds we clearly have (i), while any $x \in U' \cap U''$ gives rise to the following equation

$$0 = x + (-x).$$

Since $x \in U'$, $-x \in U''$, this is only possible in a direct sum (by (4)) if $x = 0$, i.e. (ii). Conversely, when (i) and (ii) hold, every element $x \in V$ can be written

$$x = x' + x'' \ (x' \in U',\ x'' \in U''), \tag{7}$$

and if we also have $x = y' + y''(y' \in U', y'' \in U'')$, then $x' - y' = y'' - x'' = z$ say. Here $x' - y' \in U'$, $y'' - x'' \in U''$, so $z \in U' \cap U''$ and by (ii) $z = 0$, i.e. $x' = y'$, $x'' = y''$. This shows that (7) is unique, whence (6) holds. ■

When V is expressed in the form (6), U'' is said to be a *complement* of U' in V. E.g., 3-dimensional space, regarded as a vector space, may be expressed as the direct sum of a plane and a line (say the (x, y)-plane and the z-axis).

It is an important fact that such a complement always exists, even though it will not in general be unique.

PROPOSITION 2 *Let V be any finite-dimensional vector space. Then any subspace of V has a complement.*

Proof. Let U be a subspace of V for which we want to find a complement. We know that V has finite dimension, n, say. Hence any $n+1$ vectors of U are linearly dependent and so we can find a basis for U, say $u_1, \ldots, u_r$. This is a linearly independent set of vectors in V, which can be completed to a basis of V, say $u_1, \ldots, u_n$. By definition, every element $x \in V$ can be expressed as $\sum \alpha_i u_i$, with uniquely determined scalar coefficients α_i, but this means that $V = U \oplus U'$, where U' is the subspace spanned by $u_{r+1}, \ldots, u_n$. Hence U' is a complement of U in V. ∎

Of course any subspace will in general have many complements, because there are many ways of completing the basis of U to a basis of V. E.g., if V is the plane and U a line through the origin (a 1-dimensional subspace), any line other than U will, with U itself, span V.

Going back to the proof of Prop. 2, we see that $\dim V = n$, $\dim U = r$ and these two numbers are equal iff $u_1, \ldots, u_r$ already form a basis for V. But this holds precisely when $U = V$, so we have the

COROLLARY *If V is any finite-dimensional space and U a subspace, then $\dim U \leqslant \dim V$ with equality if and only if $U = V$.* ∎

Whereas most of the results proved so far still hold for infinite-dimensional spaces, this last corollary does not; any infinite-dimensional space has a proper subspace of the same dimension, just as an infinite set has a proper subset which is still infinite, of the same cardinal number.

As an application of Prop. 2 we shall obtain a formula relating the dimension of the sum of two subspaces to the sum of their dimensions.

THEOREM 3 *Let V be a finite-dimensional vector space and U', U'' any subspaces. Then*

$$\dim (U' + U'') = \dim U' + \dim U'' - \dim (U' \cap U''). \tag{8}$$

Proof. Write $D = U' \cap U''$; this is a subspace of U' and of U'', so we may take a complement W' of D in U' and a complement W'' of D in U''. We assert

$$U' + U'' = W' \oplus W'' \oplus D. \tag{9}$$

Clearly any vector in $U' + U''$ is a sum of vectors in W', W'' and D. To prove directness in (9), assume that $y' + y'' + z = 0$, where $y' \in W'$, $y'' \in W''$ and $z \in D$. Then $y' = -(y'' + z)$; on the one hand $y' \in W' \subseteq U'$ and, on the other hand, $-(y'' + z) \in W'' + D = U''$, hence $y' \in U' \cap U''$, i.e. $y' \in D$. But $y' \in W'$ and, by definition of W', $W' \cap D = \mathbf{0}$, hence $y' = 0$. Similarly $y'' = 0$, and it follows that $z = 0$; this proves (9).

Now $\dim U' = \dim W' + \dim D$ and similarly for $\dim U''$, hence by (9)

$$\dim (U'+U'') = \dim W' + \dim W'' + \dim D$$
$$= \dim U' + \dim U'' - \dim D, \text{ i.e. (8)} \quad \blacksquare$$

To illustrate Th. 3, two planes which together span 3-space intersect in a line. It would be difficult to give a more general example without going into more than 3 dimensions and hence lose the advantage of spatial intuition. But there is another way of illustrating this theorem, which is both simpler and more general. We draw a diagram in which each subspace under discussion is represented by a point and the inclusion of one space in another is indicated by a sloping line from the larger space down to the smaller. E.g., the situation of Th. 3 is represented by Fig. 2. The equal sides of the parallelogram are intended to suggest that the differences $\dim (U'+U'') - \dim U'$ and $\dim U'' - \dim D$ are equal.

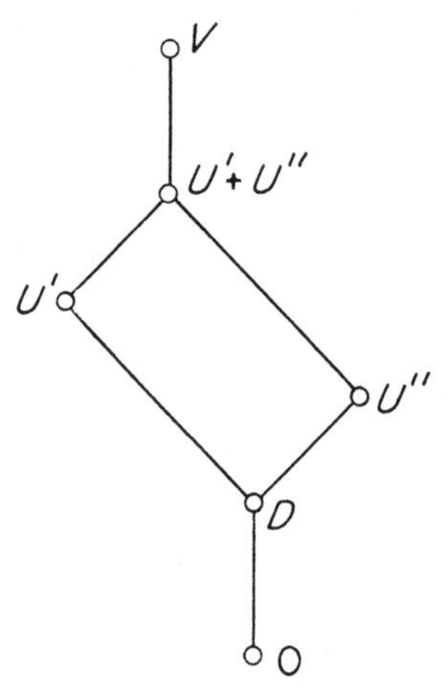

Figure 2

Exercises

(1) Find $U' \cap U''$ when U' is the subspace of $\mathbf{R}^4$ spanned by $(4, -3, 2, 0)$, $(7, 0, 5, 3)$ and U'' the subspace spanned by $(2, -5, 3, 1)$, $(5, -2, 6, 4)$, $(7, -7, 9, 5)$. Hence obtain a basis for $U'+U''$.

(2) Let V be a 10-dimensional space and U', U'' subspaces of 8 and 9 dimensions respectively. Show that there are only two possible values for the dimension of $U' \cap U''$.

(3) Let V be a finite-dimensional space and $V = U_1 + \cdots + U_k$. Show that the sum is direct iff $\dim V = \sum \dim U_i$.

(4) Given subspaces U_1, U_2, U_3 of a space V, show that $\dim (\sum U_i) = \sum \dim U_i - \sum \dim (U_i \cap U_j) + \dim (U_1 \cap U_2 \cap U_3)$.

(5) Show that a sum $\sum_1^k U_i$ of subspaces is direct iff $(U_1 + \cdots + U_{i-1}) \cap U_i = \mathbf{0}$, for $i = 2, \ldots, k$.

(6) Let V be a vector space and U_1, U_2 two subspaces. If neither of U_1, U_2 is contained in the other, find a vector in $U_1 + U_2$ which is not in either of U_1, U_2.

(7) Show that any two subspaces of the same dimension in a (finite-dimensional) space have a common complement. (Hint. Use Ex. (6).)

(8) Show that in Ex. (7) we cannot replace 2 by 3. (Hint. Take a 2-dimensional space over the field of 2 elements.)

4.5 The space of linear mappings

Let V be a vector space over a field F. If X is any non-empty set, then the set V^X of all mappings $f: X \to V$ may be defined as a vector space; if $f, g \in V^X$ and $\alpha \in F$, then $f+g$ and αf are defined as the mappings

$$f+g: x \mapsto xf + xg, \qquad \alpha f: x \mapsto \alpha(xf). \tag{1}$$

The verification that V^X forms a vector space under these operations is straightforward and may be left to the reader (cf. Ex. (7), **4.1**, for the case $V = F$).

Suppose now that instead of an abstract set X we have a vector space U. We shall be particularly interested in the mappings from U to V that are linear; they form a subset of V^U which will be denoted † by Hom (U, V), or $\text{Hom}_F(U, V)$ when we wish to stress the field F. We claim that Hom (U, V) is a subspace of V^U. To prove this, let $f, g \in \text{Hom}(U, V)$, $x, y \in U$ and $\lambda, \mu \in F$, then

$$\begin{aligned}(\lambda x + \mu y)(f+g) &= (\lambda x + \mu y)f + (\lambda x + \mu y)g = \lambda xf + \mu yf + \lambda xg + \mu yg \\ &= \lambda xf + \lambda xg + \mu yf + \mu yg = \lambda x(f+g) + \mu y(f+g).\end{aligned}$$

This shows $f+g$ to be linear; the proof for αf is similar.

Suppose that U and V are both finite-dimensional. It is natural to ask whether in this case Hom (U, V) is also finite-dimensional. We shall find that the answer is 'yes'; more precisely, if $\dim U = m$, $\dim V = n$, then $\dim \text{Hom}(U, V) = mn$. For the proof we shall need to find a basis for Hom (U, V).

Let $u_1, \ldots, u_m$ be a basis of U and $v_1, \ldots, v_n$ a basis of V. Given any linear mapping $f: U \to V$, we express the images of the us in terms of the vs:

$$u_i f = \sum a_{ir} v_r \qquad (a_{ir} \in F), \tag{2}$$

In these equations the summation is for $r = 1, \ldots, n$, while i ranges over $1, \ldots, m$. Thus (a_{ir}) is a collection of mn elements of F arranged in m rows of n elements, as we see by writing out a simple case in full, say $m = n = 2$:

$$\begin{aligned}u_1 f &= a_{11} v_1 + a_{12} v_2, \\ u_2 f &= a_{21} v_1 + a_{22} v_2.\end{aligned} \tag{3}$$

† Recall that linear mappings are homomorphisms preserving the vector space structure.

An array of mn elements of F arranged in m rows and n columns in this way is called an $m \times n$ *matrix* over F. We usually denote it by a single letter or also by the general term in parentheses; thus we write $A = (a_{ir})$. We may regard an $m \times n$ matrix as an mn-tuple of F, arranged in a certain way, namely as the array

$$A = \begin{pmatrix} a_{11} & a_{12} & \dots & a_{1n} \\ a_{21} & a_{22} & \dots & a_{2n} \\ \cdot & \cdot & \cdot & \cdot \\ a_{m1} & a_{m2} & \dots & a_{mn} \end{pmatrix}.$$

The set of all $m \times n$ matrices over F will be denoted by ${}^mF^n$ (just as the set of mn-tuples is written F^{mn}). Clearly ${}^mF^n$ is a vector space; in fact it is nothing other than F^{mn}, but with its standard basis arranged in a certain way.

The equations (2) provide a mapping from Hom (U, V) to ${}^mF^n$; it will be our aim to show that this is an isomorphism of vector spaces. In the first place we note that in (2) f is completely determined by the matrix A and that, conversely, every $A \in {}^mF^n$ defines a unique linear mapping f by the equations (2).

To prove this assertion we take $x \in U$ and write it in the form $x = \sum \xi_i u_i$. Applying f we have, by linearity,

$$xf = (\sum \xi_i u_i)f = \sum \xi_i(u_i f), \qquad \text{i.e. by (2),}$$
$$xf = \sum_{ir} \xi_i a_{ir} v_r, \tag{4}$$

where the sum is over i and r. Conversely, if $A = (a_{ir}) \in {}^mF^n$ and we define f by (4), then for any vectors $x = \sum \xi_i u_i$ and $y = \sum \eta_i u_i$ in U we have

$$(\lambda x + \mu y)f = \sum (\lambda \xi_i + \mu \eta_i) a_{ir} v_r = \lambda \sum \xi_i a_{ir} v_r + \mu \sum \eta_i a_{ir} v_r = \lambda(xf) + \mu(yf),$$

and this shows f to be linear.

Thus (2) is a bijection between Hom (U, V) and ${}^mF^n$; to show that it is an isomorphism, assume that f, g correspond to $A = (a_{ir})$ and $B = (b_{ir})$ respectively, then

$$u_i(\lambda f + \mu g) = \lambda u_i f + \mu u_i g = \sum \lambda a_{ir} v_r + \sum \mu b_{ir} v_r = \sum (\lambda a_{ir} + \mu b_{ir}) v_r,$$

hence $\lambda f + \mu g$ corresponds to $\lambda A + \mu B$ and the linearity is established. We sum up the result as

THEOREM 1 *Let U, V be vector spaces over a field F, of dimensions m, n respectively. Then*

$$\mathrm{Hom}_F (U, V) \cong {}^mF^n, \tag{5}$$

and hence $\dim \mathrm{Hom}_F (U, V) = mn$. ■

It is important to note that the isomorphism (5) depends on the choice of bases in U and V, i.e. the isomorphisms $U \cong F^m$, $V \cong F^n$. But, once these are fixed, the isomorphism (5) is uniquely determined.

Just as the vector space F^n has the standard basis $\{e_1, \ldots, e_n\}$, so the space ${}^mF^n$ has a standard basis $\{E_{ir}\}$; here E_{ir} is the matrix whose (i, r)-entry (i.e. the entry in the ith row and rth column) is 1, while all the other entries are 0. In terms of this basis the general matrix $A = (a_{ir})$ is expressed as the double sum

$$A = \sum a_{ir}E_{ir}.$$

The corresponding basis $\{e_{ir}\}$ of Hom (U, V) is given by

$$u_ie_{jr} = \begin{cases} v_r \text{ if } j = i \\ 0 \text{ if } j \neq i \end{cases} \qquad (i, j = 1, \ldots, m;\ r = 1, \ldots, n) \tag{6}$$

This formula (6) for e_{ir} may be written more briefly using the following function, which is called the *Kronecker delta*:

$$\delta_{ij} = \begin{cases} 1 & \text{if } i = j, \\ 0 & \text{if } i \neq j. \end{cases} \tag{7}$$

Using (7) we can rewrite (6) in the form

$$u_ie_{jr} = \delta_{ij}v_r.$$

We see that (relative to given bases) matrices can be used to describe linear mappings, just as rows (or columns) are used to describe vectors. Thus any computation of linear mappings can always be reduced to a computation with matrices. This is indeed the chief use of matrices: they provide us with a manageable way of discussing linear mappings. On the other hand, any explicit calculation with matrices is usually space- and time-consuming, and we therefore carry any discussion as far as possible before going into matrices, just as any argument involving vectors should always be taken as far as possible before going into components.

So far we have regarded Hom (U, V) merely as a vector space, but there is another operation on linear mappings, the composition. Given mappings $f\colon U \to V$, $g\colon V \to W$, we obtain by composition a mapping $fg\colon U \to W$ and it is easily seen that when f, g are both linear then so is fg. Moreover, the product itself is a linear function of each of its arguments, or more briefly, it is *bilinear*:

$$\begin{aligned} (\lambda f+\lambda' f')g &= \lambda fg+\lambda' f'g, \qquad & f, f'\colon U \to V, \quad g, g'\colon V \to W, \\ f(\mu g+\mu' g') &= \mu fg+\mu' fg'. & \lambda, \lambda', \mu, \mu' \in F. \end{aligned} \tag{8}$$

Moreover, given $f\colon U \to V$, $g\colon V \to W$, $h\colon W \to X$, then $(fg)h$ and $f(gh)$ are both defined, as mappings from U to X, and are equal:

$$(fg)h = f(gh). \tag{9}$$

For this is just an instance of the general associative law of mappings.

If we now express these linear mappings in terms of matrices, the product of mappings will give rise to a product of matrices, and this product will be associative, by (9) and bilinear, by (8). Let U, V, W be vector spaces with bases $u_1, \ldots, u_m$; $v_1, \ldots, v_n$; $w_1, \ldots, w_p$ respectively, and suppose that $f \in \mathrm{Hom}(U, V)$ and $g \in \mathrm{Hom}(V, W)$ are given by the equations

$$u_i f = \textstyle\sum_i a_{ir} v_r, \qquad v_r g = \textstyle\sum_k b_{rk} w_k,$$

then $fg \colon U \to W$ is the linear mapping given by the equations

$$u_i(fg) = (\textstyle\sum_r a_{ir} v_r) g = \textstyle\sum_{rk} a_{ir} b_{rk} w_k,$$

where the last sum is a double sum, over r and k. Note that the last equality depends on the linearity of g. We see that fg has the matrix $C = (c_{ik})$, where

$$c_{ik} = \textstyle\sum_r a_{ir} b_{rk}. \tag{10}$$

This equation is usually written more briefly as $C = AB$. It represents the rule for multiplying matrices (so as to correspond to the multiplication of linear mappings): We see from (10) that the product AB is defined precisely when the number of columns of A equals the number of rows of B, and the (i, k)-entry of AB is obtained by multiplying the ith row of A into the kth column of B (i.e. corresponding components of this row and column are multiplied and the results added, as indicated in (10)). The product AB has as many rows as A and as many columns as B. This relation between the rows and columns is easily remembered when it is written in the form

$$\underset{m \times n}{A} \; \underset{n \times p}{B} = \underset{m \times p}{C}.$$

By expressing (8) and (9) in terms of coordinates we obtain the following laws for matrices: Let $A, A' \in {}^mF^n$, $B, B' \in {}^nF^p$, $C \in {}^pF^q$ and $\lambda, \lambda', \mu, \mu' \in F$, then

$$(\lambda A + \lambda' A')B = \lambda AB + \lambda' A'B,$$
$$A(\mu B + \mu' B') = \mu AB + \mu' AB',$$
$$(AB)C = A(BC).$$

Consider now the special case when $U = V$; we are then dealing with linear mappings of V into itself. They are also called *endomorphisms* of V and instead of $\mathrm{Hom}(V, V)$ we write $\mathrm{End}(V)$ or sometimes $\mathrm{End}_F(V)$. The endomorphisms of V can be added and multiplied, with the identity mapping on V as the neutral element for multiplication; thus, altogether, we have the following rules:

A. 1 End (V) *is a vector space over* F.

A. 2 End (V) *is a monoid under multiplication.*

A. 3 *Multiplication is bilinear*, i.e. (8) *holds.*

A. 1–3 are expressed by saying that End (V) is a *linear algebra over* F, or briefly, an *F-algebra*; this is a particular case of a ring, a topic to which we shall return in Ch. **6**.

Let us see what becomes of the isomorphism of Th. 1 in this case. To each endomorphism f of V there corresponds an $n \times n$ matrix $A = (a_{ij})$, defined by the equations

$$v_i f = \textstyle\sum a_{ij} v_j. \tag{11}$$

Of course we must use the same basis of V on both sides of this equation. By means of this correspondence the set of $n \times n$ matrices becomes an F-algebra; instead of "F^n" we shall denote this set by $\mathfrak{M}_n(F)$ or simply F_n (to be distinguished from F^n). Sometimes n is called the *order* of the matrices in F_n. From (11) we see that the identity mapping is represented by the *unit matrix*

$$I = \begin{pmatrix} 1 & 0 & 0 & \dots & 0 & 0 \\ 0 & 1 & 0 & \dots & 0 & 0 \\ . & . & . & . & . & . \\ 0 & 0 & 0 & \dots & 0 & 1 \end{pmatrix}.$$

This is the matrix with 1s down the main diagonal and 0s elsewhere; using the Kronecker delta it may be written more briefly as $I = (\delta_{ij})$. Since it corresponds to the identity mapping, it is neutral under multiplication, a fact which is easily checked directly. The rules of matrix calculation were recognized and made explicit by Cayley in the 1860s, though they had been used earlier in working with determinants, which go back to the 17th century (Leibniz).

If we compare the rules of F_n with the description of fields in Ch. **2**, we see that F_n fails to be a field in two respects: (i) the multiplication is not commutative and (ii) there are elements besides 0 which fail to have an inverse. Both (i) and (ii) may be illustrated by quite simple examples. Let $A = \begin{pmatrix} 1 & 0 \\ 0 & 0 \end{pmatrix}$, $B = \begin{pmatrix} 0 & 1 \\ 0 & 0 \end{pmatrix}$, then $AB = \begin{pmatrix} 0 & 1 \\ 0 & 0 \end{pmatrix}$ while $BA = \begin{pmatrix} 0 & 0 \\ 0 & 0 \end{pmatrix} = 0$; thus $AB \neq BA$. Moreover neither A nor B can have an inverse, for if A had an inverse A^{-1} say, then $B = B(AA^{-1}) = (BA)A^{-1} = 0$, a contradiction. Of course there are nevertheless many matrices that do have an inverse, e.g. I and, more generally, any matrix with non-zero entries along the main diagonal and zeros elsewhere: $(\alpha_i \delta_{ij})$, where $\alpha_i \neq 0$. We shall meet many other examples of invertible matrices later.

In using matrices we shall frequently want to interpret the rows of the matrix as vectors, e.g. in (2) the rows of $A = (a_{ir})$ represent the vectors $u_i f$ in coordinate form. At other times we may wish to interpret the columns as vectors and anything proved about the rows of a general matrix will, for reasons of symmetry, also apply to the columns. To make this explicit, we introduce the operation of *transposition* on matrices. If A is an $m \times n$ matrix, its *transpose* A^{T}, is the $n \times m$ matrix obtained by writing the columns of A as rows. Thus the (i, r)-entries of A and A^{T} are related by the equation

$$(A^{\mathrm{T}})_{ir} = a_{ri}.$$

To give a concrete example, if $A = \begin{pmatrix} a_{11} & a_{12} & a_{13} \\ a_{21} & a_{22} & a_{23} \end{pmatrix}$, then

$$A^{\mathrm{T}} = \begin{pmatrix} a_{11} & a_{21} \\ a_{12} & a_{22} \\ a_{13} & a_{23} \end{pmatrix}.$$

Transposition is related to addition and multiplication of matrices by the following rules, again valid whenever all the sums and products are defined:

$$(A+B)^{\mathrm{T}} = A^{\mathrm{T}}+B^{\mathrm{T}}, \tag{12}$$

$$(AB)^{\mathrm{T}} = B^{\mathrm{T}}A^{\mathrm{T}}, \qquad I^{\mathrm{T}} = I, \tag{13}$$

$$A^{\mathrm{TT}} = A. \tag{14}$$

These rules are easily checked, by comparing corresponding entries on both sides; we shall do this for the first equation in (13) and leave the others to the reader. The left-hand side gives as (i, k)-entry

$$[(AB)^{\mathrm{T}}]_{ik} = (AB)_{ki} = \sum a_{kr}b_{ri},$$

while the right-hand side gives

$$[B^{\mathrm{T}}A^{\mathrm{T}}]_{ik} = \sum (B^{\mathrm{T}})_{ir}(A^{\mathrm{T}})_{rk} = \sum b_{ri}a_{kr} = \sum a_{kr}b_{ri}$$

and this shows that the two sides agree.

Finally suppose that A is invertible; this means that A is a square matrix with inverse A^{-1} say, satisfying

$$AA^{-1} = A^{-1}A = I.$$

If we transpose the matrices in these equations and apply (13), we find that $(A^{-1})^{\mathrm{T}}A^{\mathrm{T}} = A^{\mathrm{T}}(A^{-1})^{\mathrm{T}} = I^{\mathrm{T}} = I$. This shows that A^{T} is invertible, with inverse $(A^{-1})^{\mathrm{T}}$. Thus if A is invertible, then so is A^{T}, with inverse

$$(A^{\mathrm{T}})^{-1} = (A^{-1})^{\mathrm{T}}. \tag{15}$$

Exercises

(1) Give a geometrical interpretation of the linear mapping

$$R_\theta = \begin{pmatrix} \cos\theta & -\sin\theta \\ \sin\theta & \cos\theta \end{pmatrix}.$$

Verify that R_θ is invertible and find its inverse.

(2) Interpret the linear mappings

$$\alpha I = \begin{pmatrix} \alpha & 0 \\ 0 & \alpha \end{pmatrix}, \qquad D(\alpha, \beta) = \begin{pmatrix} \alpha & 0 \\ 0 & \beta \end{pmatrix}, \qquad S_c = \begin{pmatrix} 1 & c \\ 0 & 1 \end{pmatrix}.$$

(3) Prove that every linear mapping of $\mathbf{R}^2$ can be written in the form $S_c \,.\, D(\alpha, \beta)R_\theta$.

(4) Interpret the linear mapping $A = \begin{pmatrix} a & -b \\ b & a \end{pmatrix}$ $(a, b \in \mathbf{R})$, by expressing it as rR_θ, for suitable r and θ. Show that the matrices of this form are isomorphic to the field of complex numbers under the mapping $A \leftrightarrow a+bi$.

(5) Put $I = \begin{pmatrix} 1 & 0 \\ 0 & 1 \end{pmatrix}$, $J = \begin{pmatrix} i & 0 \\ 0 & -i \end{pmatrix}$, $K = \begin{pmatrix} 0 & 1 \\ -1 & 0 \end{pmatrix}$, $L = \begin{pmatrix} 0 & i \\ i & 0 \end{pmatrix}$; where $i = \sqrt{-1}$. With this notation show that for any real a, b, c, d, $(aI+bJ+cK+dL)(aI-bJ-cK-$

$dL) = (a^2+b^2+c^2+d^2)I$. Deduce that $aI+bJ+cK+dL = 0$ iff $a = b = c = d = 0$ and hence that every nonzero matrix of this form has an inverse (this is Hamilton's algebra of quaternions, a linear algebra admitting division, but non-commutative).

(6) Using the definition of matrix addition and multiplication, give a direct proof of the laws satisfied by these operations.

(7) Find the products AB and BA in each of the following cases:

(i) $A = \begin{pmatrix} 0 & 3 \\ 4 & 5 \end{pmatrix}$, $B = \begin{pmatrix} 2 & -1 \\ 3 & 2 \end{pmatrix}$, (ii) $A = \begin{pmatrix} 1 & -2 & 5 \\ 3 & 0 & 4 \end{pmatrix}$, $B = \begin{pmatrix} 2 & 1 \\ 3 & 6 \\ 1 & 5 \end{pmatrix}$,

(iii) $A = \begin{pmatrix} 2 & -1 \\ 3 & 2 \end{pmatrix}$, $B = \begin{pmatrix} 1 & -4 \\ 12 & 1 \end{pmatrix}$.

(8) Evaluate A^2, where $A = \begin{pmatrix} 2 & -5 \\ 3 & 1 \end{pmatrix}$, and find scalars α, β such that $\alpha I+\beta A+A^2 = 0$. Use your result to find A^{-1}. Treat the matrix $A = \begin{pmatrix} a & b \\ c & d \end{pmatrix}$ in the same way. What is the condition for A to be invertible?

(9) Show that any matrix that commutes with every diagonal matrix must be diagonal. What form does a matrix take which commutes with every matrix?

(10) If A is any square matrix such that $A^3 = 0$, show that $I-A$ has an inverse of the form $I+\alpha A+\beta A^2$. Find α, β.

(11) A *permutation matrix* is a square matrix with a 1 in each row and column and zeros elsewhere. Verify that each permutation matrix arises by permuting the rows (or also by permuting the columns) of the unit matrix. If P is a permutation matrix, show that $P^{-1} = P^{\mathrm{T}}$.

(12) Is there a linear mapping from $\mathbf{R}^3$ to $\mathbf{R}^2$ with $(5, 0, 3) \mapsto (1, 0)$, $(3, -2, 1) \mapsto (0, 1)$, $(1, 1, 2) \mapsto (1, -1)$. Is there such a mapping when $\mathbf{R}$ is replaced by $\mathbf{F}_p$?

(13) Find a linear mapping from $\mathbf{R}^3$ to $\mathbf{R}^2$ such that $(1, 2, -1) \mapsto (1, 0)$, $(2, 1, 4) \mapsto (0, 1)$. Show that there is more than one such mapping, but for all of them the image of $(1, 1, 1)$ is the same. What happens when $\mathbf{R}$ is replaced by $\mathbf{F}_3$?

(14) If $f: U \to V$ is a surjective linear mapping between finite-dimensional spaces, show that $\dim U \geqslant \dim V$. Find g such that $gf = 1$, and show that $fg = 1$ iff $\dim U = \dim V$.

(15) If F is a finite field with q elements, show that $\mathrm{End}\,(F^n)$ has q^{n^2} elements. Find the number of automorphisms of F^2 when $q = 2$.

4.6 Change of basis

We now consider what happens when we change the basis of a vector space, and particularly how this affects the matrix of a linear mapping. Let U be a vector space with a basis $u_1, \ldots, u_m$ and take a new basis $u'_1, \ldots, u'_m$ in U. These bases are related by equations of the form

$$u'_i = \sum p_{ij} u_j \qquad p_{ij} \in F, \tag{1}$$

$$u_j = \sum \check{p}_{jk} u'_k \qquad \check{p}_{jk} \in F. \tag{2}$$

If we substitute from (2) in (1) we find $u'_i = \sum p_{ij}\check{p}_{jk}u'_k$ and, since the u'_i are linearly independent, we may equate coefficients in this equation:

$$\sum p_{ij}\check{p}_{jk} = \delta_{ik}, \tag{3}$$

where the δ_{ik} is the Kronecker delta introduced earlier. Similarly if we substitute from (1) in (2) and equate coefficients, we obtain in the same way as before (after renaming the suffixes),

$$\sum \check{p}_{jk} p_{ki} = \delta_{ji}. \tag{4}$$

If we put $P = (p_{ij})$, $\check{P} = (\check{p}_{ij})$, then the last two sets of equations may be written in matrix form as

$$P\check{P} = I, \qquad \check{P}P = I. \tag{5}$$

Hence $\check{P} = P^{-1}$ and we see that the passage from one basis to another is accomplished by applying an invertible matrix.

We observe that (1), (2) can themselves be written in matrix form if we introduce the column matrix of vectors $\mathbf{u} = (u_1, \ldots, u_m)^{\mathrm{T}}$ and $\mathbf{u}' = (u'_1, \ldots, u'_m)^{\mathrm{T}}$. We shall use boldface type for rows or columns of vectors. Any linear combination of $u_1, \ldots, u_m$, say $\sum \alpha_i u_i$ can then be written in matrix form as $\alpha\mathbf{u}$, where $\alpha = (\alpha_1, \ldots, \alpha_m)$. Thus the fact that $\mathbf{u}$ consists of linearly independent vectors is expressed by the condition

$$\alpha\mathbf{u} = 0 \Rightarrow \alpha = 0 \quad \text{for any } \alpha \in F^m. \tag{6}$$

We note that (6) holds even when α is replaced by a matrix A (with m columns); for we can take each row of A separately and apply (6).

Now (1) and (2) can be rewritten as

$$\mathbf{u}' = P\mathbf{u}, \qquad \mathbf{u} = \check{P}\mathbf{u}'.$$

By substituting we find $\mathbf{u} = \check{P}P\mathbf{u}$, i.e. $(I-\check{P}P)\mathbf{u} = 0$, hence $\check{P}P = I$; similarly $P\check{P} = I$, i.e. (5).

Suppose now that P is any invertible $m \times m$ matrix and let $\mathbf{u} = (u_1, \ldots, u_m)^{\mathrm{T}}$ be any basis of U. Then $\mathbf{u}' = P\mathbf{u}$ is again a basis; for the equation $\mathbf{u} = P^{-1}P\mathbf{u} = P^{-1}\mathbf{u}'$ shows that the subspace spanned by the u'_i contains all the u_i and hence coincides with U. Thus $\mathbf{u}'$ is a spanning set of m elements of U, i.e. a basis. Taken together, these results show that in terms of any one basis $\mathbf{u}$ of U we can express all the others in the form $P\mathbf{u}$, where P is invertible of order m. If we look more closely we find that it was not necessary to assume P invertible but merely that P' exists satisfying $P'P = I$. This shows that for $m \times m$ matrices over a field, one-sided inverses are in fact two-sided; this will be proved in detail in **4.7**.

Let $x \in U$, then in terms of a basis $\mathbf{u} = (u_1, \ldots, u_m)^{\mathrm{T}}$ we have $x = \sum \xi_i u_i$ or in matrix notation,

$$x = \xi\mathbf{u}, \qquad \text{where } \xi = (\xi_1, \ldots, \xi_m).$$

In terms of a new basis $\mathbf{u}'$ we have $x = \sum \xi_i' u_i'$ or, again in terms of matrices,

$$x = \xi'\mathbf{u}', \qquad \text{where } \xi' = (\xi_1', \ldots, \xi_m').$$

Now $\mathbf{u}' = P\mathbf{u}$, hence $x = \xi\mathbf{u} = \xi'\mathbf{u}' = \xi' P\mathbf{u}$; by linear independence we find

$$\xi = \xi' P, \qquad \xi' = \xi P^{-1}.$$

In words: if the new basis is related to the old by a matrix P, then the *old* coordinates are related to the *new* ones by the same matrix P; the new coordinates will be related to the old ones by the inverse matrix P^{-1}. This reversal is expressed by saying: the coordinates transform *contragrediently* to the basis. It is nothing other than an expression of the well-known fact that if e.g. we *halve* the length of the unit of measurement, all our measurements will be *twice* as large.

We now consider how the matrix of a linear mapping changes when the bases in both spaces are transformed. Let U have bases $\mathbf{u}, \mathbf{u}'$ and let V have bases $\mathbf{v}, \mathbf{v}'$, where

$$\mathbf{u}' = P\mathbf{u}, \qquad \mathbf{v}' = Q\mathbf{v}.$$

A linear mapping $f: U \to V$ has the matrix A say, relative to $\mathbf{u}, \mathbf{v}$ and the matrix A' relative to $\mathbf{u}', \mathbf{v}'$. Thus

$$\mathbf{u}'f = A'\mathbf{v}' \tag{7}$$

and $\mathbf{u}f = A\mathbf{v}$, hence $\mathbf{u}'f = (P\mathbf{u})f = P(\mathbf{u}f) = PA\mathbf{v} = PAQ^{-1}\mathbf{v}'$. A comparison with (7) shows that

$$A' = PAQ^{-1}. \tag{8}$$

This formula describes how the matrix of a linear mapping changes under a change of basis. For example, let $f: (\xi_1, \xi_2, \xi_3) \mapsto (\xi_1, \xi_2)$ be a mapping from F^3 to F^2 and introduce new bases $(1, -1, 0)$, $(1, 0, -1)$, $(1, 1, 1)$ and $(4, 3)$, $(3, 2)$ respectively (in terms of the standard basis). Then

$$A = \begin{pmatrix} 1 & 0 \\ 0 & 1 \\ 0 & 0 \end{pmatrix} \quad P = \begin{pmatrix} 1 & -1 & 0 \\ 1 & 0 & -1 \\ 1 & 1 & 1 \end{pmatrix} \quad Q = \begin{pmatrix} 4 & 3 \\ 3 & 2 \end{pmatrix} \quad Q^{-1} = \begin{pmatrix} -2 & 3 \\ 3 & -4 \end{pmatrix}$$

hence

$$A' = \begin{pmatrix} 1 & -1 & 0 \\ 1 & 0 & -1 \\ 1 & 1 & 1 \end{pmatrix}\begin{pmatrix} 1 & 0 \\ 0 & 1 \\ 0 & 0 \end{pmatrix}\begin{pmatrix} -2 & 3 \\ 3 & -4 \end{pmatrix} = \begin{pmatrix} -5 & 7 \\ -2 & 3 \\ 1 & -1 \end{pmatrix}.$$

The fact that the matrix of a linear mapping depends on the choice of basis can be exploited to reduce the matrix to a particularly simple form. We shall meet examples of this process in **4.7** and in Ch. **11**.

Exercises

(1) If the matrix of an endomorphism of U is A, show that after change of basis $\mathbf{u}' = P\mathbf{u}$ it is PAP^{-1}. If U is 3-dimensional and f is the linear mapping $u_1 \mapsto 2u_1 + u_2 + 3u_3$, $u_2 \mapsto u_1 - 2u_3$, $u_3 \mapsto 4u_1 + 3u_2 + u_3$, write down its matrix relative to the basis u_1, u_2, u_3 and the basis $\alpha_1 u_1, \alpha_2 u_2, \alpha_3 u_3$, where $\alpha_i \neq 0$.

(2) Show that if the elements of a basis are written in a different order, the corresponding change of coordinates is accomplished by a permutation matrix.

(3) Using the formulae $e^{ix} = \cos x + i \sin x$, $e^{-ix} = \cos x - i \sin x$, describe the matrix of transformation from the basis e^{ix}, e^{ix} to the basis $\cos x, \sin x$. Find also the inverse of the matrix.

(4) Show that an $n \times n$ matrix over F is invertible iff its columns form a basis of F^n.

(5) Describe the matrix of transformation from $v_1, \ldots, v_n$ to $v_1 - v_2, v_1 - v_3, \ldots, v_1 - v_n, \sum v_i$ over $\mathbf{R}$. If $\mathbf{R}$ is replaced by $\mathbf{F}_p$, what restriction on p is needed for this to be a basis?

4.7 The rank

With any linear mapping $f: U \to V$ we can associate two subspaces of U and V respectively, the *kernel* and *image* of f:

$$\ker f = \{x \in U \mid xf = 0\},$$
$$\operatorname{im} f = \{y \in V \mid y = xf \text{ for some } x \in U\}.$$

The linearity of f ensures that both $\ker f$ and $\operatorname{im} f$ are subspaces of U and V respectively: if $x, x' \in \ker f$ and $\lambda, \lambda' \in F$, then $(\lambda x + \lambda' x')f = \lambda(xf) + \lambda'(x'f) = 0$, hence $\ker f$ is a subspace of U. The proof that $\operatorname{im} f$ is a subspace of V is entirely similar. When U and V are finite-dimensional, we define two numbers associated with f, the *rank* $\rho(f)$ and the *nullity* $\nu(f)$, by the equations

$$\rho(f) = \dim \operatorname{im} f, \qquad \nu(f) = \dim \ker f.$$

Of these two numbers the rank is usually regarded as the basic invariant; the nullity is of secondary importance because it is related to the rank by a simple formula. This is given in the next theorem, which also provides a simple normal form for the matrix of f:

THEOREM 1 *If $f: U \to V$ is any linear mapping between finite-dimensional spaces, then*

$$\rho(f) + \nu(f) = \dim U, \tag{1}$$

and for a suitable choice of bases in U and V, f has the matrix

$$E_r = \begin{pmatrix} I_r & 0 \\ 0 & 0 \end{pmatrix} \qquad (r = \rho(f)), \tag{2}$$

where I_r is the unit matrix of order r and the 0s in (2) stand for zero matrices of the appropriate sizes, i.e. $r \times n - r$, $n - r \times r$ and $n - r \times n - r$ respectively.

Proof. Let us take a basis of $\ker f$ and complete it to a basis of U. The result is a basis $u_1, \ldots, u_m$ of U, where $u_{r+1}, \ldots, u_m$ is a basis of $\ker f$. We claim that the set $\{u_1 f, \ldots, u_r f\}$ is linearly independent. For if

$$\textstyle\sum_1^r \alpha_i(u_i f) = 0,$$

then $\sum_1^r \alpha_i u_i \in \ker f$, say $\sum_1^r \alpha_i u_i = \sum_{r+1}^m \beta_j u_j$ and by the linear independence of $u_1, \ldots, u_m$ we see that $\alpha_1 = \cdots = \alpha_r = \beta_{r+1} = \cdots = \beta_m = 0$. This shows that $u_1 f, \ldots, u_r f$ are linearly independent, as claimed.

Now $\operatorname{im} f$ is spanned by $u_1 f, \ldots, u_m f$ and the first r of these vectors are linearly independent, while the remaining $m-r$ are 0. Therefore $u_1 f, \ldots, u_r f$ form a basis of $\operatorname{im} f$; in particular this shows that $\rho(f) = r$. If we recall that $u_{r+1}, \ldots, u_m$ was a basis of $\ker f$, we find $\nu(f) = m-r$, hence $\rho(f)+\nu(f) = m = \dim U$, i.e. (1).

To get a matrix representation for f let us choose a basis $v_1, \ldots, v_n$ for V by taking $v_i = u_i f\,(i = 1, \ldots, r)$ and completing these r vectors in any way to a basis of V. Referred to these bases, f is given by

$$u_i f = \begin{cases} v_i & \text{if } i = 1, \ldots, r, \\ 0 & \text{if } i > r. \end{cases}$$

Thus the matrix for f is $A = (a_{ij})$, where

$$a_{ij} = \begin{cases} 1 & \text{if } i = j \leqslant r, \\ 0 & \text{otherwise.} \end{cases}$$

Hence $A = E_r$ as claimed. ■

How can we find the rank of a linear mapping f? Let the matrix of f, referred to any bases in U and V, be A; then the rows of A represent the images of the basis of U, referred to the basis of V. Thus if A has r linearly independent rows on which every other row of A is linearly dependent, then r is the rank of f. This number r, representing the number of rows in a maximal linearly independent set of rows of A, is called the *row rank* of A and is written $\rho(A)$. For example, the row rank of

$$\begin{pmatrix} 1 & -3 & 6 & 0 \\ 0 & 0 & 0 & 0 \\ 2 & -6 & 12 & 0 \end{pmatrix}$$

is 1; the row rank of the matrix E_r in (2) is r. In precisely the same way one can define the *column rank* of A as the number of columns in a maximal linearly independent set of columns of A, but, as we shall see below, these two ranks are always equal for matrices over a field. Practical methods of computing the rank of a matrix will be given in the next chapter.

Before proving the equality of row and column rank we shall deduce a reduction theorem for matrices from Th. 1, which is interesting in its own right and will be of use later. Let A be any $m \times n$ matrix; this defines a linear mapping $f: F^m \to F^n$ when referred to standard bases, say. If we now apply

Th. 1 to find bases for which the matrix of f has the form (2) and observe the formula (8), **4.6**, for the change of basis, we obtain the

COROLLARY *Given any $m \times n$ matrix A over a field F, there exist invertible matrices P, Q of orders m, n respectively, such that*

$$PAQ^{-1} = E_r, \tag{3}$$

where $r = \rho(A)$ and E_r is as in (2).

The fact that $r = \rho(A)$ follows because A and PAQ^{-1} both have the same row rank, namely $\rho(f)$, and $\rho(E_r) = r$. ■

To show that row rank equals column rank, we note that the column rank of A is the row rank of the transposed matrix A^T, so we need only prove

$$\rho(A^T) = \rho(A). \tag{4}$$

Let P, Q be as in the Cor. then $PAQ^{-1} = E_r$; transposing this equation, we find

$$(Q^{-1})^T A^T P^T = E_r^T.$$

Now E_r^T like E_r has rank r and $(Q^{-1})^T$, P^T are invertible, hence $\rho(A) = \rho(E_r) = \rho(E_r^T) = \rho(A^T)$, i.e. (4). From now on we shall simply speak of the *rank* of A.

Only for square matrices do we speak of nullity: if A is a square matrix of order n, its *nullity* $\nu(A)$ is defined as $n - \rho(A)$. By Th. 1 this agrees with the nullity of the linear mapping defined by A, as well as that of the linear mapping defined by A^T.

The case of zero nullity is particularly important. Some conditions for this to happen are given in the next result. For brevity we shall instead of ${}^1F^n$ and ${}^nF^1$ write F^n and ${}^nF$ respectively; thus F^n is the space of all rows and ${}^nF$ the space of all columns with n components.

PROPOSITION 2 *For any square matrix A of order n over a field the following four conditions are equivalent*:

(a) *The rows of A are linearly independent*: $uA = 0 \Rightarrow u = 0$ *for all* $u \in F^n$.

(a*) *The columns of A are linearly independent*: $Ax = 0 \Rightarrow x = 0$, *for all* $x \in {}^nF$.

(b) *A has a right inverse*: $AB = I$ *for some* $B \in F_n$.

(b*) *A has a left inverse*: $BA = I$ *for some* $B \in F_n$.

In Ch. **7** we shall see that a ⇔ a* holds more generally for matrices over any commutative ring. When (a) and (a*) hold, A is said to be *non-singular*, otherwise *singular*. Thus Prop. 2 shows that a square matrix over a field is invertible precisely when it is non-singular.

Proof. (a) ⇔ (a*) because A and A^T have the same nullity. If (b) holds, we can deduce (a): let $\mathbf{u}A = 0$, then $\mathbf{u} = \mathbf{u}AB = 0$. Thus (b) ⇒ (a) and likewise (b*) ⇒ (a*). Now assume (a), i.e. A has nullity 0, then $E_r = PAQ^{-1}$ has nullity 0, and so $r = n$, i.e. $PAQ^{-1} = I$, whence $A = P^{-1}Q$ has a two-sided inverse. ∎

Our final result in this section gives an estimate for the rank and nullity of a sum or product of linear mappings.

PROPOSITION 3 (i) *Let f_1, f_2 be linear mappings from U to V, where* $\dim U = m$, *then*

$$|\rho(f_1)-\rho(f_2)| \leqslant \rho(f_1+f_2) \leqslant \rho(f_1)+\rho(f_2), \tag{5}$$

$$\nu(f_1)+\nu(f_2)-m \leqslant \nu(f_1+f_2) \leqslant m-|\nu(f_1)-\nu(f_2)|. \tag{6}$$

(ii) *Let $f: U \to V$, $g: V \to W$*, $\dim V = n$, *then*

$$\rho(f)+\rho(g)-n \leqslant \rho(fg) \leqslant \min\{\rho(f), \rho(g)\}, \tag{7}$$

$$\nu(fg) \leqslant \nu(f)+\nu(g). \tag{8}$$

The inequality (8) is sometimes called *Sylvester's law of nullity.*

Proof. (i) For any $x \in U$, $x(f_1+f_2) = xf_1+xf_2$, hence $\operatorname{im}(f_1+f_2) \subseteq \operatorname{im} f_1 + \operatorname{im} f_2$ and therefore

$$\rho(f_1+f_2) \leqslant \rho(f_1)+\rho(f_2). \tag{9}$$

This proves the second half of (5); of course the inequality is generally strict, e.g. when $f_1 = -f_2 \neq 0$. To prove the other part of (5) we replace f_1, f_2 by $f_1+f_2, -f_2$ in (9) and note that $\rho(f) = \rho(-f)$: we get $\rho(f_1) \leqslant \rho(f_1+f_2)+\rho(f_2)$, hence

$$\rho(f_1)-\rho(f_2) \leqslant \rho(f_1+f_2).$$

This proves (5) when $\rho(f_1) \geqslant \rho(f_2)$; the other case follows by symmetry. Now we can use the formula for the nullity: $\rho(f)+\nu(f) = m$, to get (6).

(ii) We have $Ufg \subseteq Vg$, hence $\rho(fg) \leqslant \rho(g)$. Further, a basis of Uf is mapped to a spanning set of Ufg, hence $\rho(fg) \leqslant \rho(f)$; altogether this shows that $\rho(fg) \leqslant \min\{\rho(f), \rho(g)\}$, i.e. the second half of (7). Now the first half of (7) is equivalent to (8), in view of the nullity formulae for f, g and fg. Therefore it will be enough to prove (8):

Let V' be any subspace of V and g' the mapping from V' to W defined by the rule: $xg' = xg$ for all $x \in V'$. This is the restriction of g to the subspace V' and is easily seen to be a linear mapping from V' to W. Clearly $\ker g' \subseteq \ker g$, hence by the nullity formula (1), $\dim V' - \dim V'g' = \nu(g') \leqslant \nu(g)$, i.e.

$$\dim V' - \dim V'g \leqslant \nu(g). \tag{10}$$

Now

$$\begin{aligned} \nu(fg) &= \dim U - \dim Ufg \\ &= (\dim U - \dim Uf) + (\dim Uf - \dim Ufg) \\ &\leqslant \nu(f)+\nu(g) \end{aligned}$$

by (10) and the nullity formula for f. This proves (8). ∎

Exercises

(1) Describe the general linear mapping of rank 1; show that any linear mapping of rank r can be written as a sum of r linear mappings of rank 1.

(2) Show that an $n \times n$ matrix has rank $\leqslant r$ iff it can be written as the product of an $n \times r$ matrix and an $r \times n$ matrix.

(3) Let A be a matrix of rank r. Show that, by suitably permuting the rows of A and likewise the columns, we can obtain a matrix in which the submatrix formed by the first r rows and columns is non-singular.

(4) Let A be a square matrix of order n. If some power of A is zero, show that $A^n = 0$.

(5) Given finite-dimensional vector spaces U, V, W and linear mappings $f: U \to V$, $g: U \to W$, if $\ker f \subseteq \ker g$, show that there exists $h: V \to W$ such that $g = fh$. Is the condition on the kernels necessary?

(6) Prove that an injective endomorphism of a finite-dimensional vector space is an automorphism and deduce another proof that a square matrix with linearly independent columns is invertible. Which implications in Prop. 2 remain true in an infinite-dimensional space? (Cf. Ex. (7).)

(7) Let f be the endomorphism of $F^{(N)}$ defined by $e_i \mapsto e_{i+1}$ $(i \in N)$. Show that f is injective but not surjective. If g maps e_i to e_{i-1} for $i > 1$, and e_1 to 0, show that g is surjective but not injective. Verify that $fg = 1$, $gf \neq 1$.

(8) If $A, B \in {}^mF^n$, show that there exist invertible matrices P, Q of orders m, n respectively such that $PAQ = B$ iff $\rho(A) = \rho(B)$.

(9) An endomorphism f of a finite-dimensional vector space V is such that $\operatorname{im} f \subseteq \ker f$. Show that $\rho(f) \leqslant 1/2 \dim V$ and that for a suitable basis of V, the matrix of f has the form

$$\begin{pmatrix} 0 & C \\ 0 & 0 \end{pmatrix}.$$

(10) Let A be an $n \times n$ matrix of the form $A = \begin{pmatrix} P & 0 \\ Q & R \end{pmatrix}$, where the zero matrix in the NE corner is $r \times s$ and $r+s > n$. Show that A is singular.

(11) Let $A \in {}^mF^n$ and $B \in {}^nF^p$. If $A = (A_1\ A_2)$ where A_i is $m \times n_i$ and $B = \begin{pmatrix} B_1 \\ B_2 \end{pmatrix}$ where B_i is $n_i \times p$, show that

$$AB = (A_1B_1 + A_2B_2).$$

(12) Given $A \in F_m$, $B \in F_n$ show that the mapping $X \mapsto AXB$ is an endomorphism of ${}^mF^n$. Under what conditions is it an automorphism?

(13)* Let U, V, W, X be finite-dimensional vector spaces and $\alpha: U \to V$, $\beta: V \to W$, $\gamma: W \to X$ linear mappings. Verify that $\operatorname{im} \alpha\beta$ is a subspace of $\operatorname{im} \beta$, and if W_0 is a complement of $\operatorname{im} \alpha\beta$ in $\operatorname{im} \beta$, so that $\operatorname{im} \beta = \operatorname{im} \alpha\beta \oplus W_0$, show that $\operatorname{im} \beta\gamma = \operatorname{im} \alpha\beta\gamma + W_0\gamma$. Deduce Frobenius's inequality:

$$\rho(\alpha\beta)+\rho(\beta\gamma) \leqslant \rho(\beta)+\rho(\alpha\beta\gamma),$$

and hence obtain another proof of Prop. 3 (ii).

(14) Show that the stabilizer of E_r (in $({}^mF^n)^2$) under the action $A \mapsto PAQ^{-1}$ consists of all pairs $\begin{pmatrix} P_1 & P_2 \\ 0 & P_4 \end{pmatrix}, \begin{pmatrix} P_1 & 0 \\ Q_3 & Q_4 \end{pmatrix}$, where P_1, P_4, Q_4 are invertible of orders r, $m-r$, $n-r$ respectively, P_2 is $r \times m-r$ and Q_3 is $n-r \times r$.

4.8 Category and functor

Much of algebra consists in the study of sets with a given structure and of the mappings preserving this structure. Thus we may be dealing with groups; the mappings preserving the group structure are the homomorphisms. When we turn to linear spaces, the appropriate mappings are the linear mappings and we shall meet many other examples later. Surprisingly, a useful discussion of this situation can be given within a general framework which assumes only very little about the mappings in each case. All we need is that they are closed under composition and include the identity mapping. But we need not even assume them to be mappings. Thus the definition takes the following form.

A *category* $\mathscr{A}$ is a collection of objects; for each pair of $\mathscr{A}$-objects A, B there is a set $\operatorname{Hom}_{\mathscr{A}}(A, B)$ or $\operatorname{Hom}(A, B)$ whose members are called the *$\mathscr{A}$-morphisms* (or just *morphisms*) from A to B. Instead of $f \in \operatorname{Hom}(A, B)$ we also write $f: A \to B$ or $A \xrightarrow{f} B$. For any three objects A, B, C there is a mapping

$$\operatorname{Hom}(A, B) \times \operatorname{Hom}(B, C) \to \operatorname{Hom}(A, C).$$

Thus with each pair of morphisms $f: A \to B$, $g: B \to C$ a morphism $p: A \to C$ is associated. We call p the *composite* or *product* of f and g and write $p = fg$. These morphisms satisfy the following axioms:

C. 1 $\operatorname{Hom}(A, B) \cap \operatorname{Hom}(A', B') = \varnothing$ *unless* $A' = A$ *and* $B' = B$. *Thus each morphism* f *determines a unique pair of objects* A, B *such that* $f \in \operatorname{Hom}(A, B)$; *$A$ is called the 'source' and B the 'target' of f.*

C. 2 *If* $f: A \to B$, $g: B \to C$, $h: C \to D$, *so that* $(fg)h$ *and* $f(gh)$ *are both defined, then they are equal*:

$$(fg)h = f(gh). \tag{1}$$

C. 3 *With each object A a morphism* $1_A: A \to A$ *is associated such that for any* $f: A \to B$, $g: C \to A$,

$$1_A f = f, \qquad g \,.\, 1_A = g. \tag{2}$$

In words, 1_A acts as left neutral for morphisms with source A and as right neutral for morphisms with target A. As in monoids, one verifies that 1_A is uniquely determined by the equations (2). 1_A is called the *identity morphism* on A.

A morphism $f: A \to B$ is called an *isomorphism* if there exists an *inverse morphism* $f': B \to A$ with the property $ff' = 1_A$, $f'f = 1_B$. Again such an inverse, if it exists, is easily seen to be unique and it may again be denoted by f^{-1}. An isomorphism with source and target both A is called an *automorphism* of A.

Examples of categories. (i) There is a category 'Sets' whose objects are sets and whose morphisms are mappings between sets, with the usual composition of mappings. Here 1_A is the identity mapping of the set A. Strictly speaking, a morphism in this category is a triple (A, B, f) consisting of two sets A, B and a mapping f from A to B. For example, if a mapping f from A to B is given which maps A into a proper subset B' of B, then (A, B, f) and (A, B', f) count as different morphisms, even though as mappings they are the same. It is understood that similar remarks apply to the other examples that follow.

(ii) Groups form the objects of a category in which the morphisms are homomorphisms. Likewise monoids and their homomorphisms form a category.

(iii) Given a field F, there is a category Vect_F whose objects are vector spaces over F and whose morphisms are linear mappings between vector spaces.

(iv) Let F be a field and consider a category whose objects are pairs $(V, \mathbf{v})$ consisting of a finite-dimensional vector space V over F and a basis $\mathbf{v}$ of V. The morphisms from $(U, \mathbf{u})$ to $(V, \mathbf{v})$ are to be linear mappings from U to V which map the basis $\mathbf{u}$ (as an ordered family) bijectively to $\mathbf{v}$. This is a category in which all morphisms are isomorphisms.

(v) Let M be a monoid; we can regard M as a category with a single object, whose morphisms are the elements of M. If all morphisms are isomorphisms, the monoid is a group.

(vi) A partially ordered set S may be regarded as a category whose objects are the elements of S, and Hom (a, b) has a single element or is empty according to whether $a \leqslant b$ or not.

From these examples we see that generally, though not always, the morphisms turn out to be mappings and we shall illustrate the composition of morphisms by diagrams as for mappings. Thus given $f: A \to B$, $g: B \to C$, $p: A \to C$, we have the diagram shown and we say that this diagram commutes if $p = fg$.

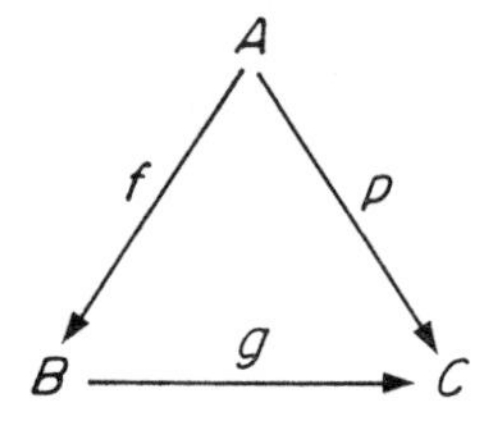

Similarly, given f, g as before and $h: C \to D$, we have the second diagram shown in which the two triangles commute, by definition; now the associative law (1) may be expressed by saying that the square commutes.

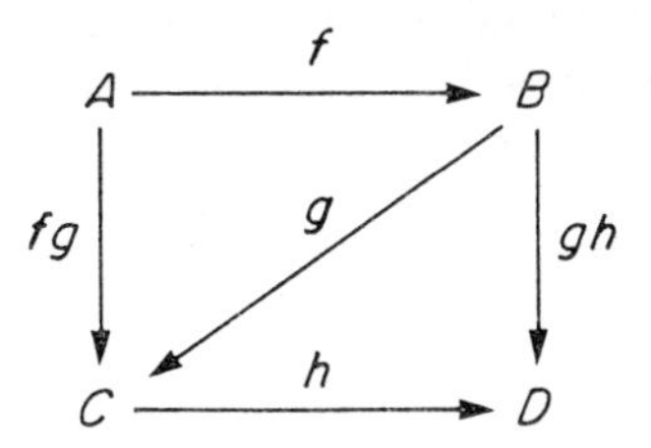

Given categories $\mathscr{A}$ and $\mathscr{B}$, we shall say that $\mathscr{B}$ is a *subcategory* of $\mathscr{A}$ if each $\mathscr{B}$-object is an $\mathscr{A}$-object, each $\mathscr{B}$-morphism is an $\mathscr{A}$-morphism and composition of morphisms is the same in $\mathscr{A}$ and $\mathscr{B}$. Thus for any two $\mathscr{B}$-objects X and Y we have

$$\mathrm{Hom}_{\mathscr{B}}(X, Y) \subseteq \mathrm{Hom}_{\mathscr{A}}(X, Y). \tag{3}$$

If equality holds in (3) for all X and Y, we call $\mathscr{B}$ a *full* subcategory of $\mathscr{A}$. Whereas for a general subcategory we must specify both objects and morphisms, for a full subcategory we need only specify the objects. For example, groups form a full subcategory of monoids; every monoid-homomorphism between groups is actually a group-homomorphism. Monoids form a subcategory of sets with an associative multiplication, with multiplication-preserving mappings as morphisms; but this subcategory is not full, because not every multiplication-preserving mapping between monoids is a monoid-homomorphism (it also has to preserve the neutral element).

The idea of a 'structure-preserving mapping' can be extended to categories themselves, where it takes the following form. Let $\mathscr{A}$, $\mathscr{B}$ be categories, then a *functor* from $\mathscr{A}$ to $\mathscr{B}$ is a function which assigns to each $\mathscr{A}$-object A a $\mathscr{B}$-object AF and to each $\mathscr{A}$-morphism $f: A \to B$ a $\mathscr{B}$-morphism $fF: AF \to BF$ such that

F. 1 *If fg is defined in $\mathscr{A}$, then $fF . gF$ is defined in $\mathscr{B}$ and*

$$fF . gF = (fg)F.$$

F. 2 $1_A F = 1_{AF}$ *for each $\mathscr{A}$-object A.*

Besides the functors just defined there is another kind of functor, which reverses the multiplication. This is defined as a function F which assigns to each $\mathscr{A}$-object A a $\mathscr{B}$-object AF and to each $\mathscr{A}$-morphism $f: A \to B$ a $\mathscr{B}$-morphism $fF: BF \to AF$ (note the reversed order) such that **F.** 2 above holds, but in place of **F.** 1,

F. 1* *If fg is defined in $\mathscr{A}$, $gF . fF$ is defined in $\mathscr{B}$ and*

$$gF . fF = (fg)F.$$

Functors satisfying **F.** 1*–2 are called *contravariant*, in contrast to the kind satisfying **F.** 1–2, which are *covariant*.

From every category $\mathscr{A}$ we can form another category, called its *opposite* and denoted by $\mathscr{A}^{op}$, by 'reversing all the arrows'. In detail, $\mathscr{A}^{op}$ has the same objects as $\mathscr{A}$, but to each morphism $f: X \to Y$ in $\mathscr{A}$ corresponds a morphism $f': Y \to X$ in $\mathscr{A}^{op}$, so that $f'g'$ is defined whenever gf is defined, and then

$$(gf)' = f'g'.$$

Now a contravariant functor from $\mathscr{A}$ to $\mathscr{B}$ may also be described as a covariant functor from $\mathscr{A}^{op}$ to $\mathscr{B}$, or as a covariant functor from $\mathscr{A}$ to $\mathscr{B}^{op}$.

Examples of functors. (i) There is a functor U from groups to sets, which associates with each group G the underlying set on which it is defined and regards each homomorphism $f: G \to H$ as a mapping of sets. U is a covariant functor, called the *forgetful functor* (U 'forgets' the group structure).

(ii) If $\mathscr{B}$ is a subcategory of $\mathscr{A}$, the inclusion of $\mathscr{B}$ in $\mathscr{A}$ is a functor, the *inclusion functor*.

(iii) Let M, N be monoids, regarded as categories. A covariant functor from M to N is nothing other than a homomorphism of monoids.

(iv) Let $\mathscr{A}$ be any category and A a given $\mathscr{A}$-object. Then there is a functor h^A from $\mathscr{A}$ to sets, which associates with each $\mathscr{A}$-object X the set Hom (A, X) and with each morphism $f: X \to Y$ associates the mapping Hom $(A, X) \to$ Hom (A, Y) which makes $\alpha: A \to X$ correspond to $\alpha f: A \to Y$. Clearly, if $Y = X$ and $f = 1_X$, then $\alpha \mapsto \alpha 1_X$ is the identity mapping. Further, a reference to the diagram shows that $(\alpha f)g = \alpha(fg)$, i.e. $fh^A \,.\, gh^A = (fg)h^A$. The functor h^A is called the *covariant hom-functor*; it is an important example, which we shall meet again later.

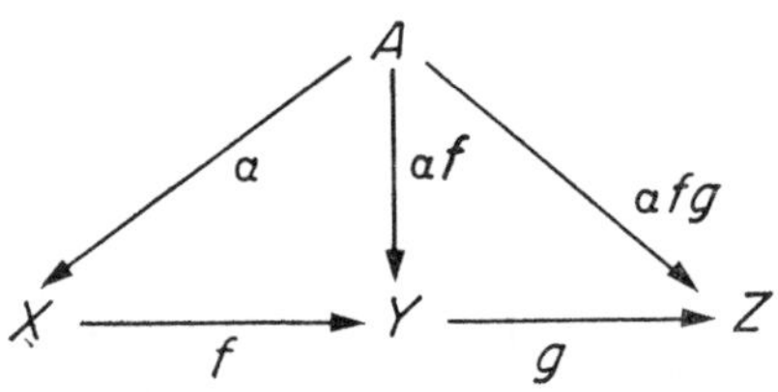

(v) Let $\mathscr{A}$ and A be as in (iv), then there is a contravariant functor h_A from $\mathscr{A}$ to sets, which associates with each $\mathscr{A}$-object X the set Hom (X, A) and with each $f: X \to Y$ the mapping from Hom (Y, A) to Hom (X, A) which makes $\beta: Y \to A$ correspond to $f\beta: X \to A$. Again it is easily checked that $1_X h_A = 1_{Xh_A}$ and, from the second diagram, we see that $(fg)\beta = f(g\beta)$, which shows

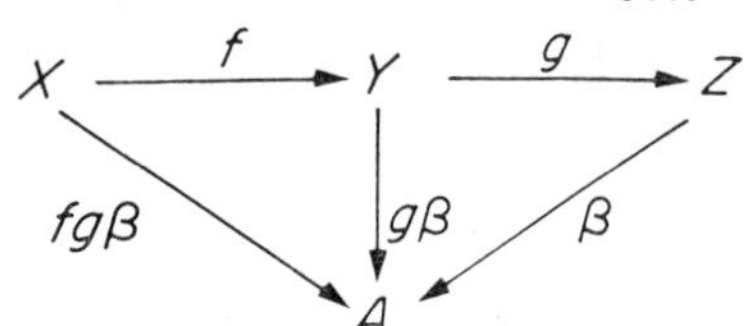

that $fg \,.\, h_A = gh_A \,.\, fh_A$, if we bear in mind that $f(g\beta)$ means: first apply g, then f (to β). This is called the *contravariant hom-functor*.

We conclude this section by describing the universal mapping property met in **4.2** in the context of category theory. In any category $\mathscr{A}$ an *initial object* is an object I such that for each $\mathscr{A}$-object X, Hom (I, X) has exactly one element. E.g., the trivial group is an initial object in groups. A category may have more than one initial object, but they are all isomorphic. For if I, I' are both initial, there exist unique $\mathscr{A}$-morphisms $\alpha: I \to I'$ and $\beta: I' \to I$ and so $\alpha\beta \in$ Hom (I, I). But Hom (I, I) has only one element, which must be the identity morphism for I, hence $\alpha\beta = 1_I$. Similarly $\beta\alpha = 1_{I'}$ and this shows α to be an isomorphism, as claimed. Observe that we have actually shown that any two initial objects are isomorphic, by a unique isomorphism.

As an illustration consider the universal mapping property for the coordinate space F^n described in **4.2**. We form the following category: The objects are mappings of the standard basis of F^n into a space. If the space is V and the mapping f, we denote the object by (V, f). We define a morphism between objects (V, f) and (V', f') as a linear mapping α: $V \to V'$ such that $f\alpha = f'$. Now the universal mapping property proved in **4.2** asserts that this category has an initial object, namely (F^n, ι), where ι is the inclusion mapping which takes each vector e_i to itself. We also say that (F^n, ι) is the solution of a universal problem.

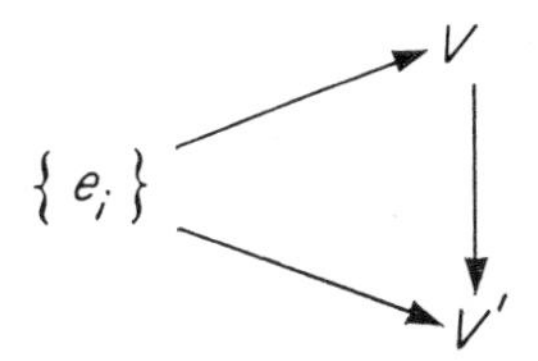

We observe that once the universal mapping property is verified, the above argument shows that (F^n, ι) is unique up to isomorphism. In the present case this is rather obvious, but later we shall meet instances where this is not so and we can then deduce the uniqueness by appealing to the general argument given above.

An object in a category $\mathscr{A}$ is called *final* if it is initial in the opposite category $\mathscr{A}^{op}$; clearly such an object, if it exists, is again unique up to isomorphism. Just as the universal mapping property can be described as the existence of a certain initial object, so there is another, dual, property which asserts the existence of a final object. For example, let S be a fixed set and consider the category whose objects are pairs of mappings from a set into S, f_i: $X \to S$ $(i = 1, 2)$, while the morphisms $(X, f_1, f_2) \to (Y, g_1, g_2)$ are mappings λ: $X \to Y$ such that $f_i = \lambda g_i$ $(i = 1, 2)$. As the reader may verify, this category has as final object the Cartesian square S^2, together with the projection mappings p_1: $(x, y) \mapsto x$, p_2: $(x, y) \mapsto y$. This fact is expressed by saying that S^2 with its projection mappings is universal for pairs of mappings into S. Thus there are two kinds of universality, characterized as initial and final objects in certain categories.

Exercises

(1) If an ordered set is interpreted as a category (as in the text), show that least and greatest elements become initial and final objects respectively. Describe a categorical notion generalizing minimal and maximal elements.

(2) In any category an object that is both initial and final is called a *zero object* and any morphism that can be taken via the zero object is called a *zero morphism*. Show that in any category with zero object there is for each pair of objects A, B a unique zero morphism from A to B. Show that the category of groups has a zero object.

(3) Let Vect_F^* be the category whose objects are pairs $(V, \mathbf{v})$ consisting of a finite-dimensional vector space V and a basis $\mathbf{v} = (v_1, \ldots, v_n)$ of V, while the morphisms $(V, \mathbf{v}) \to (V', \mathbf{v}')$ are linear mappings $f: V \to V'$. Let Mat_F be the category whose objects are non-negative integers, while the morphisms $m \to n$ are $m \times n$ matrices over F. Show that there is a functor from Vect_F^* to Mat_F which associates with each $(V, \mathbf{v})$ the number $\dim V$ and with each morphism f the matrix referred to the given bases. Show that Vect_F^* has a zero object; what does it correspond to in Mat_F?

(4) Let Group* be the category whose objects are pairs (G, a) consisting of a group G and an element a of G, while the morphisms $(G, a) \to (G', a')$ are homomorphisms $f: G \to G'$ such that $af = a'$. Show that $(\mathbf{Z}, 1)$ (where $\mathbf{Z}$ is the infinite cyclic group and 1 a generator) is an initial object in Group*.

Further exercises on Chapter 4

(1) Show that the set of convergent series forms a subspace of $\mathbf{R}^N$; prove that it is not finite-dimensional.

(2) Prove that if $\{x_1, \ldots, x_r\}$ is a linearly independent family of vectors in a vector space, then the x_i are distinct.

(3) If $\mathbf{F}_p^n$ is the vector space of n-tuples over the field with p elements (where p is a prime), show that any subgroup of $\mathbf{F}_p^n$ (as abelian group) is a subspace. Show that this does not remain true if $\mathbf{F}_p$ is replaced by $\mathbf{Q}$.

(4) Show that the differentiable functions (in one variable x, say) form a linear space.

(5) On the space of all differentiable functions, show that the following are linear mappings: (i) $D: f \mapsto f'$, (ii) $S: f \mapsto \int_0^x f(t)\,dt$. Show that $SD = 1$ and deduce that S is injective and D surjective, but that D is not injective, nor S surjective.

(6) Let θ be an endomorphism of a vector space. If $\theta^n = 0$ but $\theta^{n-1} \neq 0$, show that $1, \theta, \ldots, \theta^{n-1}$ are linearly independent.

(7) A square matrix is said to be *upper* (*lower*) *triangular* if all entries below (above) the main diagonal are zero. Show that a triangular matrix is non-singular iff the product of the diagonal elements is non-zero.

(8) Show that the invertible $n \times n$ matrices over a field F form a group. This group is denoted by $\mathbf{GL}_n(F)$ and called the *general linear group*. Find the order of $\mathbf{GL}_2(\mathbf{F}_p)$.

(9) Let V be a vector space with basis $v_1, \ldots, v_n$. If $u_1 = v_1$, $u_2 = a_{21}v_1 + v_2$, $u_3 = a_{31}v_1 + a_{32}v_2 + v_3, \ldots$, show that $u_1, \ldots, u_n$ form a basis.

(10) In $\mathbf{R}^3$ find a complement U for the space spanned by $u = (1, 1, 1)$ and a complement V for the space spanned by $v = (1, -1, 1)$ such that $U \cap V$ is a complement for $\langle u, v \rangle$.

(11) If $E(x) = \begin{pmatrix} x & 1 \\ -1 & 0 \end{pmatrix}$, prove the formula $E(x)E(0)E(y) = -E(x+y)$ and deduce that $E(x)$ has an inverse, given by $E(0)E(-x)E(0)$. Show that $E(x)$ has finite (multiplicative) order when $x = 0, \pm 1$.

(12) If $A = \begin{pmatrix} 1 & 1 & 1 \\ 0 & 1 & 1 \\ 0 & 0 & 1 \end{pmatrix}$ and $A(x) = \begin{pmatrix} 1 & x & \binom{x}{2} \\ 0 & 1 & x \\ 0 & 0 & 1 \end{pmatrix}$ show that $A^n = A(n)$. Prove also that $A(r)A(s) = A(r+s)$ and deduce that $A(r/s)^s = A(r)$.

(13) On the space of differentiable functions of x, let $fP = xf$, $fQ = f'$. Show that $PQ - QP = 1$.

(14) Prove the law of nullity for square matrices in the form $\nu(AB) \leqslant \nu(A) + \nu(B)$ using Th. 1, **4.7**, Cor.

(15) Let P be a preordered set and define an equivalence on P (as in Ex. (5), Further Exercises, Ch. **1**) by writing '$x \sim y$' if '$x \leqslant y$ and $y \leqslant x$'. Show that there is a natural ordering on the set $\tilde{P}$ of equivalence blocks and that the mapping $\lambda : x \mapsto [x]$ of any element to its equivalence block is order-preserving (i.e. $x \leqslant y \Rightarrow x\lambda \leqslant y\lambda$). Show that every order-preserving mapping f from P to an ordered set S is constant on each equivalence block; deduce that there is a unique order-preserving mapping $f_1 : \tilde{P} \to S$ such that $f = \lambda f_1$ and interpret this as a universal mapping property.

(16) Check that, except for two places in **4·5** (and some exercises), the commutativity of multiplication in the underlying field was not used in this chapter.

5

Linear Equations

Generally, when dealing with a linear problem, we find that the solutions constitute a certain subspace, which we can determine, e.g. by finding a basis. In practice, what we are faced with is a system of linear equations. In this chapter we shall consider the practical methods of solving such systems. Of course we shall use the notations and results of Ch. **4** wherever possible.

5.1 Systems of linear equations

The general system of linear equations that we wish to consider has the form

$$\begin{aligned} a_{11}x_1 + \cdots + a_{1n}x_n &= b_1, \\ a_{21}x_1 + \cdots + a_{2n}x_n &= b_2, \\ \cdots\cdots\cdots\cdots & \\ a_{m1}x_1 + \cdots + a_{mn}x_n &= b_m. \end{aligned} \tag{1}$$

Here a_{ij}, b_i lie in the given field F and we are looking for values of $x_1, \ldots, x_n$ in F to satisfy the equations. We can abbreviate the system (1) as $\sum a_{ij}x_j = b_i$; usually we shall write it even more briefly in matrix form

$$Ax = b, \qquad (A \in {}^mF^n,\ b \in {}^mF), \tag{2}$$

where $A = (a_{ij})$ is the matrix of the system, $b = (b_1, \ldots, b_m)^{\mathrm{T}}$ and $x = (x_1, \ldots, x_n)^{\mathrm{T}}$. The matrix (A, b) obtained by adjoining the column b to A is called the *augmented matrix* for the system (2). If, in (2), $b = 0$, the system is said to be *homogeneous*; even when solving the general case (2) we shall often want to consider the associated homogeneous system

$$Ax = 0. \tag{3}$$

From **4.5** we see that A may be interpreted as a linear mapping from ${}^nF$ to ${}^mF$ and the solutions of (2) are the vectors in ${}^nF$ that map to b. From **4.7** we know that the solutions of (3) form a subspace of ${}^nF$, the kernel of the mapping defined by A, and it will turn out that the solutions of (2) (if any) form a coset in ${}^nF$ of the solution space of (3). The special case $m = n$ is of particular importance. Here we can interpret A as a linear mapping of ${}^nF$ into itself; either A is non-singular, then (2) has a unique solution for each value of b;

or A is singular, then (2) either has no solution at all, or has many solutions, forming a coset of the kernel.

THEOREM 1 *Let*

$$Ax = b \tag{2}$$

be a system of m equations in n unknowns. Then the following assertions are equivalent:

(i) *the system* (2) *has a solution,*

(ii) *the augmented matrix has the same rank as* A: $\rho(A, b) = \rho(A)$,

(iii) *for any* $u \in F^m$, $uA = 0 \Rightarrow ub = 0$.

When a solution exists, the general solution of (2) *has the form* $x = x_0 + x'$, *where* x' *is a particular solution of* (2) *and* x_0 *is the general solution of the associated homogeneous system*

$$Ax = 0, \tag{3}$$

depending on $n - \rho(A)$ *arbitrary constants.*

Thus to solve (2) completely we need a particular solution of (2) and the most general solution of (3).

Proof. (i) $\Rightarrow$ (iii). Let x, u satisfy $Ax = b$ and $uA = 0$, then $ub = u(Ax) = (uA)x = 0$.

(iii) $\Rightarrow$ (ii). Write $\rho(A) = r$ and denote the rows of A by $R_1, \ldots, R_m$. If $R_1, \ldots, R_r$ are linearly independent and $R_\nu = \sum_1^r \alpha_i R_i$, then $uA = 0$, where $u_i = \alpha_i$ for $i = 1, \ldots, r$, $u_\nu = -1$ and $u_j = 0$ for $j \neq 1, \ldots, r, \nu$. By (iii) it follows that $b_\nu = \sum_1^r \alpha_i b_i$ and this shows that the augmented matrix (A, b) also has rank r, i.e. (ii) holds.

To prove (ii) $\Rightarrow$ (i), let $C_1, \ldots, C_r$ be linearly independent columns of A, where $r = \rho(A)$, then every column of A is linearly dependent on them, as well as b, because $r = \rho(A, b)$. Thus b is linearly dependent on the columns of A and this is just an expression of the fact that (2) has a solution.

Now let x', x'' be any two solutions of (2), then $x_0 = x' - x''$ satisfies (3): $Ax_0 = Ax' - Ax'' = b - b = 0$. Conversely, if x' satisfies (2) and x_0 satisfies (3), then $x_0 + x'$ satisfies (2): $A(x_0 + x') = Ax_0 + Ax' = 0 + b = b$. This shows that the general solution of (2) is obtained by taking a particular solution x' and adding to it the general solution of (3).

It is clear that the solutions of (3) form a vector space, which by Th. 1, **4.7**, has dimension $n - r$. Explicitly, we can find invertible matrices P, Q such that $PAQ^{-1} = E_r$ and the equation $Ax = 0$ is equivalent to $E_r Qx = 0$. Writing $Qx = x'$, we see that the system reduces to $x_1' = \cdots = x_r' = 0$, while $x_{r+1}', \ldots, x_n'$ are arbitrary. Thus the solution space is indeed $(n-r)$-dimensional. ■

COROLLARY 1 *If* A *is a non-singular* $n \times n$ *matrix, then* $Ax = b$ *has a unique solution, whatever the vector* b.

For this means that the n columns of A are linearly independent, so $n-r = 0$ and $\rho(A) = \rho(A, b)$. In fact the unique solution is then $x = A^{-1}b$. ∎

COROLLARY 2 *If A is a square matrix then $Ax = 0$ has a non-zero solution if and only if A is singular.*

For A is singular iff the columns of A are linearly dependent. ∎

We briefly note the geometrical interpretation of the system (2). An equation of the form

$$\alpha_1 x_1 + \cdots + \alpha_n x_n = c, \tag{4}$$

where the α_i are not all 0, represents for $n = 2$ a line, for $n = 3$ a plane and generally a *hyperplane*, i.e. a coset of an $(n-1)$-dimensional subspace in n-space. Now (2) consists of m equations of the form (4), one for each of the rows of A, and the solution of (2) is the intersection of m hyperplanes. To say that the rank of (2) is r means that this intersection can be found from r of the hyperplanes, and these are independent; thus the intersection, if non-empty, is $(n-r)$-dimensional. We shall return to this point of view in **5.2** to look at particular examples.

A slightly different interpretation views (2) as the expression in m-space, of the vector b as a linear combination of the columns of A. In terms of this picture, condition (iii) of Th. 1 states that any $(m-1)$-dimensional subspace containing the vectors represented by the columns of A also contains the vector represented by b. Now the intersection of all the subspaces containing the columns of A is just the space spanned by these columns and hence we conclude that b is a linear combination of the columns of A. In essence this is another proof that (iii) $\Rightarrow$ (i).

Th. 1 has an analogue in the theory of linear inequalities which is of importance in linear programming. For any sequence of real numbers $x = (x_1, \ldots, x_n)$ let us write $x \geqslant 0$ if $x_i \geqslant 0$ for each component x_i. Then Farkas's theorem states:

Let $A \in {}^mR^n$, $b \in {}^mR$ and assume that for any $u \in R^m$,

$$uA \geqslant 0 \quad \textit{implies} \quad ub \geqslant 0, \tag{5}$$

then there exists $x \in {}^nR$ such that

$$x \geqslant 0 \quad \textit{and} \quad b = Ax. \tag{6}$$

Without giving a detailed proof we observe that this result again follows by geometrical considerations. Just as $\sum \alpha_i x_i = 0$ represents a subspace, so the inequality $\sum \alpha_i x_i \geqslant 0$ represents a 'half-space' (e.g. in 3-space this is the space on one side of a plane). Now (5) states that every half-space containing the columns of A also contains b, and it is not hard to show that the intersection of all half-spaces containing a given set S of vectors is the convex hull of this set and may be obtained as the set of all linear combinations from S

with non-negative coefficients. It follows then that b is in the convex hull of the columns of A and this is expressed by (6).

Exercises

(1) Solve the system

$$\begin{aligned} x_1+2x_2-5x_3 &= 2 \\ -3x_2+6x_3 &= 3 \\ x_2-2x_3 &= -1. \end{aligned}$$

(2) Find a condition for the system $Ax = b$ to have at most one solution.

(3) Two systems of linear equations in $x_1, \ldots, x_n$ are said to be *equivalent* if each equation in one is a linear combination of the equations in the other. Verify that this is indeed an equivalence relation. If $Ax = b$, $A'x = b'$ are equivalent, show that there exists an invertible matrix P such that $(A', b') = P(A, b)$.

(4) Use Th. 1 to show that if n vectors in F^n have their last component 0, they are linearly dependent, and deduce that any $n+1$ vectors in F^n are linearly dependent.

(5) Show that C is a coset of a subspace of V iff for any $u, v, w \in C$ and any $\lambda \in F$, $\lambda(u-v)+w \in C$.

(6) Let V be a space and U', U'' two subspaces. If a coset of U' is contained in one of U'', show that $U' \subseteq U''$.

(7) Given $x_1, \ldots, x_r \in R^n$, let $H(x_1, \ldots, x_r)$ be the set of all vectors $\sum \alpha_i x_i$ with $\alpha_i \geqslant 0$. Verify that any half-space containing each x_i also contains $H(x_1, \ldots, x_r)$. Show that $H(x_1, \ldots, x_r)$ is the precise intersection of the half-spaces containing it, in the cases $r = 1, 2$.

(8) Show that an $m \times n$ system of rank m always has a solution.

(9) If x' is a solution of $Ax = b'$ and x'' a solution of $Ax = b''$, show that $\lambda' x' + \lambda'' x''$ is a solution of $Ax = \lambda' b' + \lambda'' b''$.

5.2 Elementary operations

When we have a system of equations, Th. 1, **5.1**, tells us what sort of solution to expect but it does not tell us how to get it. Although we can in principle find a solution by following through the arguments of Ch. **4**, what we need now is a method for solving the equations in the most direct way. The method which we shall develop here can also be used to find the rank of a matrix, invert a non-singular matrix and calculate determinants (to be described later, in Ch. **7**). In essence it is just the familiar method of eliminating the unknowns, though organized in such a way as to cope with the possibility of a singular matrix.

The first question is: When can the matrix A be regarded as reduced in the sense that the solution of the system $Ax = b$ can be read off? To give an answer we need a definition. The matrix A is said to be in *echelon form* † if in each row the first non-zero element (if any) is 1 and for any such occurrence of 1, all the elements which lie below it in the same column or an earlier column are zero. Suppose that these 1s occur in the columns labelled $i_1, \ldots, i_r$ and for simplicity assume that the corresponding rows are $1, 2, \ldots, r$. Then our system takes the form

$$\begin{array}{rl} x_{i_1} + \cdots \qquad\qquad & = b_1 \\ x_{i_2} + \cdots \qquad & = b_2 \\ \cdots\cdots\cdots\cdots & \cdot \\ x_{i_r} + \cdots & = b_r \\ 0 & = b_{r+1} \\ \cdots & \\ 0 & = b_m. \end{array} \tag{1}$$

It is easy to see that a solution exists iff $b_{r+1} = \cdots = b_m = 0$ and, when this is so, we get the general solution by assigning arbitrary values to x_i for $i \neq i_1, \ldots, i_r$ and determining $x_{i_r}, \ldots, x_{i_1}$ (in that order) from the given equations. E.g., the system

$$\begin{aligned} x_1 - 4x_2 - 6x_3 - 20x_4 &= -1, \\ x_3 + 3x_4 &= 1, \end{aligned}$$

is in echelon form. To solve it we put $x_2 = \lambda$, $x_4 = \mu$, then $x_3 = 1 - 3\mu$, from the last equation, and hence $x_1 = 4\lambda + 6(1-3\mu) + 20\mu - 1 = 5 + 4\lambda + 2\mu$.

It only remains to transform an arbitrary matrix to echelon form without affecting the solution of the corresponding system. In order to do this we define the following *elementary operations* on the rows $R_1, \ldots, R_m$ say, of A:

(α) *Interchange* R_i *and* R_j: $R_i \leftrightarrow R_j$.

(β) *Replace* R_i *by* λR_i, *where* $\lambda \neq 0$: $R_i \to \lambda R_i$.

(γ) *Replace* R_i *by* $R_i + \mu R_j$ $(j \neq i)$: $R_i \to R_i + \mu R_j$.

These operations, applied to the augmented matrix (A, b), clearly do not change the solutions of the system $Ax = b$. Thus (α) amounts to rewriting the equations in a different order, (β) multiplies one equation by a non-zero scalar and (γ) adds a multiple of one equation to another. Now our system is solved by applying

THEOREM 1 *Every matrix can be transformed to echelon form by elementary operations on its rows.*

Proof. If $A = 0$, there is nothing to prove, so we assume $A \neq 0$ and make an induction on the rank of A. Consider the first non-zero column of A; omitting

† Echelon: 'Formation of troops in parallel divisions, each with its front clear of that in advance' (Concise Oxford English Dictionary).

earlier columns (which will not affect the proof) we may assume that the first column is non-zero. By elementary operations of type (α) we arrange for the first element in this column to be non-zero and by an operation of type (β) we can make it 1. Now we use operations of type (γ) to make all the other elements in the first column equal to 0. Thus if the first element in R_i is a_{i1}, then $R_i - a_{i1}R_1$ has zero as its first element. Now A has been reduced to the form

$$\begin{pmatrix} 1 & c \\ 0 & A' \end{pmatrix}, \tag{2}$$

where c is a row and A' is a matrix of smaller rank than A, because the first row of (2) cannot be linearly dependent on the rows of A'. Thus we can apply induction on the rank of A to complete the proof. ■

Since the rank of the matrix A can be determined from the set of solutions of $Ax = 0$, it follows that the rank of A is unchanged by applying elementary row operations. Now if we compare the explicit solution of (1) with Th. 1, **5.1**, we see that the rank of an echelon matrix is the number of its non-zero rows. Thus we obtain the following rule for computing the rank of a matrix:

Reduce A to echelon form by elementary row operations and count the number of non-zero rows.

We conclude this section by working through two examples to illustrate the reduction of a system to echelon form. It will be convenient to omit the unknowns and operate directly on the augmented matrix. Thus consider the system

$$\begin{aligned} x_1 + x_2 + \ x_3 &= 3 \\ 3x_1 + x_2 + 5x_3 &= 5 \\ x_1 + \qquad 2x_3 &= t. \end{aligned} \tag{3}$$

The problem is to determine t so that the system has a solution, and with this value of t solve it completely. We rewrite the system in coefficient form and indicate the elementary operations performed on the rows:

$$\left.\begin{array}{ccc|c} 1 & 1 & 1 & 3 \\ 3 & 1 & 5 & 5 \\ 1 & 0 & 2 & t \end{array}\right. \quad \begin{array}{l} \\ R_2 \to R_2 - 3R_1 \\ R_3 \to R_3 - R_1 \end{array} \quad \left.\begin{array}{ccc|c} 1 & 1 & 1 & 3 \\ 0 & -2 & 2 & -4 \\ 0 & -1 & 1 & t-3 \end{array}\right.$$

$$R_2 \to -\tfrac{1}{2}R_2 \quad \left.\begin{array}{ccc|c} 1 & 1 & 1 & 3 \\ 0 & 1 & -1 & 2 \\ 0 & -1 & 1 & t-3 \end{array}\right. \quad R_3 \to R_3 + R_2 \quad \left.\begin{array}{ccc|c} 1 & 1 & 1 & 3 \\ 0 & 1 & -1 & 2 \\ 0 & 0 & 0 & t-1 \end{array}\right.$$

This completes the reduction, since the matrix is now in echelon form. The equation corresponding to the third row reads $t-1 = 0$, so for a solution we must have $t = 1$. When this holds, the third row can be omitted and the remaining two equations determine x_1 and x_2 in terms of x_3. Put $x_3 = \lambda$, then we have $x_2 - \lambda = 2$, hence $x_2 = \lambda + 2$, and $x_1 + (\lambda + 2) + \lambda = 3$, i.e.

$x_1 = 1-2\lambda$, so the complete solution is

$$\begin{aligned} x_1 &= 1-2\lambda \\ x_2 &= 2+\lambda \\ x_3 &= \lambda. \end{aligned}$$

We also note from the final form of the system that $\rho(A) = 2$. If we rewrite the solution in the form

$$x = \begin{pmatrix} 1 \\ 2 \\ 0 \end{pmatrix} + \lambda \begin{pmatrix} -2 \\ 1 \\ 1 \end{pmatrix},$$

we see more clearly that it consists of a coset of $\mathbf{R}(-2, 1, 1)^T$. Geometrically the solution represents a line through the point $(1, 2, 0)^T$ and this is the intersection of the 3 planes whose equations are given by (3), for $t = 1$. For other values of t the planes have a common direction, but no common points; these two possibilities ($t = 1$ and $t \neq 1$) are illustrated in Fig. 3 and Fig. 4 respectively.

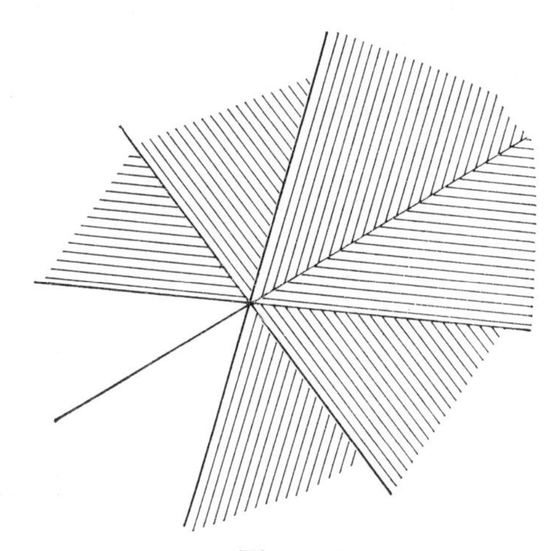

Figure 3

Next consider the system (in coefficient form)

$$\begin{array}{rrrr|r} 2 & -8 & 1 & -1 & 11 \\ 5 & -20 & -4 & -22 & 21 \\ -3 & 12 & 5 & k & -10 \end{array}$$

Here the problem is to determine k so that the system has a solution and then to solve it completely. To avoid having to divide we subtract twice the first row from the second and then interchange the first two rows before starting our reduction:

$$\begin{array}{l} \\ R_2 \to R_2 - 2R_1 \\ \\ \end{array} \quad \begin{array}{rrrr|r} 2 & -8 & 1 & -1 & 11 \\ 1 & -4 & -6 & -20 & -1 \\ -3 & 12 & 5 & k & -10 \end{array}$$

$$R_1 \leftrightarrow R_2 \quad \left(\begin{array}{cccc|c} 1 & -4 & -6 & -20 & -1 \\ 2 & -8 & 1 & -1 & 11 \\ -3 & 12 & 5 & k & -10 \end{array}\right)$$

$$\begin{array}{l} R_2 \to R_2 - 2R_1 \\ R_3 \to R_3 + 3R_1 \end{array} \quad \left(\begin{array}{cccc|c} 1 & -4 & -6 & -20 & -1 \\ 0 & 0 & 13 & 39 & 13 \\ 0 & 0 & -13 & k-60 & -13 \end{array}\right)$$

$$\begin{array}{l} R_3 \to R_3 + R_2 \\ R_2 \to \frac{1}{13} R_2 \end{array} \quad \left(\begin{array}{cccc|c} 1 & -4 & -6 & -20 & -1 \\ 0 & 0 & 1 & 3 & 1 \\ 0 & 0 & 0 & k-21 & 0 \end{array}\right)$$

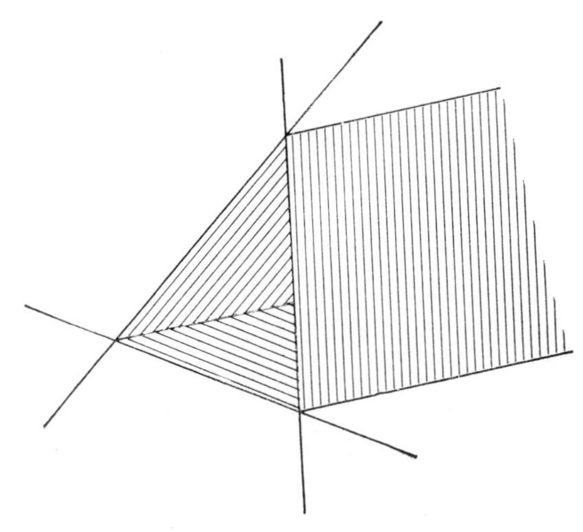

Figure 4

The last row shows that for a solution we must have $k = 21$ and, with this value, the last row can be omitted. We now obtain the solution $x_2 = \lambda$, $x_4 = \mu$, $x_3 = 1-3\mu$, $x_1 = -1+4\lambda+6(1-3\mu)+20\mu = 5+4\lambda+2\mu$, thus

$$x = \begin{pmatrix} 5 \\ 0 \\ 1 \\ 0 \end{pmatrix} + \lambda \begin{pmatrix} 4 \\ 1 \\ 0 \\ 0 \end{pmatrix} + \mu \begin{pmatrix} 2 \\ 0 \\ -3 \\ 1 \end{pmatrix}.$$

From the final form of the solution we see that the rank of A is 2.

Exercises

(1) Reduce the following matrices to echelon form. For each matrix A solve the corresponding homogeneous system $Ax = 0$:

(i) $\begin{pmatrix} 2 & 1 \\ 5 & -3 \end{pmatrix}$ (ii) $\begin{pmatrix} 1 & 3 & -1 \\ 2 & 4 & 3 \\ 1 & 5 & -6 \end{pmatrix}$ (iii) $\begin{pmatrix} 4 & 12 & -7 & 6 \\ 2 & 6 & -4 & 2 \\ 3 & 9 & -2 & 11 \end{pmatrix}$ (iv) $\begin{pmatrix} 1 & 2 & 1 & 2 \\ 3 & 5 & 3 & 5 \\ 4 & 8 & 4 & 8 \end{pmatrix}$

(2) The following schemes represent the augmented matrices of systems of equations. In each case determine t so that there is a solution and with this value of t solve the system completely.

$$\text{(i)}\ \left(\begin{array}{cc|c} 1 & -1 & 3 \\ 2 & 3 & 1 \\ 4 & 2 & t \end{array}\right) \quad \text{(ii)}\ \left(\begin{array}{ccc|c} 2 & 0 & 4 & 1 \\ 1 & 3 & 5 & 1 \\ 1 & -1 & 1 & t \end{array}\right) \quad \text{(iii)}\ \left(\begin{array}{cccc|c} 1 & 1 & 2 & 1 & 1 \\ 1 & 3 & 2 & -1 & t \\ 2 & 1 & 4 & 3 & 1 \end{array}\right)$$

(3) If a non-singular matrix is reduced to echelon form, show that the unit matrix is obtained. Illustrate this with the matrix

$$\begin{pmatrix} 1 & 5 & -1 \\ 0 & 7 & 2 \\ -1 & 1 & 3 \end{pmatrix}$$

(4) What modifications are needed if the above questions are taken over the field of p elements?

(5) Show that the elementary operation (α) may be expressed in terms of (β) and (γ).

(6) List the echelon forms for a 3×4 matrix.

(7) What elementary operations are possible on a matrix in echelon form without disturbing the echelon form?

(8) If permutations of the columns are allowed as well as elementary row operations, show that any matrix can be brought to the form

$$\begin{pmatrix} I & B \\ 0 & 0 \end{pmatrix}.$$

5.3 *PAQ*-reduction and the inversion of matrices

In Ch. **4** we saw that for every matrix A there exist invertible matrices P and Q such that

$$PAQ = E_r = \begin{pmatrix} I_r & 0 \\ 0 & 0 \end{pmatrix} \tag{1}$$

(here we have replaced Q^{-1} by Q; clearly this is immaterial). To get a practical method of reduction, let us return to the elementary operations α, β, γ, defined in **5.2**, and observe that the effect of any one of these on the rows of a matrix A is the same as multiplying A on the left by a certain *elementary matrix*, namely the matrix obtained by applying the given elementary operation to the rows of the unit matrix. Thus

(α')$R_1 \leftrightarrow R_2$ corresponds to left multiplication by $\begin{pmatrix} A & 0 \\ 0 & I \end{pmatrix}$, where $A = \begin{pmatrix} 0 & 1 \\ 1 & 0 \end{pmatrix}$,

(β') $R_1 \to \lambda R_1$ corresponds to left multiplication by $\begin{pmatrix} \lambda & 0 \\ 0 & I \end{pmatrix}$,

(γ') $R_2 \to R_2 + \mu R_1$ corresponds to left multiplication by $\begin{pmatrix} B & 0 \\ 0 & I \end{pmatrix}$, where
$B = \begin{pmatrix} 1 & 0 \\ \mu & 1 \end{pmatrix}$
and where in each case the 0s denote zero matrices of the appropriate sizes. Similarly any elementary operation on the columns of A corresponds to right multiplication by the elementary matrix obtained by performing the corresponding operation on the columns of the unit matrix.

It is quite clear that any matrix can be brought to the form (1) by elementary row and column operations: we first transform it to echelon form and then use column operations to complete the reduction. The problem is to keep track of the operations used; this we do by writing a unit $m \times m$ matrix on the left of A and carrying out the row operations on both matrices. Our reduction then goes as follows:

$$I \mid A \to P_1 \mid P_1 A \to P_2 P_1 \mid P_2 P_1 A \to \cdots \to P_r \ldots P_1 \mid P_r \ldots P_1 A,$$

where $P_r \ldots P_1 A$ is in echelon form and $P = P_r \ldots P_1$ is invertible, as product of elementary and hence invertible matrices. We now have to apply column operations on PA; to keep track of these, we write the unit matrix under PA and perform all column operations on both:

$$\frac{PA}{I} \to \frac{PAQ_1}{Q_1} \to \frac{PAQ_1Q_2}{Q_1Q_2} \to \cdots \to \frac{PAQ_1 \ldots Q_s}{Q_1 \ldots Q_s}.$$

Here the matrix PAQ, where $Q = Q_1 \ldots Q_s$, has the reduced form (1) and P, Q can be read off from the final step in the two reductions respectively. We illustrate the process by an example:

Let $A = \begin{pmatrix} 2 & -8 & 1 & -1 \\ 5 & -20 & -4 & -22 \\ -3 & 12 & 5 & 21 \end{pmatrix}$. This is the matrix of the system considered in **5.2**; we go through the reduction once more to record the matrix P.

$$\left.\begin{array}{ccc} 1 & 0 & 0 \\ 0 & 1 & 0 \\ 0 & 0 & 1 \end{array}\right| \begin{array}{cccc} 2 & -8 & 1 & -1 \\ 5 & -20 & -4 & -22 \\ -3 & 12 & 5 & 21 \end{array} \to \left.\begin{array}{ccc} 1 & 0 & 0 \\ -2 & 1 & 0 \\ 0 & 0 & 1 \end{array}\right| \begin{array}{cccc} 2 & -8 & 1 & -1 \\ 1 & -4 & -6 & -20 \\ -3 & 12 & 5 & 21 \end{array} \to$$

$$\left.\begin{array}{ccc} -2 & 1 & 0 \\ 1 & 0 & 0 \\ 0 & 0 & 1 \end{array}\right| \begin{array}{cccc} 1 & -4 & -6 & -20 \\ 2 & -8 & 1 & -1 \\ -3 & 12 & 5 & 21 \end{array} \to \left.\begin{array}{ccc} -2 & 1 & 0 \\ 5 & -2 & 0 \\ -6 & 3 & 1 \end{array}\right| \begin{array}{cccc} 1 & -4 & -6 & -20 \\ 0 & 0 & 13 & 39 \\ 0 & 0 & -13 & -39 \end{array} \to$$

$$\left.\begin{array}{ccc} -2 & 1 & 0 \\ 5/13 & -2/13 & 0 \\ -1 & 1 & 1 \end{array}\right| \begin{array}{cccc} 1 & -4 & -6 & -20 \\ 0 & 0 & 1 & 3 \\ 0 & 0 & 0 & 0 \end{array}$$

The matrix is now in echelon form and we continue the reduction by column operations. We omit the last row (consisting of zeros) as this will not affect the reduction.

$$\begin{array}{rrrr} 1 & -4 & -6 & -20 \\ 0 & 0 & 1 & 3 \\ \hline 1 & 0 & 0 & 0 \\ 0 & 1 & 0 & 0 \\ 0 & 0 & 1 & 0 \\ 0 & 0 & 0 & 1 \end{array} \to \begin{array}{rrrr} 1 & -6 & -4 & -2 \\ 0 & 1 & 0 & 0 \\ \hline 1 & 0 & 0 & 0 \\ 0 & 0 & 1 & 0 \\ 0 & 1 & 0 & -3 \\ 0 & 0 & 0 & 1 \end{array} \to \begin{array}{rrrr} 1 & 0 & 0 & 0 \\ 0 & 1 & 0 & 0 \\ \hline 1 & 6 & 4 & 2 \\ 0 & 0 & 1 & 0 \\ 0 & 1 & 0 & -3 \\ 0 & 0 & 0 & 1 \end{array}$$

Thus $PAQ = E_2$, where

$$P = \begin{pmatrix} -2 & 1 & 0 \\ 5/13 & -2/13 & 0 \\ -1 & 1 & 1 \end{pmatrix} \qquad Q = \begin{pmatrix} 1 & 6 & 4 & 2 \\ 0 & 0 & 1 & 0 \\ 0 & 1 & 0 & -3 \\ 0 & 0 & 0 & 1 \end{pmatrix}$$

Of course this reduction can be achieved in many different ways, and though the end result is always the same, the matrices P, Q are not in any way unique.

In the special case where A is a square non-singular matrix, the same process can be used to find the inverse. In this case the echelon form of A is triangular and the reduction can be completed by row operations alone. Here is an example:

$$\left.\begin{array}{rrr} 1 & 0 & 0 \\ 0 & 1 & 0 \\ 0 & 0 & 1 \end{array}\right| \begin{array}{rrr} 2 & 1 & 2 \\ 3 & 1 & 4 \\ 1 & 1 & 1 \end{array} \to \left.\begin{array}{rrr} 0 & 0 & 1 \\ 1 & 0 & 0 \\ 0 & 1 & 0 \end{array}\right| \begin{array}{rrr} 1 & 1 & 1 \\ 2 & 1 & 2 \\ 3 & 1 & 4 \end{array} \to \left.\begin{array}{rrr} 0 & 0 & 1 \\ 1 & 0 & -2 \\ 0 & 1 & -3 \end{array}\right| \begin{array}{rrr} 1 & 1 & 1 \\ 0 & -1 & 0 \\ 0 & -2 & 1 \end{array} \to$$

$$\left.\begin{array}{rrr} 0 & 0 & 1 \\ -1 & 0 & 2 \\ 0 & 1 & -3 \end{array}\right| \begin{array}{rrr} 1 & 1 & 1 \\ 0 & 1 & 0 \\ 0 & -2 & 1 \end{array} \to \left.\begin{array}{rrr} 1 & 0 & -1 \\ -1 & 0 & 2 \\ -2 & 1 & 1 \end{array}\right| \begin{array}{rrr} 1 & 0 & 1 \\ 0 & 1 & 0 \\ 0 & 0 & 1 \end{array} \to \left.\begin{array}{rrr} 3 & -1 & -2 \\ -1 & 0 & 2 \\ -2 & 1 & 1 \end{array}\right| \begin{array}{rrr} 1 & 0 & 0 \\ 0 & 1 & 0 \\ 0 & 0 & 1 \end{array}$$

Hence the inverse of $\begin{pmatrix} 2 & 1 & 2 \\ 3 & 1 & 4 \\ 1 & 1 & 1 \end{pmatrix}$ is $\begin{pmatrix} 3 & -1 & -2 \\ -1 & 0 & 2 \\ -2 & 1 & 1 \end{pmatrix}$.

Exercises

(1) Find a PAQ-reduction to the form E_r for the following matrices:

$$\text{(i)} \begin{pmatrix} 1 & 2 & 3 \\ 2 & 3 & 4 \\ 3 & 4 & 5 \end{pmatrix} \qquad \text{(ii)} \begin{pmatrix} 1 & 3 & 2 & 1 \\ 3 & 5 & 4 & 2 \\ 4 & 3 & 2 & 2 \end{pmatrix} \qquad \text{(iii)} \begin{pmatrix} 3 & 2 & -1 & 4 \\ 2 & 0 & 5 & 1 \\ 8 & 4 & 3 & 9 \\ 1 & 1 & 1 & 1 \end{pmatrix}$$

(2) Find the ranks of the following matrices:

(i) $\begin{pmatrix} 2 & -3 & 1 & -2 \\ 3 & 0 & 5 & 1 \\ -1 & 6 & 3 & 5 \end{pmatrix}$ (ii) $\begin{pmatrix} 4 & 3 & 1 & 2 \\ 3 & 1 & 7 & -1 \\ 5 & 5 & -5 & -5 \end{pmatrix}$ (iii) $\begin{pmatrix} 2 & 1 & -3 & 4 \\ 5 & 2 & -8 & 6 \\ 1 & 1 & -1 & -1 \end{pmatrix}$

(3) Invert the following matrices:

(i) $\begin{pmatrix} 3 & 1 & 2 \\ 2 & 0 & 1 \\ 4 & 3 & 4 \end{pmatrix}$ (ii) $\begin{pmatrix} -1 & 1 & 1 \\ 1 & -1 & 1 \\ 1 & 1 & -1 \end{pmatrix}$ (iii) $\begin{pmatrix} 1 & 2 & 4 \\ 1 & 3 & 9 \\ 1 & -1 & 1 \end{pmatrix}$

(iv) $\begin{pmatrix} 0 & 0 & 1 & 0 \\ 0 & 1 & 0 & 0 \\ 0 & 0 & 0 & 1 \\ 1 & 0 & 0 & 0 \end{pmatrix}$ (v) $\begin{pmatrix} 1 & 2 & -3 & 5 \\ 0 & 1 & 4 & 3 \\ 0 & 0 & 1 & -1 \\ 0 & 0 & 0 & 1 \end{pmatrix}$

(4) Show that $\begin{pmatrix} a & b \\ c & d \end{pmatrix}$ is invertible iff $ad-bc \neq 0$.

(5) Show that every invertible matrix can be written as a product of a lower triangular matrix times a permutation matrix times an upper triangular matrix. Give examples to show that none of these can always be omitted.

5.4 Block multiplication

Let A be an $m \times n$ and B an $n \times p$ matrix, then as we know, $C = AB$ is an $m \times p$ matrix. It consists of mp entries, each a sum of n products, and it may sometimes be convenient to split the matrices so as to allow these multiplications to be performed in stages. To be specific, let $n = n' + n''$ and partition A into two blocks by writing $A = (A' \; A'')$, where A' represents the first n' columns of A and A'' the remaining n'' columns. Similarly we partition B into two blocks of rows: $B = \begin{pmatrix} B' \\ B'' \end{pmatrix}$, where B' represents the first n' rows and B'' the remaining n''. Then it is easily seen that

$$AB = (A' \; A'')\begin{pmatrix} B' \\ B'' \end{pmatrix} = (A'B' + A''B''),$$

i.e. the product may be obtained by multiplying the blocks as if they were scalars. In fact this is only a slight rearrangement, though it can help to make the computation more manageable; here is an example to illustrate the point. To find the product

$$\left(\begin{array}{cc|c} 21 & 17 & 2 \\ 33 & 25 & -3 \end{array}\right)\left(\begin{array}{ccc} 10 & -18 & 49 \\ -15 & 24 & -70 \\ \hline 4 & 0 & 6 \end{array}\right) \tag{1}$$

we partition the matrices along the lines shown. We first compute

$$\begin{pmatrix} 21 & 17 \\ 33 & 25 \end{pmatrix}\begin{pmatrix} 10 & -18 & 49 \\ -15 & 24 & -70 \end{pmatrix} = \begin{pmatrix} -45 & 30 & -161 \\ -45 & 6 & -133 \end{pmatrix}$$

and then

$$\begin{pmatrix} 2 \\ -3 \end{pmatrix}(4 \quad 0 \quad 6) = \begin{pmatrix} 8 & 0 & 12 \\ -12 & 0 & -18 \end{pmatrix}$$

Hence, by addition, the product (1) is

$$\begin{pmatrix} -37 & 30 & -149 \\ -57 & 6 & -151 \end{pmatrix}$$

More generally, let us partition A into blocks in any way:

$$A = \begin{pmatrix} A_{11} & A_{12} & \dots & A_{1s} \\ A_{21} & A_{22} & \dots & A_{2s} \\ \cdot & \cdot & \cdot & \cdot \\ A_{r1} & A_{r2} & \dots & A_{rs} \end{pmatrix},$$

corresponding to a partition of m into r parts: $m = m_1 + \cdots + m_r$ and of n into s parts: $n = n_1 + \cdots + n_s$, so that $A_{\lambda\mu}$ is an $m_\lambda \times n_\mu$ matrix; similarly we partition B into blocks, taking care only to use the same partition of n (and some partition of $p: p = p_1 + \cdots + p_t$):

$$B = \begin{pmatrix} B_{11} & B_{12} & \dots & B_{1t} \\ \cdot & \cdot & \cdot & \cdot \\ B_{s1} & B_{s2} & \dots & B_{st} \end{pmatrix}$$

then $AB = C = (C_{\lambda\nu})$, where $C_{\lambda\nu} = \sum_\mu A_{\lambda\mu}B_{\mu\nu}$.

A judicious choice of blocks can often help to speed calculations; sometimes this is obvious, as when there is a large block of zeros. E.g., if

$$A = \left(\begin{array}{cc|cc} 5 & 3 & 0 & 0 \\ 2 & -1 & 0 & 0 \\ \hline 1 & -4 & 3 & 5 \end{array}\right) \qquad B = \left(\begin{array}{cc} 2 & 1 \\ 3 & -4 \\ \hline 7 & 2 \\ -1 & -5 \end{array}\right) \quad \text{then } AB = \begin{pmatrix} 19 & -7 \\ 1 & 6 \\ 6 & -2 \end{pmatrix}$$

Even when no zeros are present, partitioning may be helpful. Thus a 50×50 matrix has 2500 entries and the multiplication of two such matrices may exceed the storage capacity of the computer we are using. By partitioning each matrix into 4 blocks we can reduce the problem to a series of multiplications of pairs of matrices with 625 entries each.

An important case is that where one of the matrices can be partitioned into diagonal blocks. Let $A_1, \dots, A_r$ be any matrices, not necessarily of the same size. Their *diagonal sum* (sometimes called *direct* sum) is defined as the matrix

$$A = \begin{pmatrix} A_1 & 0 & 0 & \dots & 0 \\ 0 & A_2 & 0 & \dots & 0 \\ \cdot & \cdot & \cdot & \cdot & \cdot \\ 0 & 0 & 0 & \dots & A_r \end{pmatrix} \tag{2}$$

This is also abbreviated by writing $A = \operatorname{diag}(A_1, \ldots, A_r)$. Clearly if A_i is an $m_i \times n_i$ matrix, then A is an $m \times n$ matrix, where $m = \sum m_i$, $n = \sum n_i$. This notion is used in particular when each A_i is a scalar (i.e. a 1×1 matrix).

Exercises

(1) Find the following products by appropriate partitioning:

$$\text{(i)} \begin{pmatrix} 2 & 1 & 0 & 0 \\ 0 & 3 & 0 & 0 \end{pmatrix} \begin{pmatrix} 1 & 0 \\ 0 & 1 \\ 5 & 3 \\ 1 & 4 \end{pmatrix} \qquad \text{(ii)} \begin{pmatrix} 1 & 0 & 5 \\ 2 & 0 & -3 \\ 4 & 1 & 2 \end{pmatrix} \begin{pmatrix} 1 & -2 \\ 27 & 59 \\ 4 & 3 \end{pmatrix}$$

(2) Find $\begin{pmatrix} 3 & 1 & 0 & 0 & 0 \\ 0 & 2 & -1 & 0 & 0 \\ 0 & 0 & 1 & 0 & 0 \\ 0 & 0 & 0 & 4 & 0 \\ 0 & 0 & 0 & 1 & -2 \end{pmatrix}^3$

(3) If $A = \operatorname{diag}(A_1, \ldots, A_r)$, $B = \operatorname{diag}(B_1, \ldots, B_r)$, where for each $i = 1, \ldots, r$, A_i and B_i are square matrices of the same order, show that $AB = \operatorname{diag}(A_1B_1, \ldots, A_rB_r)$.

(4) Show that $\begin{pmatrix} A & B \\ C & D \end{pmatrix}^{\mathrm{T}} = \begin{pmatrix} A^{\mathrm{T}} & C^{\mathrm{T}} \\ B^{\mathrm{T}} & D^{\mathrm{T}} \end{pmatrix}$ and generalize the result to arbitrary partitions.

(5) Show that the set of matrices $A = (a_{ij})$ of order n such that $a_{ij} = 0$ for $i \leqslant r < j$ (for some fixed r) is closed under addition and multiplication. If A is the matrix of an endomorphism θ of a space V (relative to a given basis of V), interpret the restriction on the coefficients as a condition on θ.

Further exercises on Chapter 5

(1) Solve the following system of equations for all t: (i) over $\mathbf{R}$, (ii) over $\mathbf{F}_3$, (iii) over $\mathbf{F}_7$ (only the coefficients are written):

$$\begin{array}{rrrr|r} 1 & 4 & -2 & 3 & t \\ 3 & 5 & 0 & 2 & 5 \\ 0 & 7 & -6 & 7 & 13 \end{array}$$

(2) Solve for all a, b, c:

$$\begin{array}{rrr|r} 4bc & ac & -2ab & 0 \\ 5bc & 3ac & -4ab & -abc \\ 3bc & 2ac & -ab & 4abc \end{array}$$

(3) Find all sets of values of λ, μ for which the system

$$\left.\begin{array}{ccc|c} (\lambda+\lambda\mu-\mu^2) & 1+\mu & 1 & 0 \\ 0 & \lambda^2-\lambda & \lambda-\mu & 0 \\ \lambda & 1+\lambda & 1 & 0 \end{array}\right.$$

has a non-trivial solution; for these values solve the system completely.

(4) Find the inverses of the following matrices:

(i) $\begin{pmatrix} 17 & -15 & -2 \\ 7 & -7 & -1 \\ 5 & -6 & -1 \end{pmatrix}$ (ii) $\begin{pmatrix} 2 & 4 & 1 \\ 1 & 1 & 1 \\ 2 & 3 & 1 \end{pmatrix}$ (iii) $\begin{pmatrix} -2 & -1 & -2 \\ -4 & -1 & -3 \\ -1 & -1 & -1 \end{pmatrix}$

(5) Find a PAQ-reduction for

$$\begin{pmatrix} 4 & 12 & -7 & 6 \\ 1 & 3 & -2 & 1 \\ 3 & 9 & -2 & 11 \end{pmatrix}$$

(6) Find the rank of the matrix $\begin{pmatrix} 0 & t & 2 \\ t & 0 & 0 \\ 0 & 2t & t^2 \end{pmatrix}$ over $\mathbf{R}$, and over $\mathbf{F}_2$.

(7) If A is an $m \times n$ matrix and B an $n \times m$ matrix (over some field) such that $AB = I$, show that $m \leqslant n$ with equality iff $BA = I$.

(8) Let A be a square matrix of order n. Find matrices B, C, D of order n such that $\begin{pmatrix} A & B \\ C & D \end{pmatrix}^2 = 0$. (Hint. Try $n = 1$ first.)

(9) Let A, B be square matrices of orders m, n respectively, and let C be an $m \times n$ matrix. Show that $\begin{pmatrix} A & C \\ 0 & B \end{pmatrix}$ is invertible iff A and B are and that its inverse is then $\begin{pmatrix} A^{-1} & -A^{-1}CB^{-1} \\ 0 & B^{-1} \end{pmatrix}$.

(10)* Find the inverse of a matrix X by partitioning it as $\begin{pmatrix} A & B \\ C & D \end{pmatrix}$ and writing it as a product $\begin{pmatrix} A & 0 \\ C & I \end{pmatrix}\begin{pmatrix} I & P \\ 0 & Q \end{pmatrix}$, when possible. Apply this method to find the inverse of $\begin{pmatrix} 3 & 2 & 1 \\ 4 & 3 & 1 \\ -1 & 2 & -2 \end{pmatrix}$.

6

Rings and fields

6.1 Definition and examples

In Ch. **2** we saw that the set of integers **Z** admits three of the four basic operations of arithmetic and satisfies the laws **Z**. 1–5 stated there. We shall encounter many systems having these features in common with **Z** and it will be useful to have a general definition.

By a *ring* we understand a set R with two binary operations, $x+y$, called *addition*, and xy, called *multiplication*, such that

R. 1 *R is an abelian group under addition.*

R. 2 *R is a monoid under multiplication.*

R. 3 *Addition and multiplication are related by the distributive laws*:

$$(x+y)z = xz+yz, \qquad x(y+z) = xy+xz.$$

The neutral element for addition is called *zero* and written 0, while the neutral element for multiplication is called *one* or *unit-element* and written 1, or sometimes 1_R, if the ring is to be emphasized. The additive inverse of x is written $-x$ and the operation $x-y$, defined as $x+(-y)$, is called *subtraction*, as usual. As in the special case discussed in Ch. **2** we have again the generalized distributive law:

$$(\sum x_i)(\sum y_j) = \sum_{ij} x_i y_j.$$

The commutativity of multiplication is not assumed (that is why we needed two versions of the distributive law in **R**. 3). If commutativity holds, i.e. if

$$xy = yx \quad \text{for all } x, y \in R,$$

the ring R is said to be *commutative*. Of course, addition is always commutative, by definition; as a matter of fact, even if one did not postulate commutativity of addition, it would follow from the other axioms (cf. Ex. (1)).

Two rings are said to be *isomorphic* if there is an *isomorphism* between them, i.e. a bijection which preserves all the operations. Isomorphic rings (like isomorphic groups) may be regarded as abstractly equivalent; intrinsically they are the same, although it may be necessary to distinguish between them in their relation to other rings containing them (just as congruent triangles may differ in their position in the surrounding space).

In any ring R,

$$0x = x0 = 0 \qquad \text{for all } x \in R. \tag{1}$$

For $x0 = x(0+0) = x0+x0$, hence by subtraction, $x0 = 0$ and similarly $0x = 0$. If R consists of a single element, then necessarily $1 = 0$, but this is the only case where this happens. For suppose that $1 = 0$, then for any $x \in R$, $x = x1 = x0 = 0$, by (1), hence R has only one element. The one-element ring, called the *trivial ring*, is usually excluded from consideration as a degenerate case.

Let R be any ring; an element $a \in R$ is said to be *invertible* or a *unit* (not to be confused with unit-element) if it has an inverse in the multiplicative monoid of R, i.e. if there exists $a' \in R$ such that

$$aa' = a'a = 1. \tag{2}$$

We know from Ch. **3** that a' is then unique and, as in any monoid, we shall denote it by a^{-1}. By (1), 0 can never be invertible (unless $1 = 0$); the most we can ask is that every non-zero element be invertible. Rings where this is true are called *division rings* or *skew fields*; a commutative division ring is also called a *field*.†

Given rings R and S, we say that S is a *subring* of R if S is contained in R and forms a ring under the same operations as R (and with the same 0 and 1). For example, the integers form a subring of the field of rational numbers. If S is merely isomorphic to a subring, S_1 say, of R, we say that S can be *embedded* in R and the mapping from S to S_1 is called the *embedding*. In this case, if we identify each element of S_1 with the element of S to which it corresponds, we obtain a ring R_1 which actually contains S as a subring and which is isomorphic to R. This process is described as 'identifying S with the subring S_1 of R'. Thus if S can be embedded in R and the problem under consideration allows us to replace R by an isomorphic ring, then we may take S to be a subring of R by identifying it with its image under the embedding.

Given a ring R and a subset X of R, we observe that any subring of R that contains X also contains all finite sums $\sum a_i p_i$, where $a_i \in \mathbf{Z}$ and p_i is a product (possibly empty) of elements of X. Hence the set S of all such sums is contained in all subrings containing X. Now it is easily checked that S is itself a subring containing X; thus it is the least subring of R containing X. It is called the subring of R *generated* by X; in particular, if this subring is R itself, X is called a *generating set* of R.

Let us now list some examples of rings, including those considered earlier:

(i) The integers $\mathbf{Z}$.

(ii) The integers mod m: $\mathbf{Z}/m$ (for any $m \in \mathbf{N}$).

(iii) Any field, e.g. $\mathbf{Q}$, $\mathbf{R}$, $\mathbf{C}$. Each of these is a subfield of the next.

(iv) Hamilton's algebra of quaternions (Ex. (5), **4.5**) is an example of a skew field.

† Here is a case (rare but not unknown even in non-technical English) where the range of a noun is restricted by *omitting* a qualifying adjective: a field is a particular kind of skew field, just as tea is a particular kind of leaf tea.

(v) Let S be a set and B the set of all its subsets. On B define addition and multiplication by the rules:

$$X+Y = (X \cap Y') \cup (X' \cap Y), \qquad XY = X \cap Y,$$

where X' denotes the complement of X in S. Then B is a ring with the empty set as 0 and S as 1, as is easily verified. This is an example of a *Boolean algebra*, to be discussed in Vol. 2.

(vi) Let A be any abelian group, written additively, and define $R = \mathbf{Z} \times A$ as a ring by the rules:

$$(m, a)+(n, b) = (m+n, a+b), \qquad (m, a)(n, b) = (mn, mb+na).$$

Then R is a commutative ring with unit-element $(1, 0)$.

(vii) If R is any ring and n any positive integer, the $n \times n$ matrices over R form a ring $\mathfrak{M}_n(R)$ or R_n, the *full $n \times n$ matrix ring* over R. We note that if $n > 1$ (and R has more than one element) then R_n is non-commutative, even though R itself may be commutative.

The non-commutativity of R_n follows from the examples given in Ch. **4**. There we also saw that in a full matrix ring we can generally find elements a, b such that

$$ab = 0 \quad \text{but} \quad a \neq 0, b \neq 0. \tag{3}$$

We shall divide all non-zero elements in any ring into two classes. Given $a \neq 0$, if there exists $b \neq 0$ such that ab or ba is zero, then a is said to be a *zero-divisor*; otherwise a is a *non-zerodivisor*. Note that 0 is neither, by definition.

A non-trivial ring with no zero-divisors is said to be *entire*; a commutative entire ring is also called an *integral domain*.† We observe that R is entire iff $1 \neq 0$ and for any $x, y \in R$, $x, y \neq 0$ implies $xy \neq 0$. Hence we have

THEOREM 1 *A ring is entire if and only if the set of its non-zero elements forms a monoid under multiplication.* ■

COROLLARY *Any skew field, and more generally any subring of a skew field is entire.* ■

An important property of non-zerodivisors is the *cancellation law*:

If c is a non-zerodivisor in a ring R, then for any $a, b \in R$ $ca = cb$ or $ac = bc$ implies $a = b$.

For if $ca = cb$, then $c(a-b) = 0$ and hence $a-b = 0$, i.e. $a = b$, and similarly if $ac = bc$. Thus we can cancel any non-zerodivisor; in particular, in an entire ring we can cancel any non-zero element.

† The distinction between commutative and non-commutative entire rings is useful at this introductory stage. In more advanced work the distinction is less important and the term 'integral domain' is used to cover both cases.

Exercises

(1) Let R be a group (not necessarily abelian) under addition and a monoid under multiplication which is related to addition by the distributive laws. Prove that R is a ring (i.e. prove that the commutative law of addition is a consequence of the other laws.)

(2) Determine all rings on at most 5 elements.

(3) Sometimes rings are defined without 1, i.e. as additive group with an associative multiplication satisfying the distributive laws. Find all such rings without 1, on at most 5 elements.

(4) Let R be a 'ring without 1' or, rather, a ring in which the existence of 1 has not been postulated, and suppose that for any $a, b \neq 0$ in R the equation $ax = b$ has a solution. Prove that if R has more than one element, then it is a skew field (and so it has a 1 after all).

(5) Which of the following are subrings of the rings indicated: (i) $\{x \in \mathbf{Q} \mid 3x \in \mathbf{Z}\}$, (ii) $\{x \in \mathbf{Q} \mid 3^n x \in \mathbf{Z}$ for some $n \geqslant 0\}$, (iii) $\{c \in \mathbf{C} \mid c = a+bi$ where $a, b \in \mathbf{Q}\}$, (iv) $\{c \in \mathbf{C} \mid c = a+bi$ where $a, b \in \mathbf{Z}\}$, (v) $\{c \in \mathbf{C} \mid c = a+bi$ where $a \in \mathbf{Z}$, $b \in \mathbf{Q}\}$. ($i = \sqrt{-1}$)

(6) In any ring show that the subgroup of the additive group generated by the unit-element is a subring.

(7) Let R be a ring, X a subset of R and define the *centralizer* of X as $\mathbf{C}(X) = \{c \in R \mid cx = xc$ for all $x \in X\}$. Show that $\mathbf{C}(X)$ is a subring of R. Under what conditions is $X \subseteq \mathbf{C}(X)$? When is $X \subseteq \mathbf{CC}(X)$?

(8) In any ring R show that the set of units is a subgroup of the multiplicative monoid of R.

(9) Let A be an abelian group, written additively, and denote the set of endomorphisms of A by E. Define the sum and product of $\alpha, \beta \in E$ by $a(\alpha+\beta) = a\alpha + a\beta$, $a(\alpha\beta) = (a\alpha)\beta$ ($a \in A$). Show that with these operations E is a ring, with the identity mapping as 1. What is this ring when A is a finite-dimensional vector space over $\mathbf{F}_p$ (p a prime number)? What goes wrong in the above construction if we allow the group A to be non-abelian?

(10) An element c of a ring R is said to be *nilpotent* if $c^n = 0$ for some $n \geqslant 1$. Show that the ring $\mathbf{Z}/n$ has no nilpotent elements other than 0 precisely when n is squarefree (i.e. n is not divisible by the square of any number $\neq 1$).

(11) A ring is said to be *Boolean* if $x^2 = x$ for all $x \in R$. Verify that the ring of all subsets of a set S in the example (v) on p. 117 is Boolean. Show that any Boolean ring is commutative and satisfies the identity $2x = 0$. (Hint. Expand $(x+y)^2$).

(12)* Show that any ring in which each element x satisfies $x^3 = x$ is commutative and satisfies $6x = 0$.

(13) In any entire ring, if an element x satisfies $x^2 = 1$, show that $x = 1$ or $x = -1$.

(14) Let a, b be elements of a skew field such that $ab \neq 0, 1$. Show that $a-(a^{-1}+(b^{-1}-a)^{-1})^{-1} = aba$ (Hua's identity).

(15) Let S be a ring and R a subring of S, then R is said to be *inert* in S and S is an *inert extension* of R, if $c \in R$, $c = ab$ where $a, b \in S$, implies that au, $u^{-1}b \in R$ for some unit $u \in S$. Show that any inert extension of an entire ring is again entire. Show also that a subring of a skew field K which is inert in K is again a skew field.

6.2 The field of fractions of an integral domain

Even when working with integers it is often helpful to make use of the rational numbers. Does anything analogous exist for a general ring? If a ring R is to be a subring of a field, then clearly R must be an integral domain. Conversely, we shall show that every integral domain R can be embedded in a field; the construction is essentially the one which leads from the integers to the rational numbers. That special case suggests that the fractions to be constructed must have the form $ab^{-1} = a/b$ say ($a, b \in R$, $b \neq 0$), but a fraction in this form will not generally be unique:

$$a/b = a'/b' \quad \text{whenever } ab' = ba'. \tag{1}$$

Further, the addition and multiplication of rational numbers is given by the formulae:

$$a_1/b_1+a_2/b_2 = (a_1b_2+a_2b_1)/b_1b_2, \tag{2}$$

$$(a_1/b_1)(a_2/b_2) = a_1a_2/b_1b_2, \tag{3}$$

and it seems reasonable to try these formulae also in the general case. So to construct a field F of fractions for an integral domain R we take all pairs (a, b) from R with $b \neq 0$, and show that the relation (1) defined on such pairs is an equivalence which is preserved by addition and multiplication, defined by (2). It remains to show that the set of equivalence classes forms a field in which R can be embedded. This programme can be simplified a little by first leaving addition out of account. Thus we have the monoid of non-zero elements of R and we have to embed this monoid in a group. We state this step separately as

LEMMA 1 *A commutative monoid S can be embedded in a group if and only if it admits cancellation by all elements:*

$$ac = bc \quad \textit{implies} \quad a = b, \qquad \textit{for all } a, b, c \in S. \tag{4}$$

Proof. The necessity of this condition is clear, so assume that (4) holds. We shall construct fractions a/b by following the steps outlined above. Thus we consider pairs $(a, b) \in S^2$ and define a relation on these pairs by setting

$$(a, b) \sim (a', b') \Leftrightarrow ab' = ba'.$$

Clearly this is reflexive and symmetric. It is also transitive, for if $(a, b) \sim (a', b')$ and $(a', b') \sim (a'', b'')$, then $ab' = ba'$, $a'b'' = b'a''$, hence $ab'b'' =$

$ba'b'' = bb'a''$ and by (4), $ab'' = ba''$, i.e. $(a, b) \sim (a'', b'')$. Let us denote the equivalence class of (a, b) by $[a, b]$ and define a multiplication on these equivalence classes by writing

$$[a_1, b_1][a_2, b_2] = [a_1a_2, b_1b_2]. \tag{5}$$

To show that this multiplication of classes is well defined we must check that the class on the right of (5) really only depends on the *classes* of (a_1, b_1), (a_2, b_2) and not on these pairs themselves. Let us pick other representatives for these classes: if $[a_1, b_1] = [a_1', b_1']$ say, then $a_1b_1' = b_1a_1'$ and hence $a_1a_2b_1'b_2 = a_1'a_2b_1b_2$, thus the class on the right of (5) is unchanged and the same holds if we alter the second representative.

Thus the multiplication is well-defined. Clearly it is associative: $[a_1, b_1][a_2, b_2][a_3, b_3] = [a_1a_2a_3, b_1b_2b_3]$ whichever bracketing is chosen on the left. Let $Q(S)$ be the set of all equivalence classes with this multiplication; we claim that $Q(S)$ is a group. The neutral element is $[1, 1] = [a, a]$, and the inverse of $[a, b]$ is $[b, a]$, for $[a, b][b, a] = [ab, ab] = [1, 1]$. Finally S is embedded in $Q(S)$ by the mapping

$$\lambda: a \mapsto [a, 1]. \tag{6}$$

For this is clearly a homomorphism, and it is injective: if $[a, 1] = [a', 1]$, then $a1 = a'1$, i.e. $a = a'$. ■

The group $Q(S)$ constructed in the lemma is called the *group of fractions* of S. If we identify S with its image in $Q(S)$ under λ, so that $[a, 1] = a$, then the elements of $Q(S)$ can be written in the form ab^{-1} $(a, b \in S)$; for $[a, b] = [a, 1][1, b] = [a, 1][b, 1]^{-1}$. Frequently we shall also write a/b in place of ab^{-1}. In a/b, a is called the *numerator* and b the *denominator*; it should be noted that these terms are associated with the *expression* a/b and not with the element of $Q(S)$ it represents. To give an example, the rational number 3/5 can be written as a fraction with the denominator 10, viz. 6/10, and here 'having the denominator 10' is a property of the expression 6/10 and not of the rational number it represents.

A natural question arises whenever we have a construction such as the one given for $Q(S)$ in Lemma 1: How far is $Q(S)$ unique? Intuitively we would expect the construction to be unique since no arbitrary choices were made at any stage. More precisely, if Q_1, Q_2 are two groups of fractions of S and $\lambda_i: S \to Q_i$ $(i = 1, 2)$ the embeddings, we would expect to have an isomorphism $\alpha: Q_1 \to Q_2$ such that $\lambda_1\alpha = \lambda_2$, i.e. the triangle shown in the diagram commutes; if we identify S with its image in Q_i by means of λ_i,

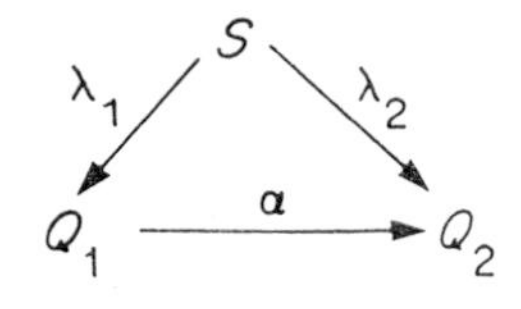

this means that there is an isomorphism between Q_1 and Q_2 which leaves S elementwise fixed. This is in fact the case; we shall deduce it by showing that $Q(S)$ satisfies the universal mapping property:

PROPOSITION 2 *Let S be a commutative monoid with cancellation and $Q(S)$ its group of fractions, with the embedding $\lambda: S \to Q(S)$ constructed in the proof of* Lemma 1. *Given any homomorphism f from S to a group G, there is a unique homomorphism f' from $Q(S)$ to G such that*

$$f = \lambda f'. \tag{7}$$

Thus the triangle shown commutes.

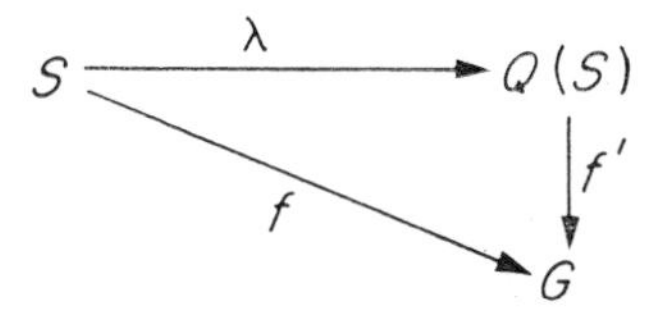

Proof. If such a homomorphism f' exists, then for any $a \in S$, we must have

$$[a, 1]f' = af,$$

and hence $[a, b]f' = (a\lambda \,.\, (b\lambda)^{-1})f' = a\lambda f' \,.\, (b\lambda f')^{-1} = af \,.\, (bf)^{-1}$. This shows that there can be at most one mapping f' satisfying (7), namely,

$$[a, b]f' = (af)(bf)^{-1}. \tag{8}$$

To show that f' is well-defined, we must check that the right-hand side depends only on the class of (a, b) and not on a, b themselves. Let (a', b') be another representative for $[a, b]$, then $ab' = ba'$, hence $af \,.\, b'f = bf \,.\, a'f$, and so $af \,.\, (bf)^{-1} = a'f \,.\, (b'f)^{-1}$. Thus if in (8) we replace a by a' and b by b' the right-hand side is unaffected. We therefore have a mapping f' satisfying (7) and this is easily seen to be a homomorphism. ■

The uniqueness of $Q(S)$ now follows from the general remark in **4.8**; but we shall give another proof here which shows a little more: we prove that whenever S is embedded in a group, the subgroup generated by the image of S is isomorphic to $Q(S)$; this will establish the uniqueness of $Q(S)$. Let f be an embedding of S in a group G, i.e. an injective homomorphism, and denote by G_1 the subgroup generated by the image of S. The image of $Q(S)$ under f' is a subgroup of G containing Sf and hence containing G_1. But by the form (8) of the mapping f', the image of $Q(S)$ is contained in the subgroup G_1, thus f' maps $Q(S)$ onto G_1. We claim that f' is injective and so provides an isomorphism between $Q(S)$ and G_1.

Suppose that $[a, b]f' = 1$, then $af \,.\, (bf)^{-1} = 1$, i.e. $af = bf$ and hence $a = b$, since f is injective; it follows that $[a, b] = 1$. This shows that $xf' = 1$ implies $x = 1$. Now assume that $xf' = yf'$, then $(xy^{-1})f' = (xf')(yf')^{-1} = 1$, hence by what has been proved, $xy^{-1} = 1$ i.e. $x = y$. Thus f' is indeed injective and we have shown that $G_1 \cong Q(S)$, i.e. the subgroup generated by the image of S (under any embedding in a group) is isomorphic to $Q(S)$.

We now apply our construction to the case of rings: Let R be an integral domain and denote by R^* the set of its non-zero elements. Then R^* is a commutative monoid satisfying cancellation (cf. **6.1**), so we can apply Lemma 1 and construct its group of fractions Q say. We claim that the set $F = Q \cup \{0\}$ can be defined as a field in such a way that R can be embedded in F. The elements of Q can all be expressed in the form a/b, where $a, b \in R^*$; correspondingly we shall use $0/a$, where $a \in R^*$, to express 0. We observe that any two elements of F can be brought to a common denominator: given a_1/b_1, a_2/b_2, we have $a_1/b_1 = a_1b_2/b_1b_2$, $a_2/b_2 = a_2b_1/b_1b_2$. Here $b_1, b_2 \in R^*$ and hence $b_1b_2 \in R^*$. Thus it will be enough to define the addition of fractions with a common denominator:

$$a/c + b/c = (a+b)/c. \tag{9}$$

This is well-defined, for if $a/c = a'/c'$ and $b/c = b'/c'$, then $ac' = ca'$, $bc' = cb'$ and hence $(a+b)c' = c(a'+b')$. The associative and commutative laws are easily checked, the neutral for addition is $0/1$ and the mapping $a \mapsto a/1$ of R into F is easily seen to be an embedding. As in the case of monoids one shows that the least subfield containing a given integral domain is unique up to isomorphism, thus we have

THEOREM 3 *Any integral domain R can be embedded in a field, and the least field containing R as subring is unique up to isomorphism.* ■

The field constructed here is called the *field of fractions* or *quotient field* of R; its elements can all be written in the form ab^{-1} $(a, b \in R, b \neq 0)$. Of course if R is a field to begin with, the field of fractions is just R itself.

In the special case $R = \mathbf{Z}$ the field of fractions is the rational field $\mathbf{Q}$. In this case, although a given rational number can be written in many forms, e.g. $3/5 = 6/10 = -9/-15$ etc., there is a certain *reduced form* for each fraction which is unique. Let us call a fraction m/n $(m, n \in \mathbf{Z}, n \neq 0)$ *reduced* if m, n are coprime and $n > 0$. Every fraction can be written in reduced form, for if we are given a/b $(a, b \in \mathbf{Z}, b \neq 0)$, let d be the highest common factor of a and b, then $a = da_0$, $b = db_0$ where a_0, b_0 are coprime. Since $b \neq 0$, we can ensure that $b_0 > 0$, by changing the sign of d if necessary. Now $a/b = a_0/b_0$ and the latter is reduced. This reduced form is unique, for if $a_0/b_0 = a_1/b_1$ where a_1/b_1 is also reduced, then $a_0b_1 = b_0a_1$. Here b_0 is prime to a_0, hence $b_0 \mid b_1$ and similarly $b_1 \mid b_0$, hence $b_0 = b_1$ because both are positive. Therefore also $a_0 = a_1$ and we obtain

PROPOSITION 4 *Every element of $\mathbf{Q}$ has a unique reduced form a/b, where a, b are coprime and $b > 0$.* ■

For a general integral domain there may be no such natural way of defining a reduced form in its field of fractions. However, this does not of course affect the construction of the field of fractions itself.

There is one important case where the forming of fractions is unnecessary:

THEOREM 5 *Every finite integral domain is a field.*

Proof. Let R be a finite integral domain, then R^*, the set of non-zero elements of R, is a finite monoid with cancellation and hence (by Prop. 2, **3.2**) a group; so R is a field. ■

The attentive reader will have noticed that this proof does not depend on the commutativity of R and so could have been stated for entire rings. But the gain in generality would have been spurious for, as we shall see in Vol. 2, every finite division ring is necessarily commutative.

Exercises

(1) Restate Lemma 1 for monoids written in additive notation and apply it to the monoid of non-negative integers under addition. What is the resulting group?

(2) Fill in the details left to the reader in the construction of the field of fractions and point out where the commutativity of the ring is used.

(3) Show that the field of fractions of an integral domain is unique up to isomorphism. Specifically, if Q_1, Q_2 are two fields of fractions of an integral domain, show that the group isomorphism between Q_1^* and Q_2^* (which exists by the remarks following Prop. 2) extends to a field isomorphism between Q_1 and Q_2.

(4) Show that any homomorphism from a skew field to a non-trivial ring is injective.

(5) Show that an injective homomorphism between integral domains can be extended in just one way to an injective homomorphism between their fields of fractions.

(6) If Th. 3 is applied to a field F, show that the resulting field of fractions is F. Deduce that any field containing a ring R contains an isomorphic copy of the field of fractions of R.

(7) Let R be the subring of the complex field $\mathbf{C}$ generated by i; describe the field of fractions of R.

(8) Let R be a commutative subring of a skew field K; show directly that the subfield of K generated by R is commutative. How can this be deduced from Prop. 2?

6.3 The characteristic

In any ring R we shall abbreviate expressions like $a+a$ by $2a$ and generally put

$$na = a+a+ \cdots +a \ (n \text{ terms}),$$

where n is any positive integer. By the general distributive law this agrees with the usual meaning of na in the case $R = \mathbf{Z}$; likewise in a general ring, for

$n = 1$ no ambiguity results since $1_R a = a$. When $n < 0$, say $n = -k$, $k > 0$, we define na to mean $-ka$, so that na is defined for all $n \in \mathbf{Z}$. It is easily seen that we have the rules

$$m(a+b) = ma+mb, \tag{1}$$

$$(a, b \in R,\ m, n \in \mathbf{Z}).$$

$$(m+n)a = ma+na, \qquad (mn)a = m(na), \tag{2}$$

Let us consider the additive order of a, i.e. the order of a in the additive group of R; this is the least positive integer k (if one exists) such that $ka = 0$. When R is entire, any two non-zero elements have the same order. For if $a \in R$ has finite order $k > 1$, then $a \neq 0$ and $ka = 0$, hence $k1_R \,.\, a = 0$. By cancelling a we find $k1_R = 0$ and hence for any $b \in R$, $kb = k1_R \,.\, b = 0$. This shows that every element of R has order dividing k. In particular, by choosing a to be of least order $k > 1$, we see that every non-zero element of R has order exactly k. Furthermore, this order k must be a prime number, for if $k = rs$, then $r1_R \,.\, s1_R = rs1_R = 0$ and since R is entire, either $r1_R = 0$ or $s1_R = 0$, i.e. k must divide r or s, and this shows k to be a prime.

In any ring R the additive order of the unit-element is usually called the *characteristic* of R; however, when this order is infinite, i.e. when no positive multiple of 1_R vanishes, we say that R has *characteristic zero*. Our conclusions can then be summarized as

THEOREM 1 *In an entire ring R the additive order of each non-zero element is the same. It is either infinite—R has characteristic zero; or a prime number p—R has characteristic p.* ■

This classification according to characteristic is particularly important for fields, thus $\mathbf{Q}$, $\mathbf{R}$, $\mathbf{C}$ all have characteristic 0, while $\mathbf{F}_p$ has characteristic p. The characteristic of a field F is written char F.

Let F be any field of characteristic 0, then the ring $\mathbf{Z}$ of integers can be embedded in F by mapping

$$n \mapsto n1_F. \tag{3}$$

By definition this represents the multiples of 1_F and their negatives; that (3) is a homomorphism is clear from the equations $m1_F + n1_F = (m+n)1_F$, $m1_F \,.\, n1_F = mn \,.\, 1_F$, $1 \,.\, 1_F = 1_F$. It is also injective, for if $m1_F = n1_F$, where $m \geqslant n$ say, then $(m-n)1_F = 0$ and hence $m-n = 0$ because F has characteristic 0. Thus $\mathbf{Z}$ is embedded in F; now F contains all rational numbers: $a = m/n$ is obtained as a solution of the equation $nx = m1_F$ in F. So every field of characteristic 0 contains a copy of the rationals; moreover, this copy of $\mathbf{Q}$ is contained in every subfield of F. For any subfield must contain 0_F and 1_F, hence all integral multiples of 1_F and so all rational multiples. This smallest subfield of F is called the *prime subfield* of F.

Next let F be a field of prime characteristic p, then the multiples of 1_F, namely $0_F, 1_F, 2_F, \ldots, (p-1)1_F$, already form a subfield isomorphic to $\mathbf{F}_p$.

This is again the least subfield of F, contained in every subfield of F, and it is again called the *prime subfield* of F. Thus we have

THEOREM 2 *Every field F contains a least subfield, the prime subfield, and this is isomorphic to* $\mathbf{Q}$ *if char* $F = 0$, *and to* $\mathbf{F}_p$ *if* char F *is a prime* p. ∎

This result still holds for skew fields, but a little more care is needed then to verify that the image of $\mathbf{Z}$ is still commutative (cf. Ex. (4)).

In an integral domain of characteristic $p \neq 0$, the binomial theorem takes on a particularly simple form. We recall that in any commutative ring we have the formula

$$(a+b)^n = \sum \binom{n}{i} a^i b^{n-i},$$

for any $a, b \in R$ and any $n \geqslant 1$. Here $\binom{n}{i} = n!/i!(n-i)!$ and when $n = p$ is a prime number, then $p \mid p!$ but for $1 < i < p$, $p \nmid i!(p-i)!$, because $i!(p-i)!$ is a product of factors less than p. It follows that $p \mid \binom{p}{i}$ for $1 < i < p$, and hence in an integral domain of characteristic p,

$$(a+b)^p = a^p + b^p. \tag{4}$$

By induction we obtain

PROPOSITION 3 *In an integral domain R of prime characteristic p,*

$$(a+b)^{p^k} = a^{p^k} + b^{p^k}, \tag{5}$$

for any $a, b \in R$ and any integer $k \geqslant 0$.

Proof. For $k = 0$ this is trivial (interpreting p^0 as 1) and for $k = 1$ it reduces to (4). Let $k > 1$ and assume that the result holds for values less than k. Then $(a+b)^{p^k} = ((a+b)^{p^{k-1}})^p = (a^{p^{k-1}} + b^{p^{k-1}})^p = a^{p^k} + b^{p^k}$, by induction, and the case $k = 1$. ∎

The reader can easily convince himself that here the assumption of commutativity cannot be dropped.

Exercises

(1) In any ring R show that (1) and (2) hold.

(2) Show that in any ring the non-zerodivisors all have the same additive order. What are the values this order can take?

(3) If in a ring R, every element has finite additive order, show that a positive integer n exists such that $nx = 0$ for all $x \in R$. What are the possible values for the least such n? Does this result remain true if we allow rings without a unit-element?

(4) For any ring R, show that $n \mapsto n \,.\, 1_R$ is a ring homomorphism of $\mathbf{Z}$ into the centre of R. When R is entire, show that the image of this mapping is either $\mathbf{Z}$ or F_p for some prime p. Deduce that Th. 2 still holds for skew fields.

(5) Show that for any integral domain R of prime characteristic p, the mapping $\varphi: x \mapsto x^p$ is an endomorphism (i.e. a homomorphism of R into itself). Verify that φ is injective and, when R is finite, that φ is an automorphism.

(6) Let R be any ring in which $(x+y)^2 = x^2+y^2$ for all $x, y \in R$. Show that R is commutative and $2x = 0$ for all $x \in R$.

6.4 Polynomials

In the study of functions the polynomial functions play an important role. They are the functions of the form

$$f = a_0 + a_1 x + \cdots + a_n x^n, \tag{1}$$

where the a_i are real (or complex) numbers. Here the role of x is twofold: it may be a variable which can take different real (or complex) values, or it may be an indeterminate, i.e. a symbol allowing us to handle expressions like (1) formally. We begin by concentrating on this formal aspect and show that for any ring R we can define a ring of polynomials in x with coefficients in R. This polynomial ring is to consist of all formal expressions (1), where now $a_i \in R$, and two expressions, f given by (1) and

$$g = b_0 + b_1 x + \cdots + b_m x^m \tag{2}$$

are equal iff $a_0 = b_0$, $a_1 = b_1, \ldots$; here we do not require $m = n$, but if $m < n$ say, then $a_{m+1} = \cdots = a_n = 0$. Addition and multiplication of polynomials are defined by the rules:

$$f+g = a_0+b_0+(a_1+b_1)x+ \cdots +(a_m+b_m)x^m+ \cdots +a_n x^n \quad \text{(if } m \leqslant n\text{, say)} \tag{3}$$

$$fg = a_0 b_0 + (a_0 b_1 + a_1 b_0)x + (a_0 b_2 + a_1 b_1 + a_2 b_0)x^2 + \cdots + a_n b_m x^{m+n}. \tag{4}$$

It is not difficult to verify that all the laws for rings are satisfied, so the polynomials in x form a ring, which will be denoted by $R[x]$. If, in (1), $n = 0$, f is called a *constant* polynomial. From (3) and (4) with $m = n = 0$ it follows that the constant polynomials add and multiply as in R and so form a subring of $R[x]$ isomorphic to R. Thus R is embedded in $R[x]$. We see that $R[x]$ is the ring generated by x and the constant polynomials, but the above method of construction (and the verifications which were omitted) are necessary to guarantee the existence of the polynomial ring. The result of this discussion may be summarized as follows:

THEOREM 1 *Let R be any ring, then there exists a polynomial ring $R[x]$ whose elements are the expressions* (1), *with operations* (3) *and* (4), *and R is embedded in $R[x]$, each element of R being represented by the corresponding constant polynomial.* ■

Any non-zero polynomial in an indeterminate x over R can be written in the form

$$f = c_0x^n + c_1x^{n-1} + \cdots + c_{n-1}x + c_n, \qquad \text{where } c_i \in R \text{ and } c_0 \neq 0.$$

We call n the *degree* of f and write $n = \deg f$; thus the elements of degree 0 are just the non-zero constants. The *leading coefficient* of f is c_0, and c_0x^n is its *leading term*. If $c_0 = 1$, f is said to be *monic*.

To obtain a relation between the degrees of f, g and $f+g$ we need only look at (3): we see that

$$\deg(f+g) \leqslant \max\{\deg f, \deg g\}, \tag{5}$$

but we must make the proviso that $f, g, f+g$ are non-zero, since the degree of the zero polynomial has not yet been defined. We could overcome this by arbitrarily defining $\deg 0$ as some fixed negative number; then (5) will hold without exception.

Next we try to relate $\deg fg$ to the degrees of f and g. For simplicity let us take R to be an entire ring. Then (4) shows that

$$\deg fg = \deg f + \deg g, \tag{6}$$

under the assumption that $f, g \neq 0$ (this already implies that $fg \neq 0$). If $f = 0$, say, (6) reduces to

$$\deg 0 = \deg 0 + \deg g.$$

Here $\deg g$ can assume any non-negative integer value, and there is clearly no integer value for $\deg 0$ to satisfy this equation, but formally we can satisfy it by setting

$$\deg 0 = -\infty. \tag{7}$$

In making this definition we are simply adopting a convention designed to ensure that (6) holds without exception; the same value for $\deg 0$ will also satisfy (5). Thus with the convention (7) the following result holds without exception:

PROPOSITION 2 *Polynomials over an entire ring (in particular a field, even skew) satisfy*

$$\deg(f+g) \leqslant \max\{\deg f, \deg g\},$$
$$\deg fg = \deg f + \deg g. \blacksquare$$

In particular if $f, g \neq 0$, then $fg \neq 0$ and we obtain the

COROLLARY *If R is an entire ring, then so is the polynomial ring $R[x]$.* ■

So far we have discussed polynomials as formal expressions of the type (1). But the basic property of a polynomial is the fact that we can substitute arbitrary values for the indeterminate. In the case of a polynomial over R these values may lie in R or, more generally, in any ring S containing R as a subring. The substitution property is then just another instance of the universal mapping property already encountered. We state this in its general form in Th. 4, restricting ourselves to commutative rings (the most important case) for simplicity and refer to the exercises for the general case. For clarity we single out as a lemma a remark needed in the proof.

LEMMA 3 *Let λ, λ' be two homomorphisms from a ring R to a ring S. If λ and λ' agree on a generating set of R, then they are equal.*

Proof. Let us write

$$R' = \{a \in R \mid a\lambda = a\lambda'\};$$

we claim that R' is a subring of R. Clearly $1_R\lambda = 1_R\lambda' = 1_S$ and, if $a, b \in R'$, then $(a+b)\lambda = a\lambda + b\lambda = a\lambda' + b\lambda' = (a+b)\lambda'$ and so $a+b \in R'$; similarly we see that $ab \in R'$ and R' is therefore a subring. By hypothesis, R' contains a generating set of R, hence $R' = R$ by the definition of generating set. This means that $a\lambda = a\lambda'$ for all $a \in R$, i.e. $\lambda = \lambda'$. ∎

THEOREM 4 *Let R be a commutative ring; then there exists a commutative ring $R[x]$ with an embedding $\mu: R \to R[x]$ and containing an element x such that for any homomorphism φ from R to a commutative ring S and any $c \in S$ there is a unique homomorphism $\varphi_c: R[x] \to S$ such that*

$$\varphi = \mu\varphi_c \quad \text{and} \quad x\varphi_c = c. \tag{8}$$

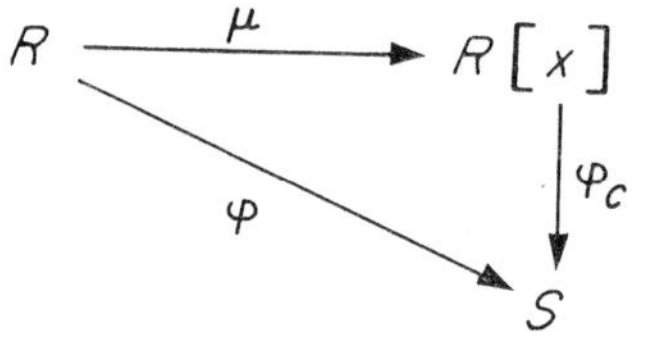

Moreover, this property determines $R[x]$ up to an isomorphism leaving R elementwise fixed.

The ring $R[x]$ in the statement is in fact the polynomial ring constructed earlier, but this theorem provides another characterization of $R[x]$, because of the uniqueness statement at the end. In detail this means, if $R[x']$ is another ring having the properties ascribed to $R[x]$, with embedding $\mu': R \to R[x']$, then there is an isomorphism $\alpha: R[x] \to R[x']$ such that $\mu' = \mu\alpha$. If we regard

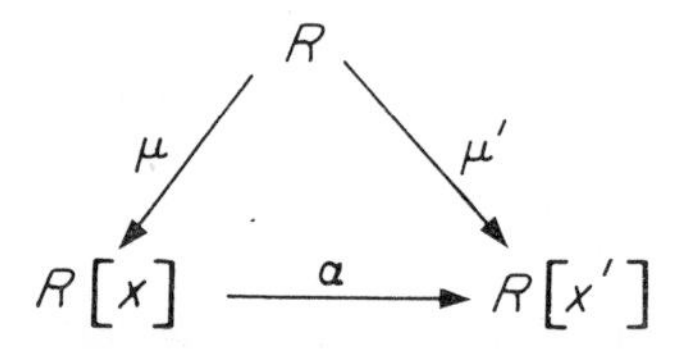

R as a subring of $R[x]$ and $R[x']$ (by identifications using μ and μ'), this means that α is an isomorphism which maps every element of R to itself, i.e. 'α leaves R elementwise fixed'.

Proof. We show that the assertion holds for the polynomial ring in x over R. Let us define φ_c by

$$\varphi_c\colon \sum a_i x^i \mapsto \sum (a_i\varphi)c^i,$$

thus φ_c substitutes c for x and applies φ to the coefficients. Since every element of $R[x]$ can be written uniquely in the form $\sum a_i x^i$, this definition is unambiguous. Clearly φ_c satisfies (8); and it is a homomorphism because (3) and (4) still hold when we replace x by c. Moreover, it is unique, by Lemma 3, because its values on R and x are prescribed, and these elements generate $R[x]$.

The uniqueness of $R[x]$ now follows by the universal mapping property, as in **4.8.** ■

It is instructive to write out the uniqueness proof (by translating the proof given in **4.8** that initial objects are isomorphic), and the reader is urged to do so.

The process of adjoining an indeterminate may be applied to any ring, in particular to the ring $R[x]$ of polynomials in x over R. Of course we must then call the new indeterminate by a name other than x, say y. In this way we obtain the ring $R[x, y]$ of polynomials in x and y; the general element is of the form

$$f = \sum a_{ij}x^i y^j \qquad (a_{ij} \in R),$$

where the sum is over some finite set of pairs (i, j) of exponents. Instead of indicating this set explicitly, one usually takes the sum over all pairs (i, j), with $a_{ij} = 0$ for all except a finite set of pairs.

More generally, we define the ring of polynomials in the indeterminates $x_1, \ldots, x_n$ inductively by the rule

$$R[x_1, \ldots, x_n] = R[x_1, \ldots, x_{n-1}][x_n].$$

In other words, we regard the elements of $S = R[x_1, \ldots, x_n]$ as polynomials in x_n with coefficients that are polynomials in $x_1, \ldots, x_{n-1}$ over R. Clearly if R is commutative, or entire, then so is S. From the definition we have the following normal form for the elements of S:

$$f = \sum a_{i_1\ldots i_n} x_1^{i_1} \ldots x_n^{i_n} \qquad (a_{i_1\ldots i_n} \in R). \tag{9}$$

Here $a_{i_1\ldots i_n}$ is uniquely determined as the coefficient of $x_1^{i_1} \ldots x_n^{i_n}$ in f; formally (9) is an infinite sum, but in fact only finitely many of the as are non-zero. Since the indeterminates commute with each other and the elements of R, the ring S depends symmetrically on $x_1, \ldots, x_n$; thus x_n plays no special role: we could equally well write f as a polynomial in x_1 with coefficients in $R[x_2, \ldots, x_n]$, or single out any other x_i.

Each product $m_{(i)} = x_1^{i_1} \ldots x_n^{i_n}$ is called a *monomial*, the corresponding term $a_{i_1\ldots i_n} m_{(i)}$ in (9) is called a *monomial term*; its *total degree* (or simply: *degree*) is $\sum i_\nu$, the *degree* in x_ν being i_ν. The *degree* of f is the maximum of the degrees of its non-zero terms; e.g. $2x_1^5x_2^3x_3 - x_1^2x_3^3 + 7x_2^6$ is of degree 5 in x_1, 6 in x_2 and 3 in x_3, the total degree being 9.

A polynomial in which all terms have the same total degree is called *homogeneous* or also a *form*. The only forms in one variable are monomial terms, but in x, y we have, e.g. the quadratic form $ax^2+hxy+by^2$. A practical test for homogeneity is the following:

The polynomial $f(x_1, \ldots, x_n)$ is homogeneous of degree k iff

$$f(tx_1, \ldots, tx_n) = t^k f(x_1, \ldots, x_n), \tag{10}$$

where t is another indeterminate. The proof is almost immediate, by a comparison of monomial terms in (10).

Sometimes it is necessary to order the monomials—a total ordering is needed even for such a mundane purpose as writing a polynomial out in full—in this case one frequently uses the *lexicographic ordering* (as in a dictionary) defined as follows: of any two monomials with the same total degree, $x_1^{i_1} \ldots x_n^{i_n}$ *precedes* $x_1^{j_1} \ldots x_n^{j_n}$ if the first non-zero difference among i_1-j_1, $i_2-j_2, \ldots, i_n-j_n$ is positive. Thus e.g., $x_1^3x_2x_3^2$ precedes $x_1^3x_3^3$ and is preceded by $x_1^3x_2^2x_3$. Now monomials of different total degrees may be compared, e.g. by taking the one of lower degree first, though different conventions may be used at different times. In any polynomial, the first monomial term (in the lexicographic ordering) among the terms of highest degree is called the *leading term.*

Exercises

(1) Given any commutative ring R and $\alpha \in R$, show that the mapping $f(x) \mapsto f(x-\alpha)$ is an automorphism of $R[x]$. Give an example to show that the requirement of commutativity cannot be omitted.

(2) Let F be a field and $\alpha \in F$. Show that the powers of $x-\alpha$ in $F[x]$ are linearly independent over F. If $\alpha_1, \ldots, \alpha_r$ are distinct elements of F, show that $(x-\alpha_1)^{-1}$, $\ldots, (x-\alpha_r)^{-1}$ are linearly independent over F; more generally, show that the powers $(x-\alpha_i)^{-j}$ $(i = 1, \ldots, r, j = 1, 2, \ldots)$ are linearly independent over F.

(3) Prove the following characterization of $R[x]$ for general rings: Given any ring R, there exists a ring $R[x]$ with an embedding $\mu: R \to R[x]$ and an element x such that for any homomorphism $\varphi: R \to S$ and any c in the centre of S, there is a unique homomorphism $\varphi_c: R[x] \to S$ such that $\varphi = \mu\varphi_c$ and $x\varphi_c = c$. Moreover, this property determines $R[x]$ up to an isomorphism leaving R elementwise fixed.

(4) Given any ring R, define the set R^N of sequences $(a_i) = (a_0, a_1, a_2, \ldots)$ as a ring by the rules:

$$(a_i)+(b_i) = (a_i+b_i), \quad (a_i)(b_i) = (c_i), \quad \text{where } c_n = \Sigma a_i b_{n-i}.$$

Verify that with these operations R^N becomes a ring and $c \mapsto (c, 0, 0, \ldots)$ is an embedding of R in R^N. If we write $x = (0, 1, 0, 0, \ldots)$, the elements of R^N may formally be written as infinite sums: $(a_i) = \sum a_i x^i$; for this reason R^N is also called the *ring of formal power series* over R and it is usually denoted by $R[[x]]$. Find an embedding of the polynomial ring $R[x]$ in $R[[x]]$.

(5) Show that the multiplication in $R[[x]]$ obeys the infinite distributive law: $(\sum a_i x^i)(\sum b_j x^j) = \sum_{ij} a_i b_j x^{i+j}$.

(6) In $R[[x]]$, show that $\sum a_i x^i$ is a unit iff a_0 is a unit in R.

(7) Show that the power series with coefficients in **C** and with positive radius of convergence form a subring of $\mathbf{C}[[x]]$.

(8) In the polynomial ring $S = R[x_1, \ldots, x_n]$ show that any permutation of $x_1, \ldots,$ x_n defines an automorphism of S. Give a definition of $R[[x_1, \ldots, x_n]]$ and prove a corresponding statement for this ring.

(9) Find an embedding of $R[[x]][y]$ in $R[y][[x]]$. Is it an isomorphism?

(10) For each $f \in R[[x]]$ define the *order* $o(f)$ of f as the least degree of the non-zero terms in f. Show that (i) $o(f-g) \geqslant \min\{o(f), o(g)\}$, (ii) $o(fg) \geqslant o(f)+o(g)$, with equality if R is entire.

(11) Prove that the formula (10) characterizes forms of degree k.

(12) Show that the number of monomials of degree k in $x_0, x_1, \ldots, x_n$ is $\binom{n+k}{n}$.

(13) By expressing $(1+x)^{m+n}$ in two ways, prove that $\sum_{\nu=0}^{k} \binom{m}{\nu}\binom{n}{k-\nu} = \binom{m+n}{k}$.

6.5 Factorization

In Ch. **2** we briefly discussed the factorization of integers; many of the properties found there hold more generally. The most useful result in this context is the fundamental theorem of arithmetic and we shall find that rings satisfying the conclusion of this theorem—unique factorization domains—display rather simple divisibility properties.

Let R be an integral domain; as in **Z** we define $b \mid a$ (for any $a, b \in R$) to mean: $a = bc$ for some $c \in R$. As before, any element u of R satisfying $u \mid 1$ is called a *unit* and two elements a, b are *associated* if $a \mid b$ and $b \mid a$; clearly this is an equivalence relation on R. If a and b are associated, then $a = bu$, $b = av$ for some $u, v \in R$; hence $a = auv$ and so either $a = b = 0$, or by cancelling a we find $uv = 1$ and hence u and v are units. Conversely, if u is a unit, a and au are associated. This proves

PROPOSITION 1 *In any integral domain two elements a, b are associated if and only if $a = bu$, where u is a unit.* ∎

By an *irreducible element* or *atom* in an integral domain R we understand a non-unit which is not a product of two non-units. Thus in **Z** the atoms are just the prime numbers and their negatives, and the fundamental theorem of arithmetic shows that every integer can be completely factorized into atoms in essentially one way. The uniqueness of factorization turns out to be a useful property, so we make the

DEFINITION. A *unique factorization domain* (UFD for short) is an integral domain in which every element not zero or a unit can be written as a product of atoms, and given two complete factorizations of the same element:

$$c = a_1 \dots a_r = b_1 \dots b_s \qquad (a_i, b_j \text{ atoms}), \tag{1}$$

then $r = s$ and after suitably renumbering the bs, a_i is associated to b_i.

With this definition **Z** is a UFD, for the unique factorization of positive integers was proved in Th. 3, **2.2**, and we obtain the factorization of negative integers by observing that -1 is a unit, so $-c$ is associated to c.

A field is trivially a UFD since every non-zero element is a unit. Later in this section we shall prove that a polynomial ring over a field is a UFD; for the moment we shall describe some of the properties of a general UFD.

In a UFD R, let $\mathscr{P}$ be a set of atoms such that every atom of R is associated to exactly one member of $\mathscr{P}$. This means that we choose one representative from each class of associated atoms, (i.e. from each class consisting of atoms of the equivalence relation defined on R by the relation of being associated). In most interesting cases $\mathscr{P}$ is infinite, but this is not essential for what follows. Every non-zero element of R can be written in just one way as

$$c = up_1^{\alpha_1} \dots p_r^{\alpha_r} \tag{2}$$

where u is a unit, $\alpha_i \geqslant 0$ and the p_i are distinct elements of $\mathscr{P}$; here we agree to disregard factors of the form $p^\circ = 1$. In fact we can even write (2) as a formally infinite product

$$c = u \prod p_i^{\alpha_i}, \tag{3}$$

where all but a finite number of the α_i are zero. In this form it is particularly simple to find the highest common factor and least common multiple. By a *highest common factor* (HCF for short) or *greatest common divisor* of two elements a, b in an integral domain R we understand an element $d \in R$ such that

(i) $d \mid a, d \mid b$,

and

(ii) $d' \mid a, d' \mid b \Rightarrow d' \mid d$.

Thus d is a common factor of a and b and is divisible by every other such factor (i.e. it is greatest among such factors). E.g., 15 and 25 have the HCF 5. The HCF is not unique: if d is an HCF, clearly so is any associate of d. Conversely, let d, d' both be HCFs of a, b then $d' \mid a$, $d' \mid b$, hence $d' \mid d$; reversing the roles of d and d' we find that $d \mid d'$, so that d' is in fact associated to d. We shall usually ignore the distinction between associates and denote any one of the HCFs of a and b by (a, b). With this convention we have the following rules for working with HCFs, which are easily checked:

(i) $(ac, bc) = (a, b)c$,
(ii) $((a, b), c) = (a, (b, c))$,
(iii) (a, b) is associated to a iff $a \mid b$,
(iv) $(a, 0) = a$.

So far the question whether HCFs always exist has been left open. We now answer this question for an important class of rings.

THEOREM 2 *In a unique factorization domain any two elements have an HCF. If*

$$a = up_1^{\alpha_1} \ldots p_r^{\alpha_r}, \qquad b = vp_1^{\beta_1} \ldots p_r^{\beta_r}, \tag{4}$$

where u, v are units and $p_1, \ldots, p_r$ are pairwise non-associated atoms, then

$$(a, b) = p_1^{\gamma_1} \ldots p_r^{\gamma_r} \qquad \text{where } \gamma_i = \min\{\alpha_i, \beta_i\}. \tag{5}$$

Proof. In a UFD, if a, b are given by (4) and $d = \prod p_i^{\delta_i}$, then $d \mid a$ iff $\delta_i \leqslant \alpha_i$ for $i = 1, \ldots, r$. It follows that the expression given by (5) is a common factor, and is divisible by every common factor. ∎

Analogously we can define the *least common multiple* (LCM) of a and b as an element m satisfying (i) $a \mid m$, $b \mid m$ and (ii) $a \mid m'$, $b \mid m' \Rightarrow m \mid m'$. Again the LCM is determined up to associates. Denoting it by $[a, b]$, we find that in a UFD, if a, b are given by (4), then

$$[a, b] = p_1^{\mu_1} \ldots p_r^{\mu_r} \qquad \text{where } \mu_i = \max\{\alpha_i, \beta_i\}. \tag{6}$$

Since for any two numbers α, β, $\min\{\alpha, \beta\} + \max\{\alpha, \beta\} = \alpha + \beta$, we see that $(a, b)[a, b]$ is associated to ab; for any atom p_i occurs in $(a, b)[a, b]$ to the power $\min\{\alpha_i, \beta_i\} + \max\{\alpha_i, \beta_i\}$ and in ab to the power $\alpha_i + \beta_i$. But this rule can be proved more generally:

PROPOSITION 3 *In any integral domain R, let $a, b \in R$ have an LCM m. Then $m = 0$ if and only if either a or b vanishes, and excluding this case, ab/m is an HCF of a, b.*

Proof. The case $a = 0$ or $b = 0$ is easily settled by the reader, so we may assume $a, b \neq 0$. Clearly ab is a common multiple of a, b, hence $m \mid ab$, say $ab = dm$; in particular this shows that $m \neq 0$. Moreover, since $a \mid m$, $b \mid m$, we have $m = ab' = ba'$ say, and so $ab = dab'$, i.e. $b = db'$; similarly $a = da'$ and this shows d to be a common factor of a and b. If d' is another common factor, put $m' = ab/d'$. Writing m' as $m' = a \,.\, b/d' = b \,.\, a/d'$, we see that $a \mid m'$, $b \mid m'$, hence $m \mid m'$, by the definition of m as LCM. Thus $m' = mc$ say, and so $md = ab = m'd' = mcd'$, i.e. $d = cd'$. Thus $d' \mid d$ and d is indeed an HCF of a and b. ∎

Although the formulae (5), (6) are easy to work with, they presuppose that we have a UFD and that a, b are completely factorized. And the complete factorization of two elements (even when this exists) can be a lengthy process. Fortunately, for many rings such as $\mathbf{Z}$ and $F[x]$ there is another more direct method, using the division algorithm already encountered in Ch. **2**, which goes back to Euclid.

Let R be an integral domain and suppose that with each $a \in R$, $a \neq 0$, a non-negative integer $\varphi(a)$ is associated such that

E. 1 *$\varphi(ab) \geqslant \varphi(a)$ for all $a, b \neq 0$ in R,*

E. 2 *for any $a, b \in R$, if $b \neq 0$, there exist $q, r \in R$ such that*

$$a = bq+r, \qquad \text{where } \varphi(r) < \varphi(b) \text{ or } r = 0. \tag{7}$$

Any domain with these properties is said to be *Euclidean*. E.g., $\mathbf{Z}$ is a Euclidean ring if we put $\varphi(a) = |a|$; this is the property we used in Ch. **2**. Generally, to verify that a ring is Euclidean it is more convenient to replace **E.** 2 by another equivalent condition:

E. 2′ *For any $a, b \in R$, if $\varphi(a) \geqslant \varphi(b)$, there exists $c \in R$ such that*

$$\varphi(a-bc) < \varphi(a) \qquad \text{or } a = bc. \tag{8}$$

To prove the equivalence of **E.** 2 and **E.** 2′, let us define $\varphi(0) = -1$; this will allow us to dispense with the second alternative in each of (7), (8). If **E.** 2 holds and $\varphi(a) \geqslant \varphi(b)$, then taking q, r as in (7), we find that $\varphi(a-bq) = \varphi(r) < \varphi(b) \leqslant \varphi(a)$, hence **E.** 2′. Conversely, assume **E.** 2′; given $a, b \in R$, $b \neq 0$, choose $q \in R$ so that $\varphi(a-bq)$ takes its least value. If $\varphi(a-bq) \geqslant \varphi(b)$, then by **E.** 2′, there exists $c \in R$ such that $\varphi(a-bq-bc) < \varphi(a-bq)$ and this contradicts the choice of q. Hence $\varphi(a-bq) < \varphi(b)$, i.e. (7) holds with $r = a-bq$. This then shows **E.** 2 to be equivalent to **E.** 2′.

Next we show that for any field F, the polynomial ring $F[x]$ is Euclidean, by verifying **E.** 1 and **E.** 2′, taking $\varphi(a) = \deg a$. We have already seen that $F[x]$ is an integral domain and **E.** 1 is clear from the definition of the degree function. To prove **E.** 2′, let $a = a_0+a_1x+\cdots+a_nx^n$, $b = b_0+b_1x+\cdots+b_mx^m$, where $a_i, b_j \in F$ and $a_n, b_m \neq 0$; suppose that $\varphi(a) \geqslant \varphi(b)$ say, so that $n \geqslant m$. Then a has the leading term a_nx^n and bx^{n-m} has the leading term b_mx^n, hence a and $ba_nb_m^{-1}x^{n-m}$ have the same leading term and $\varphi(a-ba_nb_m^{-1}x^{n-m}) < \varphi(a)$. This proves **E** 2′ and it shows that $F[x]$ is a Euclidean ring.

We now establish the property which makes the Euclidean rings easy to handle: the stepwise procedure for determining the HCF, known as the *Euclidean algorithm*:†

† By an *algorithm* one generally understands a method for calculating a certain value in a finite number of steps. It should be borne in mind that we may often be able to prove the existence of an entity (especially in analysis) without being able to construct it in a finite number of steps. Finding algorithms is particularly important for machine computation.

THEOREM 4 *In a Euclidean ring any two elements a, b have a highest common factor and this can be expressed in the form*

$$(a, b) = au+bv \qquad (u, v \in R).$$

Proof. Let $\varphi(a) \geqslant \varphi(b)$ say and apply **E.** 2 repeatedly; we claim that after a finite number of steps we reach a zero remainder:

$$\begin{aligned} a &= bq_1+r_1 & \varphi(r_1) &< \varphi(b), \\ b &= r_1q_2+r_2 & \varphi(r_2) &< \varphi(r_1), \\ &\cdots & &\cdots \\ r_{n-2} &= r_{n-1}q_n+r_n & \varphi(r_n) &< \varphi(r_{n-1}), \\ r_{n-1} &= r_nq_{n+1} & r_{n+1} &= 0. \end{aligned} \tag{9}$$

For $\varphi(b) > \varphi(r_1) > \cdots$ is a strictly decreasing sequence of non-negative integers, which must break off, and this can only happen when a remainder is zero.

From the first equation we see that r_1 has the form $ax+by$ (for some $x, y \in R$); we use induction to show that the same holds for each remainder r_i. For if $r_{i-1} = ax+by$, $r_{i-2} = ax'+by'$, then $r_i = -r_{i-1}q_i+r_{i-2} = a(x'-xq_i)+b(y'-yq_i)$. Hence we have in particular

$$r_n = au+bv \qquad \text{for some } u, v \in R. \tag{10}$$

Moreover, r_n divides r_n and r_{n-1}, hence also r_{n-2} and, by induction, $r_n \mid r_i$ for all i, until we finally get $r_n \mid b$, $r_n \mid a$ (by (9)). Thus r_n is a common factor of a and b. By (10) any factor of a and b also divides r_n, so r_n is an HCF of a, b and it has the form (10), as claimed. ■

COROLLARY *In a Euclidean ring R any two elements have an HCF and an LCM; if the HCF of a and b is d, then there exist $u, v \in R$ such that $d = au+bv$.*

For the existence of the HCF, and the expression (10) for it, has just been proved. Now let $d = au+bv$ be an HCF and put $m = ab/d$ then m is clearly a common multiple of a and b. If m' is another common multiple, then $a \mid m'$, $b \mid m'$, hence $ab \mid am'$, $ab \mid bm'$, hence $ab \mid (au+bv)m'$, i.e. $dm \mid dm'$, therefore $m \mid m'$ and m is indeed an LCM. ■

We conclude this section with a general criterion for a ring to be a UFD. Let us define a *prime* in an integral domain as an element p, not zero or a unit, such that

$$p \mid ab \Rightarrow p \mid a \quad \text{or} \quad p \mid b.$$

E.g., the primes in $\mathbf{Z}$ are prime numbers and their negatives. A prime is always an atom: if a prime p satisfies $p = ab$, then by definition, $p \mid a$ or $p \mid b$, say $p \mid a$. It follows that $a = pc = abc$ for some $c \in R$, and $a \neq 0$ (as factor of p) hence $bc = 1$ by cancellation, so b is a unit. Thus p has no factorization into two non-units and so is an atom. In UFDs the converse holds and, in fact, this property may be used to characterize them. We observe that for any prime p and any elements $a_1, \ldots, a_n \in R$, if $p \mid a_1 \ldots a_n$, then $p \mid a_i$ for some $i = 1, \ldots, n$. This follows by induction from the case $n = 2$, as in the proof of Lemma 2, **2.2.**

THEOREM 5 *An integral domain R is a UFD if and only if*

(i) *every element not zero or a unit has a complete factorization,*

(ii) *every atom is a prime.*

Proof. Clearly (i) is necessary and (ii) follows easily from unique factorization: let $a = u \prod p_i^{\alpha_i}$, $b = v \prod p_i^{\beta_i}$, then if $p_1 \mid ab$, we must have $\alpha_1 + \beta_1 \geqslant 1$, hence $\alpha_1 \geqslant 1$ or $\beta_1 \geqslant 1$, i.e. $p_1 \mid a$ or $p_1 \mid b$.

Conversely, assume (i) and (ii) and take two complete factorizations of an element

$$c = a_1 \ldots a_r = b_1 \ldots b_s. \tag{11}$$

We must show that $r = s$ and after suitably renumbering the bs, b_i is associated to a_i. We use induction on r; for $r = 1$ there is nothing to prove (because s must then also be 1), so let $r > 1$. By hypothesis a_1 is an atom, hence prime and $a_1 \mid b_1 \ldots b_s$, therefore $a_1 \mid b_j$ for some j, say for $j = 1$ (by renumbering the bs). Now b_1 is also an atom, hence a_1 and b_1 are associated, say $a_1 = b_1 u$. Dividing (11) by b_1 we get

$$ua_2 \ldots a_r = b_2 \ldots b_s.$$

By induction, $r-1 = s-1$ and we can renumber the bs so that a_i is associated to b_i for $i = 2, \ldots, r$; but this also holds for $i = 1$, hence the result. ∎

It is now an easy matter to verify that any Euclidean ring is a UFD. In the first place, if p is an atom and $p \mid ab$ but $p \nmid a$, then the HCF (a, p) is a proper factor of p and so $(a, p) = 1$, because p has no other factors. Thus $1 = au + pv$ and $b = abu + pbv$; since $p \mid ab$, it follows that $p \mid (abu + pbv) = b$ and this shows p to be prime.

To construct complete factorizations we first observe: If a is a proper factor of b, i.e. $a \mid b$ but $b \nmid a$, then $\varphi(a) < \varphi(b)$. For if $\varphi(a) \geqslant \varphi(b)$, then $\varphi(a - bc) < \varphi(a)$ for some $c \in R$, by **E.** 2′, but $b = aq$ for some $q \in R$ by hypothesis and, on substituting, we find that $\varphi(a(1 - qc)) < \varphi(a)$; by **E.** 1 this means that $qc = 1$, hence $a = aqc = bc$, which contradicts the fact that $b \nmid a$. Now let $a \in R$, where R is a Euclidean ring, and suppose that

$$a = a_1 a_2 \ldots a_k$$

where each a_i is a non-unit. Then $a_i a_{i+1} \ldots a_k$ is a proper factor of $a_{i-1} a_i a_{i+1} \ldots a_k$ and this shows that

$$\varphi(a) = \varphi(a_1 a_2 \ldots a_k) > \varphi(a_2 \ldots a_k) > \cdots > \varphi(a_k) > \varphi(1).$$

Since φ assumes only non-negative integer values, $\varphi(a) \geqslant k$ and this provides an upper bound on k, the number of non-unit factors in any factorization of a. Hence we can find a factorization of a into non-unit factors which has maximal length and this must clearly be into atomic factors. This proves

THEOREM 6 *Every Euclidean ring is a UFD.* ■

In particular, since the polynomial ring $F[x]$ over a field F is Euclidean, we obtain the

COROLLARY *For any field F, the polynomial ring F[x] is a UFD.* ■

This method cannot be applied, as it stands, to polynomial rings in several variables over a field, because such rings are no longer Euclidean. However, it is possible to prove in other ways that such polynomial rings are again UFDs (cf. Ex. 7, **6.7** and vol. 2).

Exercises

(1) Show that the quotient and remainder in the division algorithm in $F[x]$ are unique.

(2) Let R be an integral domain and suppose that with each $a \neq 0$ in R a non-negative integer $\nu(a)$ is associated satisfying **E.2**. Show how to define another function $\varphi(a)$ on R satisfying **E.1** as well as **E.2** and deduce that R is Euclidean. (Hint. Replace $\nu(a)$ by the smallest integer for which the algorithm holds.)

(3) Let $q_1, \ldots, q_{n+1}, r_1, \ldots, r_n$ be the quotients and remainders in the Euclidean algorithm applied to a, b (cf. (9)). Define $P(x) = \begin{pmatrix} x & 1 \\ 1 & 0 \end{pmatrix}$; verify that $P(x)^{-1} = \begin{pmatrix} 0 & 1 \\ 1 & -x \end{pmatrix}$ and show that

$$(a \quad b) = (r_n \quad 0)P(q_{n+1})P(q_n) \ldots P(q_1).$$

Deduce that r_n is an HCF of a and b and has the form $au+bv$.

(4) Find an HCF for f, g in the form $uf+vg$, when (i) $f = x^3-3x+2$, $g = x^3-2x^2-x+2$, (ii) $f = x^4-1$, $g = x^5+2x^3+x^2+x+1$.

(5) If f and g are coprime polynomials over a field, show that any polynomial h of degree less than $\deg f+\deg g$ is uniquely expressible in the form $h = uf+vg$, where $\deg u < \deg g$, $\deg v < \deg f$.

(6)* Show that for $n \geqslant 1$ the HCF of $nx^{n+1}-(n+1)x^n+1$ and $x^n-nx+n-1$ is $(x-1)^2$.

(7) Prove the rules (i)–(iv) of the text for HCFs.

(8) Let $a_1, \ldots, a_n$ be non-zero elements of a UFD and write $c = a_1a_2 \ldots a_n$, $A_\nu = c/a_\nu$; let d be an HCF of $A_1, \ldots, A_n$ and m an LCM of $a_1, \ldots, a_n$. Show that dm is associated to c.

(9) Let R be an integral domain in which any two elements have an HCF. Show that any two elements have an LCM. (Hint. Consider the HCF of different common multiples.)

(10) Show that the set of polynomials in $\mathbf{Z}[x]$ with even coefficient of x form a subring I of $\mathbf{Z}[x]$. Verify that 2 and $2x$ have an HCF but no LCM in I.

(11) In any integral domain, show that an associate of an atom is again an atom and that an associate of a prime is a prime.

(12) In a UFD R, let $p_1, p_2, \ldots$ be a set of representatives for the non-associated primes. Show that every element of the field of fractions of R can be written uniquely as

$$u\prod p_i^{\alpha_i} \quad (\alpha_i \in \mathbf{Z},\ u \text{ a unit in } R),$$

where all but a finite number of the α_i are 0.

(13) In a UFD R, fix a prime p and let $v(a)$ be the exponent of p in the complete factorization of a. Show that $v(a-b) \geqslant \min\{v(a), v(b)\}$, $v(ab) = v(a)+v(b)$, $v((a, b)) = \min\{v(a), v(b)\}$, $v([a, b]) = \max\{v(a), v(b)\}$. Show that these formulae still hold in the field of fractions of R, using the representation obtained in Ex. (12). When $R = \mathbf{Z}$ and $p = 3$, describe the subring of $\mathbf{Q}$ given by the condition $v(a) \geqslant 0$.

(14) Show that the ring of numbers $a+b\sqrt{-3}$ $(a, b \in \mathbf{Z})$ is not a UFD, even though every non-unit has a factorization into atoms. (Hint. Consider the factorizations $4 = 2.2 = (1+\sqrt{-3})(1-\sqrt{-3})$.)

(15) Let R be a UFD in which the units, together with 0, form a subring F. Show that F is in fact a subfield of R and if $F \neq R$, then R contains infinitely many non-associated primes.

(16) Use the fact that $\mathbf{Z}$ is a UFD to prove the Euler product formula for the ζ-function:

$$\sum n^{-s} = \prod_p (1-p^{-s})^{-1},$$

where the product is taken over all prime numbers. Both sides converge absolutely for $s > 1$, or they may be regarded as a formal identity.

6.6 The zeros of polynomials

One of the most important properties of polynomials is the substitution property which we met in Th. 4, **6.4**, in the form of the universal mapping property. This shows that when we replace x in $f \in R[x]$ by an element c of R (or of a ring containing R) then the resulting mapping

$$f = \sum a_i x^i \mapsto f(c) = \sum a_i c^i$$

is a homomorphism. If c is such that $f(c) = 0$, we say that c is a *zero* of the polynomial f, or also that c is a *root* of the equation $f = 0$.

For emphasis the polynomial f is sometimes written as $f(x)$; this cannot lead to confusion, as $f(x)$ can only mean the result of substituting x for x, i.e. f.

In order to exploit the Euclidean algorithm we shall need to take our coefficient ring to be a field. But many results have a more general validity; they are based on a form of the division algorithm which holds for arbitrary rings:

THEOREM 1 (General division algorithm) *Let R be any ring. Given f, g in the polynomial ring $R[x]$, if g is monic, there exist unique elements $q, r \in R[x]$ such that*

$$f = gq+r, \qquad \deg r < \deg g. \tag{1}$$

Proof. The existence of q and r is clear: we determine q so that $f-gq$ takes its least degree. If this is not less than $\deg g$, we can diminish it by subtracting a multiple of g, because g is monic. But that contradicts the definition of q; hence $\deg (f-gq) < \deg g$, and now (1) follows on putting $r = f-gq$.

To establish the uniqueness, assume that $f = gq+r = gq'+r'$, where r and r' have degree less than $\deg g$. Then

$$g(q-q') = r'-r. \tag{2}$$

If $q-q' \neq 0$, let its leading term be cx^k and suppose $\deg g = m$, so that $g(q-q')$ has leading term cx^{k+m}. But then

$$\deg (r'-r) < \deg g = m \leqslant k+m = \deg (g(q-q'))$$

and this is in contradiction with (2). Therefore $q' = q$ and $r' = r$. ■

The first application of this result relates the zeros of a polynomial to its factors:

THEOREM 2 (Remainder theorem) *Let R be any commutative ring, $\alpha \in R$ and $f \in R[x]$, then there exists $g \in R[x]$ such that*

$$f = (x-\alpha)g+f(\alpha), \tag{3}$$

and $(x-\alpha) \mid f$ if and only if $f(\alpha) = 0$.

For by the general division algorithm,

$$f = (x-\alpha)g+r,$$

where $\deg r < 1$, so $r \in R$. To find r we put $x = \alpha$, then $f(\alpha) = r$ and (3) follows. Now the remaining assertion follows easily. ■

Clearly the result holds for any ring R (not necessarily commutative) as long as we choose α in the centre of R; in fact, by taking care about the order of the factors, an even more general result can be obtained (Ex. (5)). But for the moment it is more useful to generalize the result in another direction:

THEOREM 3 *Let R be an integral domain and let f be a polynomial over R; if $\alpha_1, \ldots, \alpha_k$ are distinct zeros of f, then $(x-\alpha_1)\ldots(x-\alpha_k) \mid f$.*

Proof. For $k = 1$ this has just been proved, so let $k > 1$. By the induction hypothesis we have $f = (x-\alpha_2)\ldots(x-\alpha_k)g$ for some $g \in R[x]$. Now $f(\alpha_1)$

$= 0$, so $(\alpha_1-\alpha_2)\ldots(\alpha_1-\alpha_k)g(\alpha_1) = 0$; since R is an integral domain and the αs are all distinct, $g(\alpha_1) = 0$, therefore $g = (x-\alpha_1)h$, i.e. $f = (x-\alpha_1)(x-\alpha_2)\ldots(x-\alpha_k)h$ and the result follows. ■

One consequence of this result is that a polynomial with k distinct zeros (over an integral domain) must have degree at least k. From this remark we easily obtain a number of corollaries. The main importance of these results is for fields; we shall therefore state them for that case only, leaving the extensions to the reader (cf. Ex. (2)).

COROLLARY 1 *Any equation of degree n over a field has at most n distinct roots.* ■

COROLLARY 2 *If two polynomials f, g of degree at most n over a field agree in $n+1$ places, then $f = g$.*

For $f-g$ is a polynomial of degree at most n with more than n zeros. ■

COROLLARY 3 *If F is an infinite field and f, g are polynomials such that $f(\alpha) = g(\alpha)$ for all $\alpha \in F$, then $f = g$.* ■

This last result means that distinct polynomials (over an infinite field) define distinct functions. It is not true for finite fields, e.g. over the field $\mathbf{F}_p$ of p elements, x^p and x define the same function, by the Cor. to Fermat's theorem, p. 31.

Th. 3 and its corollaries have an extension to polynomials in several variables. We shall state the result in the form in which it is mostly used:

THEOREM 4 *Let F be an infinite field and f a polynomial in $x_1, \ldots, x_n$ over F. If $f(\alpha_1, \ldots, \alpha_n) = 0$ for all $\alpha_i \in F$, then $f = 0$.*

The proof is by induction on n. For $n = 1$ the result follows from the last Cor., so let $n > 1$ and write f as a polynomial in x_1 with coefficients in $F[x_2, \ldots, x_n]$:

$$f(x_1, x_2, \ldots, x_n) = a_0 + a_1x_1 + \ldots + a_kx_1^k, \qquad a_i = a_i(x_2, \ldots, x_n).$$

Now fix $\alpha_2, \ldots, \alpha_n \in F$, then $f(x_1, \alpha_2, \ldots, \alpha_n)$ vanishes at all points of F and so, by Cor. 3, is the zero polynomial, i.e.

$$a_i(\alpha_2, \ldots, \alpha_n) = 0 \qquad (i = 0, 1, \ldots, k)$$

for all $\alpha_2, \ldots, \alpha_n \in F$. By the induction hypothesis, $a_i = 0$, hence $f = 0$, as claimed. ■

As a corollary we have the principle of irrelevance of algebraic inequalities:

Let $f, g, h \in F[x_1, \ldots, x_n]$, $h \neq 0$ (where F is an infinite field) and suppose that $f(\alpha) = g(\alpha)$ for all $\alpha = (\alpha_1, \ldots, \alpha_n)$ for which $h(\alpha) \neq 0$. Then $f = g$.

For $(f-g)h$ vanishes for all $\alpha \in F^n$, hence it must be zero, but $h \neq 0$; since the polynomials form an integral domain it follows that $f = g$. ■

A subset D of the set F^n of n-tuples is called *dense* if there is a non-zero polynomial which vanishes at all points of the complement of D. With this terminology we can restate the above principle by saying that a polynomial function is completely determined by its values on a dense subset of F^n (where F is any infinite field).

A rational function φ over F corresponds to a function which is defined whenever the denominator of φ is non-zero. Thus the function $\tilde{\varphi}$ corresponding to φ is defined at all but a finite number of points. If $\varphi = f/g$, we can define the rational function $\tilde{\varphi}$ at any point that is not a zero of g and, for an infinite field, distinct fractions yield distinct functions. For if $\varphi = f/g$ and $\varphi_1 = f_1/g_1$ define the same function, then $fg_1 - f_1 g$ vanishes at all points of F except possibly the zeros of gg_1, so by the principle of irrelevance of algebraic inequalities, $fg_1 = f_1 g$, i.e. $\varphi = \varphi_1$.

Exercises

(1) Find all the zeros of x^2-1 over the ring $\mathbf{Z}/16$.

(2) Deduce from Cors. 1–3 of Th. 3 corresponding statements for integral domains in place of fields.

(3) If R is an integral domain, show that no polynomial of positive degree over R can be invertible.

(4) Let p be a prime number. Show that over $\mathbf{F}_p$ the polynomials $\prod_1^p(x-i)$ and x^p-x have p common zeros and the same leading term. Deduce that these polynomials are equal, and by equating coefficients, obtain Wilson's theorem: $(p-1)! \equiv -1 \pmod{p}$ for any prime p.

(5) Let R be a ring and for each $c \in R$ define a mapping λ_c: $R[x] \to R$ by the rule

$$\lambda_c : \sum a_i x^i \mapsto \sum a_i c^i.$$

Verify that λ_c is a homomorphism of additive groups (though not always a ring homomorphism) which reduces to the identity mapping on R. Show that for each $c \in R$ there exists $g \in R[x]$ such that

$$f = g\,.\,(x-c) + f\lambda_c.$$

Similarly define a mapping ρ_c: $\sum a_i x^i \mapsto \sum c^i a_i$ and show that for each $c \in R$ there exists $h \in R[x]$ such that

$$f = (x-c)\,.\,h + f\rho_c.$$

(6) Let $f(x_1, \ldots, x_n)$ be a polynomial over an infinite field, which vanishes for all distinct values $x_i = \alpha_i$, where $\alpha_i \neq \alpha_j$ for $i \neq j$. Show that $f = 0$.

(7) Prove that the intersection of any finite number of dense sets is dense.

(8) Let F be a field and $F(x)$ the field of rational functions. An expression f/g (where $f, g \in F[x]$) for an element of $F(x)$ is said to be *reduced* if g is monic and $(f, g) = 1$. Show that every element of $F(x)$ has a unique reduced form.

(9) Show that over an infinite field F, there is no polynomial $f \neq 0, 1$ such that $f(a)f(b) = f(a+b)$ for all $a, b \in F$.

(10) Let F be any field. Given $f, g \in F[x], g \notin F$, show that there is a unique expression

$$f = \sum g^i h_i \quad \text{where } \deg h_i < \deg g.$$

(11)* Given two polynomials f, g in x over a field F, show that there is a polynomial φ such that $f = \varphi(g)$ iff $g(x)-g(y) \mid f(x)-f(y)$ in $F[x, y]$. (Hint. Use Ex. (10) and show that $h_i(x) = h_i(y)$.)

6.7 The factorization of polynomials

We now look at the factorization of polynomials over some particular fields, the main task being to determine the atoms in $F[x]$, in this case also called the *irreducible* polynomials. We recall that the units in the ring $F[x]$ are just the non-zero constants, i.e. the polynomials of degree 0. Polynomials of degree 1 are sometimes called *linear*, those of degree 2 *quadratic*.

In the first place, it is clear that every linear polynomial must be irreducible; for it cannot be a unit and, in any factorization, only one factor can have positive degree. The next result shows when the converse holds:

THEOREM 1 *Let F be a field. Then the linear polynomials are atoms in $F[x]$, and they are the only atoms if and only if each equation of positive degree with coefficients in F has a root in F.*

Proof. Assume that every atom is linear: every polynomial of positive degree is divisible by an atom which, by hypothesis, is of the form $ax+b$, where $a, b \in F, a \neq 0$. Hence the equation $f = 0$ has the root $x = -b/a$. Conversely, if every equation has a root, let p be an atom and α a zero of p, then $p = (x-\alpha)q$, by the remainder theorem. Since p is an atom, q must be constant and so p is linear, as claimed. ∎

Fields over which every equation in one unknown has a root are said to be *algebraically closed.* For example, a well known theorem (sometimes called the fundamental theorem of algebra) asserts that the field of complex numbers is algebraically closed. This is most easily proved by complex analysis; later (in Vol. 2) we shall give a more algebraic (but longer) proof. However, from the algebraist's point of view, the fact that **C** is algebraically closed is relatively unimportant. What matters more is the result (also proved in Vol. 2) that every field has an algebraic closure, i.e. it can be embedded in a field which is algebraically closed.

The factorization of polynomials over an algebraically closed field such as **C** is very straightforward: every polynomial is a product of linear factors. Over the real numbers it is almost as good: every polynomial is a product of linear and quadratic factors (Ex. (1)). Over the rational numbers the situation is very different: there are irreducible polynomials of all degrees and for a given polynomial with rational coefficients it may not be at all easy to find its factors. We conclude this section with some remarks on factorizing polynomials with rational coefficients.

Given $f \in \mathbf{Q}[x]$, we multiply f by a common denominator of the coefficients so as to obtain a polynomial in $\mathbf{Z}[x]$. The first step is to show that we lose nothing by considering only polynomials over $\mathbf{Z}$; this will follow from the next result, which is stated in the more general context of unique factorization domains. A polynomial with integer coefficients, or, more generally, with coefficients in a UFD, is said to be *primitive* if the coefficients have no common factor other than units.

THEOREM 2 (Gauss's lemma) *The product of two primitive polynomials over a UFD is again primitive.*

Proof. Let $f = \sum a_i x^i$, $g = \sum b_j x^j$ be primitive polynomials with coefficients in a UFD R, and suppose that

$$fg = \lambda h, \tag{1}$$

where λ is a non-unit in R and $h \in R[x]$. Let π be a prime dividing λ; since f is primitive, not all its coefficients a_j are divisible by π. Let a_r be the first coefficient of f that is not divisible by π and similarly let b_s be the first coefficient of g that is not divisible by π. On comparing coefficients of x^{r+s} in (1) we find

$$\lambda c_{r+s} = a_0 b_{r+s} + a_1 b_{r+s-1} + \cdots + a_r b_s + \cdots + a_{r+s} b_0.$$

Here all terms except possibly $a_r b_s$ are divisible by π, hence that term is also divisible by π, but $\pi \nmid a_r$, $\pi \nmid b_s$ and this contradicts the fact that π is prime. Thus λ can have no non-unit factor and hence fg is primitive, as claimed. ■

Let f be any polynomial over a UFD R and let δ be a highest common factor of its coefficients, then $f = \delta f_1$, where f_1 is a primitive polynomial over R. The element δ, determined up to associates, is called the *content* of f. If f, g have contents α, β respectively, say $f = \alpha f_1$, $g = \beta g_1$, then f_1, g_1 are primitive and $fg = \alpha\beta f_1 g_1$. By Gauss's lemma $f_1 g_1$ is again primitive, so $\alpha\beta$ is the content of fg. This establishes

COROLLARY 1 *Over a UFD the content of a product of polynomials is the product of the contents.* ■

As an illustration, let $f \in \mathbf{Z}[x]$ be irreducible over $\mathbf{Z}$; we want to show that it is also irreducible over $\mathbf{Q}$, so suppose f factorized: $f = gh$, where $g, h \in \mathbf{Q}[x]$. By clearing g, h of denominators we can write this as $\alpha f = \beta g_1 h_1$, where g_1, h_1 are primitive polynomials over $\mathbf{Z}$ and $\alpha, \beta \in \mathbf{Z}$. Let γ be the content of f, then by Cor. 1 just proved, $\alpha\gamma = \beta$ (replacing γ by an associate if necessary), hence $f = \gamma g_1 h_1$ and this is a factorization of f over $\mathbf{Z}$. Since f was irreducible over $\mathbf{Z}$, it follows that f is also irreducible over $\mathbf{Q}$. In particular, by testing for linear factors we find

COROLLARY 2 *Let* $f = a_0 + a_1x + \cdots + a_nx^n \in \mathbf{Z}[x]$ *have a rational zero* α/β, *where* α, β *are coprime integers. Then* $\beta \mid a_n$ *and if* $\alpha \neq 0$, $\alpha \mid a_0$. *In particular, if* $a_n = 1$, *all rational zeros are integral.* ■

This enables us to find all rational zeros of f in a finite number of steps, by testing whether the numbers α/β, where α and β run over the factors of a_0 and a_n respectively, satisfy the equation $f = 0$. However, a polynomial may well be reducible without having linear factors, e.g. $x^4 - 4 = (x^2-2)(x^2+2)$.

There is no simple necessary and sufficient condition for irreducibility over $\mathbf{Q}$ and finding factors is largely a matter of trial and error.† This method (of trial and error) may eventually lead to a factor, but it will never tell us that there are no non-trivial factors. In this situation the following test is often useful:

THEOREM 3 (Eisenstein's criterion) *Let* $f = a_0 + a_1x + \cdots + a_nx^n \in \mathbf{Z}[x]$. *If there is a prime number p such that*

(i) $p \nmid a_n$,
(ii) $p \mid a_i$ *for* $0 \leqslant i < n$,
(iii) $p^2 \nmid a_0$,

then f is irreducible over $\mathbf{Q}$.

Proof. Suppose f is reducible over $\mathbf{Q}$ and hence over $\mathbf{Z}$, say $f = gh$, where g, h are of positive degrees and with integer coefficients: $g = b_0 + b_1x + \cdots + b_rx^r$, $h = c_0 + c_1x + \cdots + c_sx^s$, where $r+s = n = \deg f$, $r, s > 0$. We have $a_0 = b_0c_0$ and a_0 contains p to the first power, so one of b_0, c_0 but not the other, is divisible by p, say $p \mid b_0$, $p \nmid c_0$. Further, $b_rc_s = a_n$ is not divisible by p, so $p \nmid b_r$, $p \nmid c_s$. Thus the first but not the last coefficient of g is divisible by p. Let b_i be the first coefficient not divisible by p, then $i > 0$ and

$$a_i = b_ic_0 + b_{i-1}c_1 + \cdots + b_0c_i.$$

If we reduce mod p (i.e. if we regard this as a congruence mod p), this equation becomes $b_ic_0 \equiv 0 \pmod p$, but this is a contradiction, since $p \nmid b_i$, $p \nmid c_0$. It follows that f must be irreducible. ■

The usefulness of this criterion lies in the fact that p is at our disposal. E.g., to show that $x^3 - 4$ is irreducible over $\mathbf{Q}$, write $x = y+1$, then $x^3 - 4 = y^3 + 3y^2 + 3y - 3$ and this is irreducible, by Eisenstein's criterion with $p = 3$.

Exercises

(1) Assuming the fact that $\mathbf{C}$ is algebraically closed, show that every irreducible polynomial over $\mathbf{R}$ is either linear or quadratic, of the form ax^2+bx+c, where $b^2 < 4ac$. Deduce that every real equation of odd degree has a real root.

† As a matter of fact, it is possible to determine in a finite number of steps whether a given polynomial over $\mathbf{Q}$ is irreducible, but this is more of theoretical interest.

(2) Show that ax^2+bx+c is irreducible over $\mathbf{F}_p$, where p is an odd prime, iff b^2-4ac is not a square. Determine all irreducible quadratic polynomials over $\mathbf{F}_3$.

(3) Find all irreducible quadratic and cubic polynomials over $\mathbf{F}_2$.

(4) Let $f, g \in \mathbf{Q}[x]$; suppose that f is irreducible over $\mathbf{Q}$ and that f, g have a common zero in $\mathbf{C}$. Show that $f \mid g$.

(5) Prove that over any field there are infinitely many irreducible polynomials. (Hint. For finite fields use Ex. (15), **6.5**.)

(6) If n is a positive integer not divisible by 2 or 3, show that $(x+y)^n-x^n-y^n$ is divisible by $xy(x^2+xy+y^2)$.

(7) If R is a UFD, show that any atom in $R[x]$ is either an atom in R or a primitive polynomial. Use Gauss's lemma to show that a primitive polynomial is an atom iff it is irreducible over the field of fractions of R. Deduce that $R[x]$ is a UFD.

(8) If R is a UFD and F is its field of fractions, show that $R[x]$ is inert in $F[x]$. (Use Gauss's lemma.)

(9) If p is a prime number, show that $1+x+x^2+\cdots+x^{p-1}$ is irreducible over $\mathbf{Q}$.

(10) Factorize x^4+1 and x^4-4 completely over $\mathbf{Q}$.

6.8 Derivatives

The reader will know from analysis that at a multiple zero the derivative of a function vanishes. We shall use the same method to study multiple zeros in algebra, but it will of course be necessary to define derivatives formally.

Given any polynomial $f = a_0+a_1x+\cdots+a_nx^n$ over a ring R, we define its *derivative* Df or f' by the equation

$$f' = a_1+2a_2x+\cdots+na_nx^{n-1}. \tag{1}$$

The mapping $f \mapsto f'$ of $R[x]$ into itself satisfies the following rules:

D. 1 $(f+g)' = f'+g'$,

D. 2 $(af)' = af' \qquad (a \in R)$,

D. 3 $(fg)' = f'g+fg'$,

D. 4 $x' = 1$.

Conversely, f' is completely determined by **D.** 1–4 and this provides an alternative definition. In fact it is easy to show (by induction) that there cannot be more than one mapping $f \mapsto f'$ satisfying **D.** 1–4; To show that such a mapping exists is a little more delicate; as it happens, this difficulty can be circumvented by exhibiting the mapping (as in (1)).

Let f be a polynomial over a field F and suppose that K is an algebraically closed field containing F. Then over K we can split f into linear factors

$$f = c(x-\alpha_1)^{m_1} \dots (x-\alpha_r)^{m_r}, \tag{2}$$

where $\alpha_1, \dots, \alpha_r$ are distinct elements of K. We call m_i the *multiplicity* of α_i and call α_i an m_i*-fold* zero of f; in particular, α_i is a *simple* zero if $m_i = 1$ and a *multiple* zero if $m_i > 1$. As already remarked, every field is contained in an algebraically closed field, but this may not help us to find an expression (2) if the algebraic closure is not explicitly known. Our aim is to be able to test whether f has multiple zeros in some field containing F, without having to go outside F. This is achieved by

PROPOSITION 1 *Let F be any field and $f \in F[x]$, then an element α in any field containing F is a multiple zero of f if and only if $f(\alpha) = f'(\alpha) = 0$.*

Proof. By definition, α is a multiple zero of f iff $(x-\alpha)^2 \mid f$. On dividing f by $(x-\alpha)^2$ we obtain an equation of the form

$$f = (x-\alpha)^2 g + h,$$

where h is a linear polynomial. Putting $x = \alpha$, we see that $h(\alpha) = f(\alpha)$. Differentiating and putting $x = \alpha$ we find that $h'(\alpha) = f'(\alpha)$, hence

$$f = (x-\alpha)^2 g + (x-\alpha)f'(\alpha) + f(\alpha).$$

It follows that $(x-\alpha)^2 \mid f$ iff $f(\alpha) = f'(\alpha) = 0$. ∎

The point of this result is that although we need to know α in order to find whether $f(\alpha) = 0$, to find if f and f' have a common factor (such as $x-\alpha$) we do not need to know α. To test if f has any multiple zeros in any extension of F, we need only find the HCF of f and f', and this can be done, by the Euclidean algorithm, without going outside F. This HCF has positive degree precisely when the product of the squares of the differences of the zeros of f vanishes. This product, called the *discriminant* of f, can be expressed in terms of the coefficients of f. An explicit formula for the discriminant will be given in Ch. 7.

In view of Prop. 1, the polynomials with identically vanishing derivative are of particular interest. Let $f = a_0 + a_1 x + \dots + a_n x^n$ and suppose that $f' = 0$, then $ia_i = 0$ $(i = 0, 1, \dots, n)$. When F has characteristic 0, this shows that $a_i = 0$ for $i > 0$ and so f reduces to a constant. However, for a field of prime characteristic p, all we can conclude is that $a_i = 0$ for $p \nmid i$. Thus f has the form

$$f = a_0 + a_p x^p + a_{2p} x^{2p} + \dots + a_{rp} x^{rp},$$

in other words, f is now a polynomial in x^p. We record this fact for later use in

PROPOSITION 2 *Let F be any field and f a polynomial over F such that $f' = 0$. Then*

(i) *if* char $F = 0, f$ *is a constant,*

(ii) *if* char $F = p \neq 0, f$ *has the form* $g(x^p)$. ■

To obtain more information on the exact multiplicity, we need Taylor's formula. This is easily proved for polynomials, since we need not concern ourselves with the remainder, but we shall have to restrict the characteristic to be 0.

THEOREM 3 *Let F be a field of characteristic* 0. *Given any polynomial f of degree n over F and any $\alpha \in F$, we have*

$$f(x) = f(\alpha) + \frac{Df(\alpha)}{1!}(x-\alpha) + \frac{D^2 f(\alpha)}{2!}(x-\alpha)^2 + \cdots + \frac{D^n f(\alpha)}{n!}(x-\alpha)^n. \quad (3)$$

Here $D^\nu f$ indicates the result of differentiating ν times (i.e. applying D ν times in succession) and $D^\nu f(\alpha)$ is the result of substituting α for x.

Proof. We can arrange f as a polynomial in $x-\alpha$, i.e. writing $x = y+\alpha$, we have to express $f(y+\alpha)$ as a polynomial in y:

$$f(x) = \sum b_\nu (x-\alpha)^\nu. \quad (4)$$

To find b_i we differentiate i times: $D^i f = \sum \nu(\nu-1)\ldots(\nu-i+1)b_\nu(x-\alpha)^{\nu-i}$. Putting $x = \alpha$, we find $D^i f(\alpha) = i!\, b_i$ and, on substituting this value for b_i into (4), we obtain (3). ■

From (3) we see that f has α as a zero of exact multiplicity m iff $f(\alpha) = Df(\alpha) = \cdots = D^{m-1}f(\alpha) = 0$, $D^m f(\alpha) \neq 0$. But it should be kept in mind that this holds generally only when F has characteristic 0; more precisely, it holds whenever char $F = 0$ or char $F > m$.

Derivatives may be of help in locating roots, especially of equations with real coefficients, and we shall briefly describe some of these methods. We shall need the fact that the real numbers form a totally ordered field which has the following property:

If f is a polynomial in x with real coefficients and $\alpha, \beta \in \mathbf{R}$ are such that $\alpha < \beta$, $f(\alpha)f(\beta) < 0$, then there exists $\gamma \in \mathbf{R}$, such that $\alpha < \gamma < \beta$ and $f(\gamma) = 0$.

In other words, if a polynomial has different signs at the ends of an interval, it vanishes at some point of that interval. This is proved in courses on analysis (see also Vol. 2); apart from this we shall also need the well-known (and closely-related) fact that if a polynomial assumes equal values at the ends of an interval, then its derivative vanishes somewhere in the interval (Rolle's theorem), i.e. if $f(\alpha) = f(\beta)$, then there exists $\gamma \in \mathbf{R}$, $\alpha < \gamma < \beta$ such that $f'(\gamma) = 0$. From this fact we can derive the mean value theorem in the following form:

Given a polynomial f and $\alpha < \beta$, there exists γ, $\beta < \gamma < \alpha$ such that

$$f(\beta) = f(\alpha)+(\beta-\alpha)f'(\alpha)+\frac{(\beta-\alpha)^2}{2}f''(\gamma). \tag{5}$$

To prove (5) observe that for any polynomials F and G,

$$F(x)[G(\beta)-G(\alpha)]-G(x)[F(\beta)-F(\alpha)]$$

assumes the same value for $x = \alpha$ and $x = \beta$, hence the derivative vanishes at some γ between α and β, i.e.

$$F'(\gamma)[G(\beta)-G(\alpha)]-G'(\gamma)[F(\beta)-F(\alpha)] = 0. \tag{6}$$

We put $F(x) = f(x)+(\beta-x)f'(x)$, $G(x) = x^2+2(\beta-x)x = (2\beta-x)x$, then $F'(x) = (\beta-x)f''(x)$, $G'(x) = 2(\beta-x)$ and substituting in (6) we get $(\beta-\gamma)f''(\gamma)[\beta^2-\alpha^2-2(\beta-\alpha)\alpha]-2(\beta-\gamma)[f(\beta)-f(\alpha)-(\beta-\alpha)f'(\alpha)] = 0$. Dividing by $2(\beta-\gamma)$ and rearranging the terms, we obtain (5). ■

To get a first estimate for the size of the real roots of an equation, consider

$$f = a_0+a_1x+\cdots+a_nx^n \qquad (a_i \in \mathbf{R},\ a_n \neq 0),$$

and put

$$a = \max\{|a_0|, |a_1|, \ldots, |a_{n-1}|\}. \tag{7}$$

Then for any real $\lambda > 1$,

$$\begin{aligned} |a_0+a_1\lambda+\cdots+a_{n-1}\lambda^{n-1}| &\leqslant |a_0|+|a_1|\lambda+\cdots+|a_{n-1}|\lambda^{n-1} \\ &\leqslant a(1+\lambda+\cdots+\lambda^{n-1}) \\ &= a\frac{\lambda^n-1}{\lambda-1}. \end{aligned}$$

Since $\lambda > 1$, this is less than $a\lambda^n/(\lambda-1)$; on the other hand,

$$a\frac{\lambda^n}{\lambda-1} < |a_n|\lambda^n,$$

provided that $a/(\lambda-1) < |a_n|$, i.e.

$$\lambda > \frac{a}{|a_n|}+1. \tag{8}$$

Thus when (8) holds, $f(\lambda)$ has the same sign as a_n, for as we have seen, the leading term of f then outweighs the others. It follows that no real zero λ of f can satisfy (8). By applying the same argument to the polynomial $f(-x)$, we see that $f(x)$ can have no negative zero less than $-(a/|a_n|+1)$. Thus all zeros of f lie in the interval $\left[-\left(\frac{a}{|a_n|}+1\right), \frac{a}{|a_n|}+1\right]$.

We now come to a result which gives an upper bound on the number of real roots of an equation in a given interval. Given any sequence of real numbers: $\lambda_1, \ldots, \lambda_n$, then by the number of sign changes in the sequence we mean the number of times there is a change from a positive to a negative value or vice versa, ignoring zeros. E.g. the sequence 0, 1, −2, 0, −4, 3, 2, −1 has 3 changes.

THEOREM 4 (Budan–Fourier) *Let f be a real polynomial of degree n, and for any $\alpha \in \mathbf{R}$ denote by $\delta(\alpha)$ the number of sign changes in the sequence*

$$f(\alpha), f'(\alpha), \ldots, f^{(n)}(\alpha). \tag{9}$$

Then the number of zeros (counting multiplicities) of f in any interval $[\alpha, \beta]$, where α, β are not zeros of f, is at most $\delta(\alpha)-\delta(\beta)$. More precisely, the number of zeros is $\delta(\alpha)-\delta(\beta)-2r$, where $r \geqslant 0$.

Proof. $\delta(\lambda)$ changes only when λ passes through a zero of some $f^{(i)}$. To compute the change we distinguish two cases.

(i) λ is a zero of f, of multiplicity m. Then $f(\lambda) = \cdots = f^{(m-1)}(\lambda) = 0$, $f^{(m)}(\lambda) \neq 0$. Hence, by Taylor's formula,

$$f(\lambda+h) = \frac{f^{(m)}(\lambda)}{m!}h^m + \cdots$$

$$f'(\lambda+h) = \frac{f^{(m)}(\lambda)}{(m-1)!}h^{m-1} + \cdots$$

$$\cdot \quad \cdot \quad \cdot \quad \cdot \quad \cdot \quad \cdot \quad \cdot \quad \cdot$$

$$f^{(m)}(\lambda+h) = f^{(m)}(\lambda) + \cdots$$

In each line the term written down on the right is the dominant one (for small enough h) and as h passes from positive to negative values, the number of sign changes gained in the segment $f, f', \ldots, f^{(m)}$ of the sequence (9) is m, which is also the number of zeros gained.

(ii) λ is a zero, of multiplicity l, of some $f^{(i)} (i > 0)$ but not of $f^{(i-1)}$. Thus $f^{(i)}(\lambda) = \cdots = f^{(i+l-1)}(\lambda) = 0$, $f^{(i+l)}(\lambda) \neq 0$; as in (i) we gain l sign changes in the segment $f^{(i)}, \ldots, f^{(i+l)}$. Further, we have

$$f^{(i-1)}(\lambda+h) = f^{(i-1)}(\lambda) + \frac{f^{(i+l)}(\lambda)}{(l+1)!}h^{l+1} + \cdots$$

$$f^{(i)}(\lambda+h) = \frac{f^{(i+l)}(\lambda)}{l!}h^l + \cdots$$

When l is even, neither of $f^{(i-1)}, f^{(i)}$ changes sign, so the number of changes is l. When l is odd, $f^{(i)}$ changes sign, but not $f^{(i-1)}$, so there is another change and there are $l \pm 1$ in all, again an even number $\geqslant 0$. Thus in any case there is a non-negative even number of sign changes.

Combining (i) and (ii) (note that both may occur for the same λ), we obtain the result. ∎

As an illustration, let us estimate the number of positive roots of an equation. Taking f as before, assume that $a_n > 0$, say, then $f^{(i)}(x) = n(n-1) \ldots (n-i+1)a_n x^{n-i} + \cdots$, and this shows that $f^{(i)}(x) > 0$ if x is large enough. Thus by taking β suitably large, $\delta(\beta) = 0$ and there can be no zeros larger than β, by Th. 4, because for any $\beta' > \beta$, $\delta(\beta) - \delta(\beta') = 0$. On the other hand, $f^{(i)}(0)$ is a_i, apart from a positive factor, so $\delta(0)$ is the number of sign changes in the sequence $a_0, a_1, \ldots, a_n$. Hence we obtain the

COROLLARY (Harriot–Descartes Rule) *The equation*

$$a_0 + a_1 x + \cdots + a_n x^n = 0$$

cannot have more positive roots than the number of sign changes in the sequence $a_0, \ldots, a_n$ (*ignoring* 0s) *and it differs from the number of these changes by an even number.* ■

For example, the polynomial $f = x^6 + 3x^5 - 2x^4 - x^3 + 4x - 1$ has 3 sign changes and so has 1 or 3 positive zeros. Similarly $f(-x)$ has 3 sign changes (among its coefficients) and so has 1 or 3 positive zeros, hence the original polynomial f has 1 or 3 negative zeros.

We now turn to methods for the approximate calculation of roots. There are many such methods; we shall describe two of the simplest, the *regula falsi* and Newton's method, which together are very effective for practical purposes.

Let $f(x)$ be a real polynomial which has exactly one simple zero in a given interval $[\alpha, \beta]$; we can always arrange this, by dividing f by the HCF of f and f' to remove any multiple zeros. Then f changes sign exactly once in the interval, and the point γ, where the chord joining the points with coordinates $(\alpha, f(\alpha))$ and $(\beta, f(\beta))$ meets the horizontal axis, is an approximation to the zero (see Fig. 5). Using x, y as coordinates, we find the chord to be

$$y = \frac{f(\beta) - f(\alpha)}{\beta - \alpha} x + \frac{\beta f(\alpha) - \alpha f(\beta)}{\beta - \alpha},$$

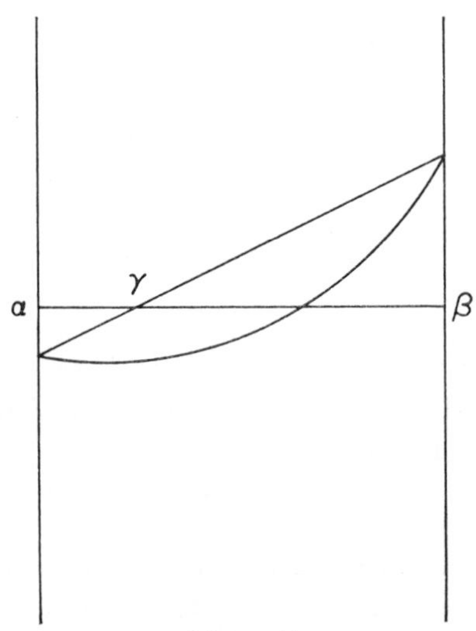

Figure 5

and γ is found by putting $y = 0$ and solving for x:

$$\gamma = \frac{\alpha f(\beta) - \beta f(\alpha)}{f(\beta) - f(\alpha)}. \tag{10}$$

This rule is called *regula falsi* (= false rule, because in the numerator each of $f(\alpha), f(\beta)$ is multiplied by the *other* coordinate). It is easily verified that γ lies between α and β. If $f(\gamma) \neq 0$, then $f(\gamma)$ has the same sign as one of $f(\alpha), f(\beta)$ and we replace the corresponding coordinate (α or β) by γ to obtain a smaller interval enclosing the zero of f. We observe that this method applies not merely to polynomials but to any continuous functions.

Newton's method is based on Taylor's formula, or rather on the form of the mean value theorem given in (5). This may also be regarded as a form of Taylor's formula with remainder. If we replace α, β by $\lambda, \lambda + h$ respectively, this takes the form

$$f(\lambda + h) = f(\lambda) + hf'(\lambda) + \frac{h^2}{2} f''(\lambda + k), \qquad \text{where } 0 < k < h. \tag{11}$$

If λ is an approximate zero and $f'(\lambda) \neq 0$, we can choose h so that $f(\lambda) + hf'(\lambda) = 0$, then $f(\lambda + h) = \frac{1}{2}h^2 f''(\lambda + k)$ and if h was small to begin with, then this is of the second order in h, so that $\lambda + h$ will be close to the actual zero. Thus under suitable conditions,

$$\lambda_1 = \lambda - \frac{f(\lambda)}{f'(\lambda)} \tag{12}$$

is a better approximation to the zero than λ.

To find these conditions, it will be helpful to draw a figure. (Of course this can only suggest a line of proof; the proof itself will be independent of any figure.) The equation

$$y = f'(\lambda)(x - \lambda) + f(\lambda)$$

represents the tangent to the curve $y = f(x)$ at $x = \lambda$ and the value λ_1 given by (12) is the point where this tangent meets the horizontal axis. As Fig. 6 shows, the values obtained by this rule are not necessarily better approximations and some care is necessary. A simple set of conditions is given by the

Newton–Fourier Rule *Let f be a real polynomial which has a single zero in the interval $[\alpha, \beta]$, and assume that for $\alpha < x < \beta$,*

$$\left| \frac{f(x)}{f'(x)} \right| < \beta - \alpha, \qquad f''(x) \neq 0. \tag{13}$$

Thus f changes sign in passing from α to β, while f'' does not. If we take λ to be α or β according as f and f'' have the same sign at α or at β, then λ_1, obtained by (12) *is a better approximation to the zero than λ.*

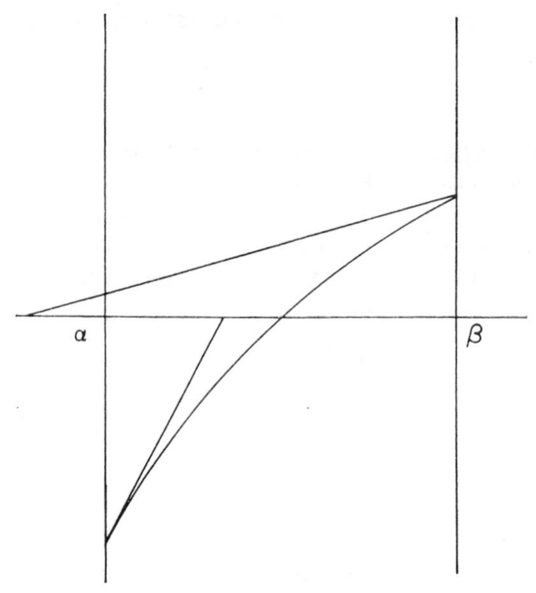

Figure 6

To establish this rule, suppose that f and f'' have the same sign at α and define h by $f(\alpha)+hf'(\alpha) = 0$. It follows from (13) that $|h| < \beta-\alpha$ and, by (11),

$$f(\alpha+h) = \frac{h^2}{2}f''(\alpha+k) \quad \text{where } 0 < k < h. \tag{14}$$

If $f(\alpha) > 0$, f decreases to 0 and so $f'(\alpha) < 0$, while if $f(\alpha) < 0, f'(\alpha) > 0$, so in either case $h > 0$. Moreover, by (14), $f(\alpha+h)$ has the same sign as $f(\alpha)$, so $\alpha+h$ lies between α and the zero we are looking for. Thus $\alpha+h$ is a better approximation than α or β. It is on the convex side of the curve (see Fig. 6) while the value obtained from the *regula falsi* is on the concave side; in this way we get a smaller interval, namely $[\alpha, \beta]$ is replaced by

$$\left[\alpha-\frac{f(\alpha)}{f'(\alpha)}, \frac{\alpha f(\beta)-\beta f(\alpha)}{f(\beta)-f(\alpha)}\right].$$

When f, f'' have the same sign at β, we determine h from the equation $f(\beta)-hf'(\beta) = 0$; as before we find that $h > 0$ and in place of (14) we find

$$f(\beta-h) = \frac{h^2}{2}f''(\beta-k).$$

Now the same argument shows that $\beta-h$ is between the true zero and β. ∎

This rule like the previous one applies not merely to polynomials but to any differentiable functions. As an illustration we calculate $\sqrt{2}$. Here

$f = x^2-2$, and there is exactly one zero in the interval [1, 2]. The *regula falsi* gives as a first approximation $\gamma_1 = 1{\cdot}33$, $f(\gamma_1) = -0{\cdot}22$. The next approximation is 1·4 (which is accurate to 0·014).

Using Newton's rule, we take the value 2 as our starting point (because f and f'' have the same sign there) and find as successive approximations 1·5, 1·417 (which is accurate to 0·003).

If we combine the two methods, the first step reduces the interval from [1, 2] to [1·33, 1·5] and the next step gives [1·35, 1·42]; here the error is within the limit 0·07; the advantage is that we can estimate the error without knowing the answer.

Exercises

(1) Show that there exists exactly one mapping satisfying **N**.1–4.

(2) Prove the formula for differentiating a function of a function: $f(g(x))' = f'(g(x))\, g'(x)$.

(3) Let R be any ring. A *derivation* of R is a mapping $\delta: R \to R$ such that $(a+b)\delta = a\delta+b\delta$, $(ab)\delta = a\delta\, .\, b+a\, .\, b\delta$ (e.g. $f \mapsto f'$ is a derivation in $F[x]$). Show that the set of *constants* of δ, defined as $C = \{a \in R \mid a\delta = 0\}$, is a subring of R.

(4) If R is a commutative ring with a derivation δ, prove Leibniz's formula: $(ab)\delta^n = \sum\binom{n}{\nu}a\delta^\nu\, .\, b\delta^{n-\nu}$. If R is an integral domain of characteristic p with a derivation δ, deduce that δ^p is again a derivation. What is this derivation in the case of the polynomial ring, if $\delta = \mathrm{d}/\mathrm{d}x$?

(5) Let F be a field of characteristic 0. Given $\alpha \in F$ and $f \in F[x]$, show that α is an m-fold zero of f iff $f(\alpha) = Df(\alpha) = \cdots = D^{m-1}f(\alpha) = 0$, $D^m(\alpha) \neq 0$.

(6)* For any ring R and $n = 0, 1, 2, \ldots$ define a mapping Δ_n on $R[x]$ by the rule $\Delta_n(\sum a_i x^i) = \sum\binom{i}{n}a_i x^{i-n}$, where x^ν is to be replaced by 0 for $\nu < 0$. Verify that

$$\Delta_n(fg) = \sum \Delta_\nu f\, .\, \Delta_{n-\nu}g.$$

Further show that, over any characteristic, f has α as m-fold zero iff $\Delta_\nu f(\alpha) = 0$ for $\nu = 0, 1, \ldots, m-1$, but $\Delta_m f(\alpha) \neq 0$. (Hint. Recall Ex. (13), **6.4**.)

(7) (i) Find the coefficient of y^2, where $y = x-1$, in $f = x^4+3x^3-5x^2+1$, (ii) find the coefficient of z, where $x = z-2$, in $f = 3x^5-2x^4-x^3+7x^2-x+3$. (Hint. Use Taylor's theorem.)

(8) Prove Euler's theorem on homogeneous functions: A polynomial f in $x_1, \ldots, x_n$ over a field of characteristic 0 is homogeneous of degree r iff $\Sigma x_i D_i f = rf$, where D_i is the derivation with respect to x_i.

(9) Solve $x^3 = 2$ (i.e. find the real root) to 2 places.

(10) Show that $x^3+5x-8 = 0$ has one real root and that it lies between 1 and 2. Find this root to 2 decimal places.

(11) Find the discriminant of ax^2+bx+c.

(12) Given $f = 2x^4+3x^3-3x^2-7x-3$, find a polynomial g with the same irreducible factors as f, but without repetitions.

(13) Show that the roots of $x^n-1 = 0$ in any field of characteristic p are distinct, provided that $p \nmid n$. Deduce that a finite field cannot be algebraically closed.

(14) Let f be a polynomial of positive degree n over a field of characteristic p. If $f' = 0$, show that $p \mid n$ and that f has at most n/p distinct zeros.

(15) Estimate the number of roots of $x^4-3x^3-x^2+8x-4 = 0$ in the interval $[-1, 1]$.

(16) Show that $3x^8-2x^5+x^4+4x^2-x-1 = 0$ has a root in $[-1, 0]$, and find it to 2 decimal places.

6.9 Symmetric and alternating functions

Let R be any ring. A polynomial in $R[x_1, \ldots, x_n]$ is said to be *symmetric* if it remains unchanged under any permutation of the indeterminates. E.g., $x_1+x_2+\cdots+x_n$, $x_1^2+x_2^2+\cdots+x_n^2$, $x_1x_2\ldots x_n$ are symmetric. Here the coefficient ring R is quite arbitrary; it need not even be commutative. However, the important case is that where R is an integral domain.

If we take the symmetric group Sym_n to act by permutations of $x_1, \ldots, x_n$, a group action on the set of monomials in $x_1, \ldots, x_n$ is defined in this way. E.g., when $n = 3$, the monomials $x_1^2x_2$, $x_1^2x_3$, $x_2^2x_1$, $x_2^2x_3$, $x_3^2x_1$, $x_3^2x_2$ constitute an orbit. In particular, monomials in a given orbit have the same degree. Clearly a polynomial in $x_1, \ldots, x_n$ is symmetric iff all the monomials in any given orbit appear with the same coefficient. This leads to an abbreviated notation for symmetric polynomials: we write down just one term from each orbit, the first in the lexicographic ordering, and prefix $\sum$. For example, $x_1+x_2+\cdots+x_n$ becomes $\sum x_1$ and $\sum_{i\neq j} x_ix_j^2$ is written $\sum x_1^2x_2$. The general symmetric polynomial is then a linear combination of terms of the form

$$\sum x_1^{\alpha_1}x_2^{\alpha_2}\ldots x_n^{\alpha_n}, \qquad \text{where } \alpha_1 \geqslant \alpha_2 \geqslant \cdots \geqslant \alpha_n \geqslant 0. \tag{1}$$

The inequalities for the αs follow because we have written down the leading term.

From the point of view of this classification the simplest symmetric functions are the functions

$$e_k = \sum x_1x_2\ldots x_k \qquad (k = 1, 2, \ldots, n). \tag{2}$$

These are the *elementary symmetric functions*; they occur as coefficients in the polynomial with zeros $-x_1, \ldots, -x_n$:

$$\prod_{i=1}^{n} (t+x_i) = t^n+e_1t^{n-1}+\cdots+e_n, \tag{3}$$

as follows by expanding the left-hand side. Clearly every polynomial in $e_1, \ldots, e_n$ yields a symmetric function in $x_1, \ldots, x_n$ when the es are replaced by their expressions in terms of the xs. The term

$$e_1^{\lambda_1} \ldots e_n^{\lambda_n} \tag{4}$$

yields a symmetric polynomial which is homogeneous of degree $\delta = \lambda_1 + 2\lambda_2 + \cdots + n\lambda_n$. This number δ is called the *weight* of the term (4). By the *weight* of a polynomial in the es one understands the largest weight of the terms occurring in it; a polynomial in the es all of whose terms have the same weight δ is said to be *isobaric* of weight δ. Thus a polynomial of weight δ in the es yields a symmetric polynomial of degree at most δ in the xs; if the given polynomial was isobaric, we get a form (i.e. a homogeneous polynomial) in the xs.

It is not hard to see that the symmetric polynomials form a subring of $R[x_1, \ldots, x_n]$; this subring is again a polynomial ring, with $e_1, \ldots, e_n$ as indeterminates:

THEOREM 1 (Fundamental theorem on symmetric functions) *Every symmetric polynomial in $x_1, \ldots, x_n$ can be written in exactly one way as a polynomial in $e_1, \ldots, e_n$. Thus the symmetric polynomials form a polynomial ring $R[e_1, \ldots, e_n]$.*

Proof. Let f be a symmetric polynomial of degree k and suppose that the leading term of degree k is $ax_1^{\alpha_1} \ldots x_n^{\alpha_n}$. Put

$$\beta_i = \alpha_i - \alpha_{i+1} \qquad (i = 1, 2, \ldots, n, \quad \text{where } \alpha_{n+1} = 0), \tag{5}$$

so that $\beta_i \geqslant 0$ by (1), and

$$\alpha_i = \beta_i + \beta_{i+1} + \cdots + \beta_n.$$

Then the term $ae_1^{\beta_1} \ldots e_n^{\beta_n}$ has weight $\beta_1 + 2\beta_2 + \cdots + n\beta_n = \alpha_1 + \alpha_2 + \cdots + \alpha_n = k$ and its leading term is

$$ax_1^{\beta_1}(x_1x_2)^{\beta_2}(x_1x_2x_3)^{\beta_3} \ldots (x_1 \ldots x_n)^{\beta_n} = ax_1^{\alpha_1} \ldots x_n^{\alpha_n},$$

hence $f_1 = f - ae_1^{\beta_1} \ldots e_n^{\beta_n}$ has a lower leading term than f. After repeating this procedure until no terms of degree k are left in f, we obtain a polynomial g, say, of lower degree than f, which differs from f by a polynomial in the es. Hence g is again symmetric and, using induction on the degree, we see that g is a polynomial in the es; hence so is f. If f can be expressed in more than one way, then by taking the difference of two such expressions we get an equation

$$\sum a_{i_1 \ldots i_n} e_1^{i_1} \ldots e_n^{i_n} = 0, \tag{6}$$

where not all the as vanish. If $ae_1^{\beta_1} \ldots e_n^{\beta_n}$ is the leading term on the left, the leading term in terms of the xs is $ax_1^{\alpha_1} \ldots x_n^{\alpha_n}$, where α_i is related to β_i by (5). But this contradicts the fact that the xs are indeterminates, hence all the coefficients in (6) must vanish and the uniqueness follows. ∎

This proof can actually be used to find the expression in terms of the es; e.g. to convert $\sum x_1^2 x_2$ we consider $e_1 e_2 = \sum x_1 \cdot \sum x_1 x_2 = \sum x_1^2 x_2 + 3 \sum x_1 x_2 x_3$, hence $\sum x_1^2 x_2 = e_1 e_2 - 3e_3$. Another method is as follows: Any symmetric polynomial can be expressed as a sum of symmetric forms, because the homogeneous components must again be symmetric. To express a symmetric form of degree k in terms of the es, we need only write down all products of weight k in the es and determine the coefficients; this can be done by giving special values to the xs. E.g., to express $\sum x_1^2 x_2$, we write this as

$$\alpha e_1^3 + \beta e_1 e_2 + \gamma e_3 = \sum x_1^2 x_2, \tag{7}$$

where we have included all products of weight 3. If we take $x_1 = 1$, $x_2 = x_3 = 0$, this reduces to $\alpha = 0$. For $x_1 = x_2 = 1$, $x_3 = 0$ we get $2\beta = 2$, so $\beta = 1$, and for $x_1 = x_2 = x_3 = 1$ we get $9 + \gamma = 6$, hence $\gamma = -3$. So we again find $\sum x_1^2 x_2 = e_1 e_2 - 3e_3$. In using this method it is particularly important not to overlook any products of the given weight, since their absence from (7) will not show up in the calculation.

Next we consider polynomials in $x_1, \ldots, x_n$ over an entire ring R, which are unchanged except for a factor in R when the variables are permuted. Suppose that f is such a polynomial, not identically zero, then for any permutation π

$$f(x_{1\pi}, x_{2\pi}, \ldots, x_{n\pi}) = \lambda_\pi f(x_1, \ldots, x_n), \qquad \text{where } \lambda_\pi \in R. \tag{8}$$

Let any $\sigma \in \mathrm{Sym}_n$ act on (8), then

$$f(x_{1\pi\sigma}, \ldots, x_{n\pi\sigma}) = \lambda_\pi f(x_{1\sigma}, \ldots, x_{n\sigma}),$$

therefore $\lambda_{\pi\sigma} f(x_1, \ldots, x_n) = \lambda_\pi \lambda_\sigma f(x_1, \ldots, x_n)$, hence (as R is entire),

$$\lambda_{\pi\sigma} = \lambda_\pi \lambda_\sigma, \qquad \lambda_1 = 1,$$

and the mapping $\lambda: \pi \mapsto \lambda_\pi$ from Sym_n to the multiplicative monoid R is a monoid homomorphism. For any transposition τ, $\tau^2 = 1$, so $(\lambda_\tau)^2 = \lambda_{\tau^2} = \lambda_1 = 1$. Thus the only possibilities are $\lambda_\tau = \pm 1$ (cf. Ex. (13), **6.1**) and as Sym_n is generated by transpositions, the image of λ is contained in $\{-1, 1\}$. Since all transpositions lie in one conjugacy class of Sym_n, they all have the same image under λ. Either this image is 1, then $\lambda_\pi = 1$ for all $\pi \in \mathrm{Sym}_n$ and f is symmetric, or char $R \neq 2$ and each transposition maps to -1, whence $\lambda_\pi = \mathrm{sgn}\,\pi$ for all $\pi \in \mathrm{Sym}_n$; in that case we have

$$f(x_{1\pi}, x_{2\pi}, \ldots, x_{n\pi}) = \mathrm{sgn}\,\pi\, f(x_1, x_2, \ldots, x_n). \tag{9}$$

A polynomial satisfying (9) (over an entire ring of characteristic $\neq 2$) is said to be *alternating*. Since Sym_n is generated by all transpositions, it is enough to require that f change sign under all transpositions.

The simplest non-vanishing alternating function is the product over all the differences $x_i - x_j$:

$$\Delta = \prod_{i>j} (x_i - x_j). \tag{10}$$

Each pair (i, j) of distinct subscripts occurs just once, so there are $\binom{n}{2} = n(n-1)/2$ factors in all. When a transposition, say (1 2) is applied, all factors on the right of (10) are permuted among themselves, except $x_2 - x_1$ which becomes $x_1 - x_2$, hence Δ changes sign. Similarly for other transpositions, so Δ is indeed alternating. Corresponding to Th. 1 we have

THEOREM 2 *Let R be an entire ring of characteristic $\neq 2$. Then any alternating polynomial in $R[x_1, \ldots, x_n]$ has the form $f = \Delta g$, where g is given by* (10) *and g is symmetric.*

Proof. $f(x_1, x_1, x_3, \ldots, x_n) = -f(x_1, x_1, x_3, \ldots, x_n)$, hence

$$2f(x_1, x_1, x_3, \ldots, x_n) = 0,$$

and here we may cancel 2; therefore f vanishes when $x_1 = x_2$. Regarding f as a polynomial in x_1 with coefficients in $R[x_2, \ldots, x_n]$ we find that $x_1 - x_2 \mid f$, by the remainder theorem. Similarly $x_1 - x_i \mid f$ for $i = 3, \ldots, n$, so $f = \prod_{i \neq 1} (x_1 - x_i) f_1$. Now f_1 is alternating in $x_2, \ldots, x_n$ and, hence, by induction on n, $f_1 = \prod_{i > j > 1} (x_i - x_j) g$ for some $g \in R[x_1, \ldots, x_n]$. Thus $f = \Delta g$ and clearly $g = f/\Delta$ is symmetric. ∎

COROLLARY *A non-zero alternating function in $x_1, \ldots, x_n$ is of degree at least $n(n-1)/2$.* ∎

Exercises

(1) Verify that the symmetric functions form a subring of $R[x_1, \ldots, x_n]$ and that the symmetric or alternating functions also form a subring.

(2) Verify the formula (3) for the elementary symmetric function defined by (2).

(3) Find the product of $\sum x_1^4 x_2^2 x_3$ and $\sum x_1^3 x_2$; check your answer by expressing both functions, and the product found, in terms of elementary symmetric functions.

(4) Express the following in terms of elementary symmetric functions: (i) $\sum x_1^2 x_2^2 x_3$, (ii) $\sum x_1^3 x_2^2$, (iii) $\sum x_1^3 x_2 x_3$, (iv) $(x_1+x_2-x_3-x_4)(x_1-x_2+x_3-x_4)(x_1-x_2-x_3+x_4)$.

(5) Show that the expression of a symmetric function of the form $\sum x_1^{\alpha_1} \ldots x_r^{\alpha_r}$ in terms of elementary symmetric functions is independent of the number of indeterminates, provided that this number is at least r.

(6) Define the power sums of $x_1, \ldots, x_n$ by $s_r = \sum x_1^r$ and show that they are related to the elementary symmetric functions by the equation

$$s_r - e_1 s_{r-1} + e_2 s_{r-2} - \cdots + (-1)^n e_n s_{r-n} = 0 \quad \text{if } r \geqslant n, \tag{11}$$

or

$$s_r - e_1 s_{r-1} + e_2 s_{r-2} - \cdots + (-1)^r e_r r = 0 \quad \text{if } n > r \geqslant 1. \tag{12}$$

(Hint. x_i satisfies the equation $\prod(t-x_\nu) = \sum(-1)^\nu e_\nu t^{n-\nu} = 0$. Multiply this equation for x_i by x_i^{r-n} and sum over i to get (11). For (12) observe that the left-hand side, when written as a polynomial in $x_1, \ldots, x_n$, say $g(x_1, \ldots, x_n)$, satisfies $g(x_1, \ldots, x_r, 0, \ldots, 0) = 0$, by (11), hence show that $g = 0$.)

(7)* Show that s_k is equal to the sum of the terms of weight k in the expression

$$\sum_r \frac{k}{r}(e_1-e_2+\cdots+(-1)^{n-1}e_n)^r \quad \text{(Waring's formula)},$$

where the summation is over all $r \in \mathbf{N}$.

(8) Express Δ^2 as a polynomial in the elementary symmetric functions, and verify that for $n = 2$ it coincides with the discriminant of $x^2-e_1x+e_2$.

(9) Show that any polynomial f in $e_1, \ldots, e_n$ (the elementary symmetric functions of $x_1, \ldots, x_n$) is isobaric of weight k iff f becomes t^kf when x_i is replaced by tx_i $(i = 1, \ldots, n)$.

(10) The complete *homogeneous symmetric function* of $\alpha_1, \ldots, \alpha_n$ of degree i, denoted by h_i, is defined as the coefficient of t^i in the product $\prod_{\nu=1}^{n}(1-\alpha_\nu t)^{-1}$, when expanded as a formal power series. If $e_1, \ldots, e_n$ are the elementary symmetric functions and $A = (a_{ij})$, where $a_{ij} = (-1)^{i+j}e_{j-i}$ and $B = (b_{ij})$, where $b_{ij} = h_{j-i}$ $(e_\nu = h_\nu = 0$ for $\nu < 0)$, show that the infinite matrices A, B are formal inverses of each other.

Further exercises on Chapter 6

(1) Show that a non-zero element a in a ring is a zero-divisor iff $aba = 0$ for some $b \neq 0$.

(2) Let R be a ring and $a, b \in R$ such that $ab+ba = 1$, $a^2b+ba^2 = a$. Show that a^2 commutes with b; show further that $2aba = a$ and deduce that a is invertible and has the inverse $2b$.

(3) Let R be a ring and define $[a, b] = ab-ba$. Show that this operation satisfies the *Jacobi-identity*:

$$[[a, b], c]+[[b, c], a]+[[c, a], b] = 0.$$

Give an example to show that [,] need not be associative.

(4) Let R be a ring and define $a+'b = a+b-1$, $a\,.'b = a+b-ab$. Show that R is a ring relative to $+'$ and $.'$ and that the mapping $a \mapsto 1-a$ is an isomorphism from the old ring to the new.

(5) For any positive integer n define a functor from rings to rings which associates with R the ring R_n of all $n\times n$ matrices.

(6) Let *Rg* be the category of rings and their homomorphisms and *Rg'* the category of 'rings without 1' (i.e. not necessarily possessing a 1) and their homomorphisms. Verify that *Rg* is a subcategory of *Rg'*, but not full. Define a correspondence E from

Rg' to Rg, by associating with the ring R a ring R^E whose additive group is $\mathbf{Z} \times R$ with multiplication

$$(m, a)(n, b) = (mn, an+mb+ab).$$

Verify that R^E is a ring with unit-element $(1, 0)$. Define the effect of E on morphisms so that it becomes a functor. What is the effect of E on a ring which already has a 1? Show that there is a Rg'-morphism $\nu_R: R \to R^E$ such that for any homomorphism f: $R \to S$, the diagram

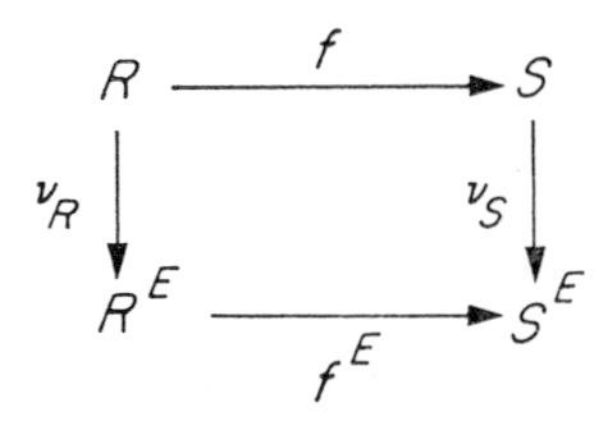

commutes.

(7) Show that any automorphism of an integral domain can be extended in just one way to an automorphism of the field of fractions. Prove that the only automorphism of **Q** is the identity.

(8) Show that an element a of **R** is non-negative iff the equation $x^2 = a$ has a solution in **R**. Deduce that the only automorphism of **R** is the identity.

(9) Show that if the numerators and denominators of the different expressions for a given rational number are plotted as coordinates in the plane, the resulting points lie on a straight line.

(10) In any ring R, if c is a nilpotent element, show that $1-c$ is a unit.

(11) Let A, B be 2×2 matrices over a field (possibly skew) such that

$$AB = \begin{pmatrix} 0 & c \\ 0 & 0 \end{pmatrix}, \quad \text{where } c \neq 0. \tag{1}$$

Show that either a_{22} or b_{11} must be 0. Deduce that a ring with matrices A, B satisfying (1), where a_{22} and b_{11} are non-zero, cannot be embedded in a skew field (Malcev has constructed entire rings with these properties, showing that not every entire ring can be embedded in a skew field).

(12) Let R be an integral domain with quotient field K. For $a, b \in K$ define $a \mid b$ to mean: $bR \subseteq aR$. Extend the definitions of HCF and LCM to elements of K and prove the rules (i)–(iv) on page 133.

(13) Let F be a field and $F[[x]]$ the ring of formal power series. Find all atoms in $F[[x]]$ and show that this is a UFD.

(14) Show that every polynomial over a field can be written in the form $f(x^2)+xg(x^2)$, where f, g are polynomials. Find the remainder in the division of $f(x^2)+xg(x^2)$ by $x^2-\alpha$.

(15)* Let p be a prime and m, n positive integers. If $(m, n) = d$, show that $(x^{p^m}-x, x^{p^n}-x) = x^{p^d}-x$.

(16) Show that a monic polynomial of degree n over a field has at most 2^n distinct monic factors.

(17) In the field of rational functions $f(x)$, if a_1/b_1 and a_2/b_2 are fractions in reduced form, under what conditions is the expression $(a_1b_2+b_1a_2)/b_1b_2$ reduced?

(18) Show that the matrices $\begin{pmatrix} a & b \\ 2b & a \end{pmatrix}$, where $a, b \in \mathbf{F}_5$, form a field of 25 elements.

(19) Factorize x^5-1 over **Q**, **R** and **C**.

(20)* Let R be a commutative ring and let $f, g \in R[x]$ be such that $fg = 0$. Writing $f = \sum a_i x^i$ $(a_i \in R)$, show that either $a_i g = 0$ for all i, or i can be chosen so that $a_i g$ has least degree δ, and then $0 \leqslant \delta < \deg g$. By examining these cases in turn, show that if $f \neq 0$, then $cf = 0$ for some $c \in R$, $c \neq 0$. (McCoy)

(21)* Let F be a field and define multiplication on $F[x]$ by the rule $f \circ g = h$, where $h(x) = f(g(x))$ (i.e. substitution). Show that $f[x]$ is a monoid under this multiplication; what is the neutral element? Using this multiplication and ordinary addition, is $F[x]$ a ring? If F has characteristic p and I consists of those polynomials $\sum a_i x^i$ for which $a_i = 0$ unless $i = p^\nu$ for some $\nu \geqslant 0$, show that I with substitution as multiplication is a ring.

(22) Let R be an integral domain with norm function $x \mapsto N(x)$ such that $N(x) \in \mathbf{Z}$, $N(xy) = N(x)N(y)$. Show that N can be extended in just one way to the quotient field K of R so as to preserve the multiplicativity. Show that R is Euclidean relative to N iff for each $u \in K$ there exists $a \in R$ such that $N(u-a) < 1$.

(23)* Let d be a negative integer and consider the set $I = \{a+b\sqrt{d} \mid a, b \in \mathbf{Z}\}$. Show that this is a subring of **C**. For each $\alpha = a+b\sqrt{d}$ define $\alpha^* = a-b\sqrt{d}$, and show that $N(\alpha) = \alpha\alpha^*$ is a positive integer unless $\alpha = 0$. Show also that $N(\alpha\beta) = N(\alpha)N(\beta)$ and find the units in I. What happens in this example if for d we use a positive number which is not a perfect square, or one that is a perfect square?

(24) Show that the complex numbers of the form $\alpha = a+bi$ $(a, b \in \mathbf{Z}, i = \sqrt{-1})$ form a ring $\mathbf{Z}[i]$ which is Euclidean relative to the norm defined in Ex. (23) ($\mathbf{Z}[i]$ is called the ring of *Gaussian integers*.)

(25) Show that the polynomial functions over an infinite field form an integral domain; give a counter-example for the finite field $\mathbf{F}_p$.

(26) Find the product of all simple factors of each of the following: (i) $x^6-2x^5+6x^4-8x^3+10x^2-6x+3$, (ii) $4x^4+8x^3+x^2+3x+1$.

(27) Prove that a symmetric rational function in $x_1, \ldots, x_n$ over a field can be written as f/g, where f, g are symmetric polynomials. (Hint. Use the fact that $F[x_1, \ldots, x_n]$ is a UFD; cf. Ex. (7) **6.7**.)

(28) Let f be a monic polynomial of degree n in x over a field F. If in some field containing F, f splits into distinct linear factors: $f = \prod(x-\alpha_i)$, then for any symmetric polynomial $g(x_1, \ldots, x_n)$, $g(\alpha_1, \ldots, \alpha_n) \in F$. Give an example to show that this may not hold for alternating polynomials.

(29)* Let α be a real number which satisfies an equation $f = 0$ of degree n, with integer coefficients but no rational roots. Show that for any integers r, s (where $s > 0$), $|f(r/s)| \geqslant 1/s^n$. If α lies in an interval containing no other zero of f and $|f'|$ is bounded by a constant M in this interval, show that

$$|\alpha-\frac{r}{s}| > \min\{1/Ms^n, 1\}.$$

Deduce that if α is a real irrational number such that for any positive c there exist integers r, s ($s > 0$) such that $|\alpha-r/s| \leqslant c/s^n$, then α satisfies no equation with integer coefficients of degree less than n. Show that $\sum 10^{-n!}$ satisfies no equation with integer coefficients. (Numbers of this type are called *Liouville numbers*, after their discoverer; they are examples of transcendental numbers, cf. Vol. 2.)

(30) Let R be a commutative ring. Given any $2n$ elements $a_1, \ldots, a_n, b_1, \ldots, b_n$ where the as are distinct, let $p_i(x)$ be the product of the $x-a_j$ for all $j \neq i$. Verify that

$$f = \sum b_i \frac{p_i(x)}{p_i(a_i)} \tag{2}$$

is a polynomial of degree at most $n-1$ such that $f(a_i) = b_i$ ($i = 1, \ldots, n$). Show that f is the only polynomial (of degree $<n$) with this property. (2) is known as the *Lagrange interpolation formula.*

7

Determinants

Given a square matrix A, we often want to know whether or not A is singular. In Ch. **4** we learnt a practical way of finding this out, but it is often useful to have a function of the entries of A whose vanishing characterizes the singularity of A. Let us for a moment look at 2×2 matrices. If the columns of A are $u = (u_1, u_2)^T$, $v = (v_1, v_2)^T$, then the singularity of A is expressed by the fact that

$$\begin{aligned} \lambda u_1 + \mu v_1 &= 0, \\ \lambda u_2 + \mu v_2 &= 0, \end{aligned} \tag{1}$$

for some scalars λ, μ, not both zero. Eliminating μ and λ in turn we find that $\lambda(u_1 v_2 - u_2 v_1) = 0$, $\mu(u_1 v_2 - u_2 v_1) = 0$ and, hence,

$$u_1 v_2 - u_2 v_1 = 0. \tag{2}$$

Conversely, assume (2); if $u \neq 0$, say $u_2 \neq 0$, then we can satisfy (1) by taking $\lambda = v_2$, $\mu = -u_2$, while if $u = 0$, (1) holds with $\lambda = 1$, $\mu = 0$. Hence (2) is necessary and sufficient for u and v to be linearly dependent.

We also note that over **R** the left-hand side of (2) can be interpreted geometrically as the area of the parallelogram spanned by u and v. For if in polar coordinates, $u = (r_1 \cos \theta_1, r_1 \sin \theta_1)$, $v = (r_2 \cos \theta_2, r_2 \sin \theta_2)$, then $u_1 v_2 - u_2 v_1 = r_1 r_2(\cos \theta_1 \sin \theta_2 - \sin \theta_1 \cos \theta_2) = r_1 r_2 \sin(\theta_2 - \theta_1)$. Similarly, for three vectors in 3-space there is a function which represents the volume of the 3-cell spanned by them, and whose vanishing indicates their linear dependence.

We shall want to define quite generally a function of n vectors in n-space to serve the same purpose. To do this we first list the algebraic properties needed to determine such a function (the determinant) and derive some consequences that will be needed later. In some of the proofs the coefficients will be allowed to lie in any commutative ring, not necessarily a field; this makes things no harder and is often useful.

7.1 Definition and basic properties

Let U, V, W be vector spaces over the same field F and consider a mapping $f(u, v)$ from the Cartesian product $U \times V$ to W:

$$f: U \times V \to W.$$

This mapping is said to be *bilinear* if for each $v \in V$, the mapping $u \mapsto f(u, v)$ is linear, and likewise for each $u \in U$, the mapping $v \mapsto f(u, v)$ is linear. Thus we require

$$f(\lambda u + \lambda' u', v) = \lambda f(u, v) + \lambda' f(u', v) \qquad u, u' \in U,\ v, v' \in V,$$
$$f(u, \lambda v + \lambda' v') = \lambda f(u, v) + \lambda' f(u, v') \qquad \lambda, \lambda' \in F.$$

More generally, given spaces $U_1, \ldots, U_n, W$ over F, a mapping $f\colon U_1 \times \ldots \times U_n \to W$ is said to be *multilinear* if it is linear in each argument when the others are kept fixed, i.e. if for $i = 1, \ldots, n$ and $u_j \in U_j$ the mapping

$$x \mapsto f(u_1, \ldots, u_{i-1}, x, u_{i+1}, \ldots, u_n)$$

of U_i into W is linear.

When the spaces considered are finite-dimensional, then by taking a basis in each, we may take them to consist of rows (or columns) over the field F. But there is now no reason why F should be a field; the definition of linearity makes good sense over any ring, although we shall here confine ourselves to commutative rings. Thus we may regard the above definitions of bilinearity and multilinearity as applying to functions on rows (or columns) over a commutative ring R, with values in R. For example, the product $x_1 x_2 \ldots x_n$ of any n elements in R is a multilinear function of its arguments, by the distributive law (here the commutativity of R is actually needed).

In order to distinguish between singular and non-singular matrices we shall define the determinant of a matrix as a function of its entries with certain properties. Of course it will then be necessary to show that such functions exist; we shall also show that they are uniquely determined by these properties (this axiomatic approach goes back to Weierstrass).

DEFINITION Let A be an $n \times n$ matrix over a commutative ring R. Then a function $d(A)$ of the n^2 entries of A is called a *determinant* (of *order* n) if it satisfies the following conditions:

D. 1 *d is a linear function of the columns of A.*

D. 2 *$d(A) = 0$ if there are two columns of A which are the same.*

D. 3 $d(I) = 1$.

Let us denote the columns of A by $a_1, \ldots, a_n$, where $a_j = (a_{1j}, a_{2j}, \ldots, a_{nj})^{\mathrm{T}}$, and write $d(A)$ as $d(a_1, a_2, \ldots, a_n)$, to indicate its dependence on the columns. The columns of the unit matrix I will be written as $e_1, \ldots, e_n$ as usual.

In detail **D.** 1 means

$$d(\lambda a_1 + \lambda' a_1', a_2, \ldots, a_n) = \lambda d(a_1, a_2, \ldots, a_n) + \lambda' d(a_1', a_2, \ldots, a_n),$$

and similarly for the other columns. This assertion must be carefully distinguished from the notion of linearity in all the coefficients. This is not postulated and will usually be false, i.e. in general

$$d(A+B) \neq d(A) + d(B).$$

But **D**. 1 may be used to expand $d(A+B)$ as a sum of 2^n determinants.

We first show that there can be at most one function with the properties **D**. 1–3. For this we shall need the following consequence of **D**. 1–2.

D. 4 *If the columns of A are permuted in any way, d(A) is multiplied by the sign of the corresponding permutation.*

We express this by saying that $d(A)$ is an *alternating* function of the columns of A; for the case of a polynomial over an integral domain of characteristic not 2 this agrees with the definition given in **6.9**, p. 156. To prove **D**. 4 it will be enough to show that $d(A)$ changes sign when two columns are interchanged. Let us take two columns, say the ith and the jth; the remaining columns of A will be kept fixed throughout the discussion, so we can abbreviate $d(a_1, \ldots, a_n)$ by $f(a_i, a_j)$. Now, by **D**. 1, f is bilinear, hence for any columns u and v,

$$f(u+v, u+v) = f(u, u)+f(u, v)+f(v, u)+f(v, v). \tag{3}$$

Further, by **D**. 2, f vanishes when both arguments are the same, hence (3) reduces to $0 = f(u, v)+f(v, u)$, therefore $f(a_j, a_i) = -f(a_i, a_j)$ as asserted.

To prove that d is uniquely determined by **D**. 1–3 we shall find an explicit formula in terms of the entries of A. Let us define functions $\varepsilon(i_1, \ldots, i_n)$, where $i_1, \ldots, i_n$ are n arguments ranging over the set $\{1, 2, \ldots, n\}$, by the rule

$$\varepsilon(i_1, \ldots, i_n) = \begin{cases} 1 & \text{if } (i_1, \ldots, i_n) \text{ is an even permutation of } (1, \ldots, n) \\ -1 & \text{if } (i_1, \ldots, i_n) \text{ is an odd permutation of } (1, \ldots, n) \\ 0 & \text{otherwise.} \end{cases} \tag{4}$$

E.g., for $n = 3$, $\varepsilon(1\ 2\ 3) = \varepsilon(2\ 3\ 1) = \varepsilon(3\ 1\ 2) = 1$, $\varepsilon(2\ 1\ 3) = \varepsilon(1\ 3\ 2) = \varepsilon(3\ 2\ 1) = -1$, $\varepsilon(1\ 1\ 2) = \varepsilon(1\ 3\ 1) = \cdots = 0$.

In order to evaluate $d(A)$ we observe that the jth column of $A = (a_{ij})$ may be written as

$$a_j = \sum e_i a_{ij}.$$

Hence

$$\begin{aligned} d(a_1, a_2, \ldots, a_n) &= d(\sum e_{i_1} a_{i_1 1}, \sum e_{i_2} a_{i_2 2}, \ldots, \sum e_{i_n} a_{i_n n}) \\ &= \sum_{i_1 i_2 \ldots i_n} a_{i_1 1} a_{i_2 2} \ldots a_{i_n n} d(e_{i_1}, e_{i_2}, \ldots, e_{i_n}), \end{aligned} \tag{5}$$

by linearity, where the summation is an n-fold one, from 1 to n for each of $i_1, i_2, \ldots, i_n$. If $i_1, \ldots, i_n$ is a permutation of $1, \ldots, n$ then by **D**. 4,

$$d(e_{i_1}, e_{i_2}, \ldots, e_{i_n}) = \pm d(e_1, e_2, \ldots, e_n), \tag{6}$$

with + or − according to whether the permutation was even or odd. But, by **D**. 3, $d(e_1, \ldots, e_n) = 1$ and, if $i_1, \ldots, i_n$ is not a permutation of $1, \ldots, n$, there must be a repetition and so the left-hand side of (6) then vanishes, by **D**. 2. Therefore in all cases,

$$d(e_{i_1}, e_{i_2}, \ldots, e_{i_n}) = \varepsilon(i_1, i_2, \ldots, i_n),$$

and inserting this value in (5) we find

$$d(a_1, a_2, \dots, a_n) = \sum \varepsilon(i_1, i_2, \dots, i_n) a_{i_1 1} a_{i_2 2} \dots a_{i_n n}. \tag{7}$$

This formula for $d(A)$ shows that the determinant is unique, if it exists. To prove that it exists we need only verify that the right-hand side of (7) satisfies **D.** 1–3.

Each term on the right of (7) contains just one entry of the first column a_1, hence the whole expression is linear in a_1, and similarly it is linear in the other columns. To prove **D.** 2, assume that $a_1 = a_2$ say, and fix a particular term on the right of (7). If $i_1 = i_2$ in this term, the ε-factor vanishes and there is no contribution. If $i_1 > i_2$, then there is a corresponding term with i_1 and i_2 interchanged, with the opposite sign, because the permutations $(i_1, i_2, \dots, i_n)$ and $(i_2, i_1, i_3, \dots, i_n)$ of $(1, 2, \dots, n)$ have opposite parity. Thus we have

$$d(a_1, a_1, a_3, \dots, a_n) = \sum_{i_1 > i_2} \varepsilon(i_1, i_2, \dots, i_n)(a_{i_1 1} a_{i_2 1} - a_{i_2 1} a_{i_1 1}) a_{i_3 3} \dots a_{i_n n}, \tag{8}$$

and this vanishes because $a_{i_1 1} a_{i_2 1} = a_{i_1 1} a_{i_2 1}$. A similar argument applies if any other two columns of A agree. Thus **D.** 2 holds. Finally **D.** 3 is clear and this shows that the determinant exists and is given by (7).

The number of terms on the right of (7) is $n!$, e.g. for $n = 2$: $d(A) = a_{11}a_{22} - a_{21}a_{12}$, $n = 3$: $d(A) = a_{11}a_{22}a_{33} + a_{21}a_{32}a_{13} + a_{31}a_{12}a_{23} - a_{21}a_{12}a_{33} - a_{11}a_{32}a_{23} - a_{31}a_{22}a_{13}$.

We go on to list some important properties of determinants:

D. 5 *The determinant is multiplied by λ if all the elements of a column are multiplied by λ.*

This follows immediately from the linearity.

D. 6 *The determinant of A remains unchanged if a multiple of one column is added to another.*

For the proof we write again $f(a_i, a_j)$ for $d(a_1, \dots, a_n)$, then we must show that $f(u, v + \lambda u) = f(u, v)$. But by **D.** 1–2,

$$f(u, v + \lambda u) = f(u, v) + \lambda f(u, u) = f(u, v).$$

This property is most useful in evaluating determinants; and it is usually easier to simplify matrices by using it than by applying (7) as it stands.

The analogues of **D.** 1–6 for rows also hold, in view of the symmetry expressed by

D. 7 $d(A^{\mathrm{T}}) = d(A)$.

To prove this we need only verify that $d(A^{\mathrm{T}})$, as a function of A, satisfies the conditions **D.** 1–3, i.e. we must show that the right-hand side of (7) satisfies **D.** 1–3, with columns replaced by rows. Clearly this expression is linear in the elements of each row: each term contains just one factor a_{1i}

with first suffix 1. To verify **D.** 2 let us rewrite (7) by putting the first suffixes in order. Suppose that

$$\begin{pmatrix} 1\,2\,\ldots\,n \\ i_1 i_2 \ldots i_n \end{pmatrix} = \begin{pmatrix} j_1 j_2 \ldots j_n \\ 1\;\,2\;\ldots\;n \end{pmatrix};$$

this means that $(i_1, \ldots, i_n)$ and $(j_1, \ldots, j_n)$ as permutations of $(1, 2, \ldots, n)$ are inverse to each other and hence have the same sign, i.e. $\varepsilon(i_1, \ldots, i_n) = \varepsilon(j_1, \ldots, j_n)$. Therefore (7) can also be written

$$d(a_1, \ldots, a_n) = \sum \varepsilon(j_1, \ldots, j_n) a_{1j_1} a_{2j_2} \ldots a_{nj_n}, \tag{9}$$

where $j_1, \ldots, j_n$ ranges over all the permutations of $1, 2, \ldots, n$. Now the same argument as before shows that $d = 0$ when two rows are the same. Finally $d(I^T) = 1$, because $I^T = I$, and this completes the proof of **D.** 7.

Conditions **D.** 1–7 enable us to compute the change in the determinant when any elementary row or column operation is applied to the matrix. As we know, any matrix over a field can be brought to diagonal form by such operations and we can then evaluate the determinant by using

D. 8 *If $A = (a_{ij})$ is a square matrix whose elements below the main diagonal are zero (i.e. $a_{ij} = 0$ for $i > j$), then*

$$d(A) = a_{11} a_{22} \ldots a_{nn}. \tag{10}$$

The same formula holds if all elements above the main diagonal are zero (i.e. $a_{ij} = 0$ for $i < j$).

To prove (10) consider the formula (7); this certainly includes a term $a_{11} a_{22} \ldots a_{nn}$; we must show that it is the only term. In the first place $a_{i1} = 0$ for $i > 1$, hence $i_1 = 1$. Secondly $a_{i2} = 0$ for $i > 2$, hence $i_2 = 1$ or 2, but if $i_2 = 1$, we have a factor $\varepsilon(1, 1, i_3, \ldots)$ which is zero, so there is no contribution unless $i_2 = 2$. Continuing in this way, we find that there is only one choice at each stage, so $d(A)$ has only one term, as in (10). The other case can now be deduced by transposition.

We illustrate **D.** 1–8 by computing some determinants. The determinant of a matrix A is generally denoted by $\det(A)$ or $|A|$; the latter notation is used particularly when the matrix is written out in full. Below we use the notation of Ch. **5** to indicate elementary operations:

$$\begin{vmatrix} 2 & 1 & 4 \\ -3 & 5 & 6 \\ 7 & 2 & 5 \end{vmatrix} \begin{matrix} \\ C_3 \to C_3 - 2C_1 \\ C_1 \to C_1 - 2C_2 \end{matrix} \begin{vmatrix} 0 & 1 & 0 \\ -13 & 5 & 12 \\ 3 & 2 & -9 \end{vmatrix} \begin{matrix} \\ C_3 \to C_3 + 3C_1 \\ \end{matrix} \begin{vmatrix} 0 & 1 & 0 \\ -13 & 5 & -27 \\ 3 & 2 & 0 \end{vmatrix}$$

$$\begin{matrix} \\ C_2 \leftrightarrow C_1 \\ C_2 \to -C_2 \end{matrix} \begin{vmatrix} 1 & 0 & 0 \\ 5 & 13 & -27 \\ 2 & -3 & 0 \end{vmatrix} \begin{matrix} \\ R_2 \leftrightarrow R_3 \\ R_3 \to -R_3 \end{matrix} \begin{vmatrix} 1 & 0 & 0 \\ 2 & -3 & 0 \\ -5 & -13 & 27 \end{vmatrix} = -81.$$

Secondly, consider the determinant (called after Vandermonde):

$$\Delta(x, y, z) = \begin{vmatrix} 1 & 1 & 1 \\ x & y & z \\ x^2 & y^2 & z^2 \end{vmatrix}, \tag{11}$$

This is an alternating function of the columns and so is alternating in x, y, z. By Th. 2, **6.9**,

$$\Delta(x, y, z) = (y-x)(z-x)(z-y)g, \tag{12}$$

where g is symmetric in x, y, z. Each term in the expansion of (11) has degree 3, hence by comparing degrees in (12) we see that g is a constant. By comparing the coefficients of a given term, such as yz^2 (corresponding to the main diagonal in (11)), we find that $g = 1$ and so the Vandermonde determinant is just the product of the differences of the variables. The same argument establishes the corresponding formula for the nth order Vandermonde determinant:

$$\begin{vmatrix} 1 & 1 & \ldots & 1 \\ x_1 & x_2 & \ldots & x_n \\ x_1^2 & x_2^2 & \ldots & x_n^2 \\ \cdot & \cdot & \cdot & \cdot \\ x_1^{n-1} & x_2^{n-1} & \ldots & x_n^{n-1} \end{vmatrix} = \prod_{i>j} (x_i - x_j).$$

We conclude this section with a further remarkable property of determinants, the *multiplication rule*:

D. 9 *For any two square matrices of the same order,*

$$\det AB = \det A \,.\, \det B.$$

To prove this rule, denote the columns of B by $b_1, \ldots, b_n$, then the columns of AB are $Ab_1, \ldots, Ab_n$, and these are linear combinations of the columns of A, namely $Ab_1 = \sum a_i b_{i1}$ etc. Hence

$$\begin{aligned} d(\textstyle\sum a_{i_1} b_{i_1 1}, \sum a_{i_2} b_{i_2 2}, \ldots, \sum a_{i_n} b_{i_n n}) &= \textstyle\sum d(a_{i_1}, a_{i_2}, \ldots, a_{i_n}) b_{i_1 1} \ldots b_{i_n n} \\ &= \textstyle\sum d(a_1, a_2, \ldots, a_n)\varepsilon(i_1, \ldots, i_n) b_{i_1 1} \ldots b_{i_n n} \\ &= d(A)d(B). \end{aligned}$$

Many other notations for determinants exist and are sometimes useful. We shall occasionally make use of Jacobi's notation which consists in writing down the entries in the main diagonal and leaving the others to be inferred; e.g. the Vandermonde determinant in this notation becomes $|1, x_2, x_3^2, \ldots, x_n^{n-1}|$.

Exercises

(1) Find $\varepsilon(1\ 3\ 4\ 5\ 2)$, $\varepsilon(2\ 5\ 1\ 4\ 3)$, $\varepsilon(3\ 1\ 6\ 4\ 2\ 5)$, $\varepsilon(4\ 1\ 3\ 4\ 5)$.

(2) Evaluate:

$$\begin{vmatrix} 3 & 7 \\ 2 & 8 \end{vmatrix}, \quad \begin{vmatrix} 4 & 2 \\ -1 & 0 \end{vmatrix}, \quad \begin{vmatrix} \cos\alpha & -\sin\alpha \\ \sin\alpha & \cos\alpha \end{vmatrix}, \quad \begin{vmatrix} x & 1 \\ -1 & 0 \end{vmatrix}, \quad \begin{vmatrix} t-1 & -2 \\ 1 & t-4 \end{vmatrix}, \quad \begin{vmatrix} 7 & 2 & 4 \\ 3 & -4 & 5 \\ 1 & 3 & -2 \end{vmatrix},$$

$$\begin{vmatrix} 3 & 1 & 5 \\ 2 & 0 & 0 \\ 4 & 9 & 6 \end{vmatrix}, \quad \begin{vmatrix} 3 & 4 & 5 \\ 6 & 7 & 8 \\ 9 & 10 & 11 \end{vmatrix}, \quad \begin{vmatrix} 0 & c & b \\ c & 0 & a \\ b & a & 0 \end{vmatrix}, \quad \begin{vmatrix} (b+c)^2 & ab & ac \\ ab & (c+a)^2 & bc \\ ac & bc & (a+b)^2 \end{vmatrix},$$

$$\begin{vmatrix} 1 & \cos(\alpha-\beta) & \cos(\alpha-\gamma) \\ \cos(\beta-\alpha) & 1 & \cos(\beta-\gamma) \\ \cos(\gamma-\alpha) & \cos(\gamma-\beta) & 1 \end{vmatrix}, \quad \begin{vmatrix} a^3 & b^3 & c^3 \\ (a+\lambda)^3 & (b+\lambda)^3 & (c+\lambda)^3 \\ (2a+\lambda)^3 & (2b+\lambda)^3 & (2c+\lambda)^3 \end{vmatrix},$$

$$\begin{vmatrix} 3 & 1 & 2 & 4 \\ 5 & 2 & 3 & 6 \\ 4 & 1 & 3 & 6 \\ 5 & 1 & 4 & 8 \end{vmatrix}, \quad \begin{vmatrix} 1 & 1 & 1 & -1 \\ 1 & 1 & -1 & 1 \\ 1 & -1 & 1 & 1 \\ -1 & 1 & 1 & 1 \end{vmatrix}, \quad \begin{vmatrix} 0 & 1 & 1 & 1 \\ 1 & 0 & c^2 & b^2 \\ 1 & c^2 & 0 & a^2 \\ 1 & b^2 & a^2 & 0 \end{vmatrix}.$$

(3) Show that over **R** the area of a triangle with vertices $a = (a_1, a_2)$, $b = (b_1, b_2)$, $c = (c_1, c_2)$ is given by

$$\tfrac{1}{2}\begin{vmatrix} 1 & a_1 & a_2 \\ 1 & b_1 & b_2 \\ 1 & c_1 & c_2 \end{vmatrix}.$$

Find an interpretation for the sign of this determinant.

(4) For any $n \times n$ matrix A and any scalar λ, show that $\det(\lambda A) = \lambda^n \det(A)$.

(5) For any square matrix A and any invertible matrix B of the same order as A, show that $\det(B^{-1}AB) = \det(A)$.

(6) Let A be a square matrix over an integral domain. If $A^r = 0$ for some $r \geqslant 1$, show that $\det(A) = 0$.

(7)* If f is a monic polynomial of degree n, which can be written as a product of linear factors, say $f = \prod(x-\alpha_i)$, show that

$$\begin{vmatrix} \alpha_1 & x & x & \dots & \\ x & \alpha_2 & x & \dots & \\ & \dots & & \dots & \\ x & x & x \dots x & \alpha_n \end{vmatrix} = (-1)^n(f - xf').$$

(8) Let $A = (a_{ij})$ be an $n \times n$ matrix with zeros above the secondary diagonal (from SW to NE), thus $a_{ij} = 0$ for $i+j \leqslant n$. Find $\det A$.

(9) Let $A_n = (a_{ij})$ be an $n \times n$ matrix such that $a_{ii} = t_i$, $a_{i\,i+1} = 1$, $a_{i+1\,i} = -1$, and all remaining as vanish. Show that

$$\det(A_n) = \det(A_{n-1})t_n + \det(A_{n-2}).$$

($\det A_n$ occurs in the formation of continued fractions and is called a *continuant*.)

(10) Show that

$$\begin{vmatrix} a & b & c \\ b & c & a \\ c & a & b \end{vmatrix} = -(a+b+c)(a^2+b^2+c^2-bc-ca-ab).$$

Write down and evaluate the corresponding determinant of order 4. (These determinants are called *circulants*.)

(11) Show that

$$\begin{vmatrix} 1 & x & x^3 \\ 1 & y & y^3 \\ 1 & z & z^3 \end{vmatrix} = \Delta(x, y, z)(x+y+z),$$

where $\Delta(x, y, z)$ stands for the product of the differences.

(12) Let $\alpha_1, \ldots, \alpha_n$ be any n positive integers, then $|x_1^{\alpha_1}, x_2^{\alpha_2}, \ldots, x_n^{\alpha_n}|$ is zero if two αs are the same and otherwise equals $\Delta(x_1, \ldots, x_n)f(x_1, \ldots, x_n)$ where f is a symmetric function of the xs, of weight $\sum(\alpha_i+1-i)$. Find f for each of the following values of the αs: (1, 2, 4, 5), (0, 1, 2, 5), (0, 1, 1, 3), (0, 1, 3, 4). (Because of their construction as the quotient of two alternating functions the fs are called *bialternants*.)

7.2 Expansion of a determinant

Let us return to the formula (7) of **7.1** for det A. This is a polynomial in the n^2 entries of A. We fix one of these elements, say a_{12}, and denote by A_{12} the coefficient of a_{12} in det A; A_{12} is called the *cofactor* of a_{12}. It is a polynomial of degree $n-1$ in the entries of A which contains no elements from the first row or the second column of A (i.e. the row and column containing a_{12}). In an analogous way we can define the cofactor A_{ij} of any other entry a_{ij} of A.

Let us now fix a row of A, say the *first* row. Since each term in (7) of **7.1** contains exactly one element from the first row, every such term will occur just once in one of the products $a_{1j}A_{1j}$ $(j = 1, \ldots, n)$ and we can therefore express the determinant as

$$\det A = a_{11}A_{11}+a_{12}A_{12}+ \cdots +a_{1n}A_{1n},$$

with the cofactors as coefficients. This is called the *expansion of* det A *by its first row*. Of course we can expand det A by any other row in this way; the result is

$$a_{i1}A_{i1}+a_{i2}A_{i2}+ \cdots +a_{in}A_{in} = \det A \qquad (i = 1, \ldots, n). \tag{1}$$

For example, the expansion of a third order determinant by its first row is

$$a_{11}(a_{22}a_{33}-a_{32}a_{23})+a_{12}(-a_{21}a_{33}+a_{31}a_{23})+a_{13}(a_{21}a_{32}-a_{31}a_{22}).$$

We notice that the cofactors in this expansion have the form of second order determinants. In general, we shall find that the cofactors in the expansion of an nth order determinant can be expressed as determinants of order $n-1$.

We can also expand det A by one of its columns, using the fact that each term of (7) of **7.1** contains just one element from each column. Thus the expansion by the jth column is

$$a_{1j}A_{1j}+a_{2j}A_{2j}+ \cdots +a_{nj}A_{nj} = \det A \qquad (j = 1, \ldots, n), \tag{2}$$

where the coefficients are again the cofactors.

In order to determine the cofactors, consider first A_{11}. This is the coefficient of a_{11} in det A, i.e.

$$A_{11} = \sum \varepsilon(1, i_2, \ldots, i_n)a_{i_2 2} \ldots a_{i_n n},$$

where $i_2, \ldots, i_n$ runs over all permutations of $2, \ldots, n$. The ε-symbol in this expression is equal to $\varepsilon(i_2, \ldots, i_n)$, so

$$A_{11} = \sum \varepsilon(i_2, \ldots, i_n) a_{i_2 2} \ldots a_{i_n n}.$$

The expression on the right is a determinant of order $n-1$:

$$A_{11} = \begin{vmatrix} a_{22} & a_{23} & \ldots & a_{2n} \\ a_{32} & a_{33} & \ldots & a_{3n} \\ \cdot & \cdot & \cdot & \cdot \\ a_{n2} & a_{n3} & \ldots & a_{nn} \end{vmatrix}. \tag{3}$$

Thus A_{11} is equal to the $(n-1)$th order determinant obtained by omitting the first row and the first column in A. To evaluate the general cofactor we first perform some row and column interchanges to bring the corresponding matrix entry to the (1, 1)-position, so that we can apply (3). Thus to find A_{ij} we move the ith row up past $\text{row}_{i-1}, \ldots, \text{row}_1$ so that after $i-1$ interchanges it becomes the new top row, and then move the jth column to the left past the first $j-1$ columns in turn so that it becomes the leftmost column. We now have a matrix which is related to A by $i-1$ interchanges of rows and $j-1$ interchanges of columns. These $i-1+j-1$ interchanges produce $i+j-2$ changes of sign, or what is the same, $i+j$ changes, so the value of the new determinant is $(-1)^{i+j} \det A$. The cofactor of a_{ij} in this determinant—in which a_{ij} is now the (1, 1)-entry—is the $(n-1)$th order determinant obtained by omitting the new first row and column, i.e. the ith row and jth column from A. Writing α_{ij} for this determinant, we have the formula

$$A_{ij} = (-1)^{i+j} \alpha_{ij}. \tag{4}$$

The $(n-1)$th order determinants α_{ij} are called the *minors* of $\det A$, and (4) expresses the cofactors in terms of the minors, with the sign determined by the chessboard rule:

$$\begin{vmatrix} + & - & + & - & \ldots \\ - & + & - & + & \ldots \\ + & - & + & - & \ldots \\ \cdot & \cdot & \cdot & \cdot & \cdot \end{vmatrix}$$

The expansions found for $\det A$ can also be used to express the inverse matrix A^{-1}, when this exists. Let us define the *adjugate* of A as the transpose of the matrix of cofactors

$$\operatorname{adj} A = \begin{vmatrix} A_{11} & A_{21} & \ldots & A_{n1} \\ A_{12} & A_{22} & \ldots & A_{n2} \\ \cdot & \cdot & \cdot & \cdot \\ A_{1n} & A_{2n} & \ldots & A_{nn} \end{vmatrix}$$

and consider the product $A \, . \operatorname{adj} A$. The (1, 1)-entry is

$$a_{11}A_{11} + a_{12}A_{12} + \cdots + a_{1n}A_{1n}$$

and this equals det A, by (1). Similarly (1) shows that every element on the main diagonal of A . adj A has the value det A. Next consider the (2, 1)-entry:

$$a_{21}A_{11}+a_{22}A_{12}+\cdots+a_{2n}A_{1n}.$$

This represents the expansion by the first row of the determinant

$$\begin{vmatrix} a_{21} & a_{22} & a_{23} & \cdots & a_{2n} \\ a_{21} & a_{22} & a_{23} & \cdots & a_{2n} \\ a_{31} & a_{32} & a_{33} & \cdots & a_{3n} \\ \cdot & \cdot & \cdot & \cdot & \cdot \end{vmatrix}$$

which has the same 2nd, 3rd, . . ., nth row as A (because the cofactors A_{11}, $A_{12}, \ldots, A_{1n}$ involve only elements from these rows), but its first row equals its second row and so its value is 0, by **D**. 2 and **D**. 7. Generally,

$$a_{i1}A_{j1}+a_{i2}A_{j2}+\cdots+a_{in}A_{jn}$$

represents the expansion of a determinant whose jth row equals the ith row of A while the other rows are as in A. Hence if $i \neq j$, the determinant has two equal rows, namely the ith and the jth, and is therefore zero. This shows that all the elements off the main diagonal in A . adj A are 0, so that

$$A \,.\, \text{adj}\, A = \det A \,.\, I. \tag{5}$$

Similarly, using (2) for the diagonal elements and **D**. 2 for the off-diagonal elements, we find

$$(\text{adj}\, A) \,.\, A = \det A \,.\, I. \tag{6}$$

These formulae hold for any square matrix A over a commutative ring R. Now suppose that det A is invertible in R and form the matrix $B = (\det A)^{-1}$ adj A. Equations (5) and (6) show that $AB = BA = I$, hence any matrix A with invertible determinant is invertible and its inverse is given by

$$A^{-1} = (\det A)^{-1} \,.\, \text{adj}\, A. \tag{7}$$

Conversely, if A is invertible, it has an inverse and, by the multiplication rule **D**. 8,

$$1 = \det (A \,.\, A^{-1}) = \det A \,.\, \det A^{-1},$$

which shows that det A is invertible. Thus we have proved

THEOREM 1 · *A square matrix over a commutative ring R is invertible if and only if its determinant is a unit in R. In particular, over a field the matrix is invertible if and only if its determinant is non-zero.* ■

The formula (7) for the inverse may also be used to give an explicit expression for the solution of a system of equations with an invertible matrix. By (7) the solution of the system

$$Ax = b \tag{8}$$

is

$$x = A^{-1}b = (\det A)^{-1}(\text{adj}\, A)b,$$

or written out in full,

$$x_i = (\det A)^{-1}(A_{1i}b_1 + A_{2i}b_2 + \cdots + A_{ni}b_n).$$

The sum on the right represents the expansion by the ith row of the determinant of the matrix

$$A^{(i)} = (a_1, \ldots, a_{i-1}, b, a_{i+1}, \ldots, a_n), \tag{9}$$

which is obtained by replacing the ith column in A by b. Hence the solution of (8) may be written as the quotient of two determinants:

$$x_i = \frac{\det A^{(i)}}{\det A} \quad (i = 1, \ldots, n), \tag{10}$$

where the matrix $A^{(i)}$ is defined by (9). This formula for solving the system (8) is known as *Cramer's rule*. Of course it must be borne in mind that it applies only when the matrix A is invertible; even then, if we only want the solution of (8) it is usually quicker to find it by direct elimination as in Ch. **5**, rather than compute determinants.

Exercises

(1) Find the inverses of the matrices in Ex. (3) **5.3** by calculating the minors, the cofactors and the adjugate.

(2) Let A, B be invertible $n \times n$ matrices and let $x_i = |B^{(i)}|/|B|$ be Cramer's formula for the solution of $Bx = c$. Prove that

$$\frac{|AB|}{|B|} = \frac{|AB^{(i)}|}{|B^{(i)}|} \quad \text{for } i = 1, \ldots, n.$$

Deduce that the left-hand side is independent of B and hence obtain another proof of the multiplication rule (due to Kronecker).

(3) Express the rank of adj A in terms of the rank r of A. (Hint. Distinguish the cases $r = n$, $r = n-1$, $r < n$.)

(4) Show (i) adj $(A^{\mathrm{T}}) = (\text{adj } A)^{\mathrm{T}}$, (ii) adj $(\lambda A) = \lambda^{n-1}$ adj A, (iii) adj AB = adj B . adj A.

(5) Show (i) det (adj A) = $(\det A)^{n-1}$, (ii) adj (adj A) = $(\det A)^{n-2}A$. (Hint. Assume first that $\det A \neq 0$.)

7.3 The determinantal rank

Our objective in introducing determinants was to have a test for linear dependence. We now show that this is indeed achieved; for simplicity we confine our attention to vector spaces over fields, the case of prime importance (but see also Ex. (4)). As in **7.1** we denote the determinant of n column vectors $u_1, \ldots, u_n$ by $\det(u_1, \ldots, u_n)$.

THEOREM 1 *Let F be a field. A family of n vectors in F^n, $u_1, \ldots, u_n$ say, is linearly dependent if and only if* $\det(u_1, \ldots, u_n) = 0$.

Proof. Let A be the matrix with columns $u_1, \ldots, u_n$. Then the us are linearly dependent iff there exist $\lambda_1, \ldots, \lambda_n \in F$, not all 0, such that $\sum u_i\lambda_i = 0$, i.e. $A\lambda = 0$, where $\lambda = (\lambda_1, \ldots, \lambda_n)^T \neq 0$. But this expresses the fact that A is singular, which is the case precisely when $\det A = 0$. ∎

We can also use determinants to describe the rank of a general matrix (not necessarily square). Let A be an $m \times n$ matrix; if we single out r rows and r columns of A (for some $r \leqslant \min\{m, n\}$), the determinant of the $r \times r$ matrix at the intersection of these rows and columns is called a *minor* of *order* r in A.

THEOREM 2 *An $m \times n$ matrix A over a field has rank at least r if and only if it has a non-zero minor of order r.*

Proof. Suppose that A has rank at least r and permute the rows so that the first r rows are linearly independent. Then the $r \times n$ matrix consisting of the first r rows of A has rank r, hence we can find r linearly independent columns in it. They will constitute a non-singular submatrix and its determinant is the required minor of order r. If A has rank less than r, then any r rows are linearly dependent, hence any $r \times r$ submatrix of A is singular and so any rth order minor is zero. ∎

COROLLARY *A matrix has rank r if and only if there is a non-zero minor of order r but every minor of higher order than r is zero.* ∎

Exercises

(1) If a change of basis in an n-dimensional vector space is given by $u' = Pu$, show that

$$\det(u_1', \ldots, u_n') = (\det P) \,.\, \det(u_1 \ldots, u_n).$$

(2) How many rth order minors can be formed from an $m \times n$ matrix?

(3) If every rth order minor of a matrix A is zero, show that every minor of order greater than r is also zero. Deduce the following sharper form of Th. 2, Cor.: A matrix has rank r if there is a non-zero minor of order r but not one of order $r+1$.

(4) Over any ring R, a family of column vectors $u_1, \ldots, u_r$ is called *linearly dependent* if there exist $\lambda_1, \ldots, \lambda_r$ in R, not all 0, such that $\Sigma u_i\lambda_i = 0$. Show that n column vectors $u_1, \ldots, u_n$ with n components over a commutative ring R are linearly dependent iff $\det(u_1, \ldots, u_n)$ is a zero-divisor (or zero) in R.

(5) The equation of a plane in 3-space may be written as

$$a_1x_1 + a_2x_2 + a_3x_3 + a = 0.$$

Show that the plane through the points $p = (p_1, p_2, p_3)$, $q = (q_1, q_2, q_3)$ and $r = (r_1, r_2, r_3)$ is

$$\begin{vmatrix} x_1 & x_2 & x_3 & 1 \\ p_1 & p_2 & p_3 & 1 \\ q_1 & q_2 & q_3 & 1 \\ r_1 & r_2 & r_3 & 1 \end{vmatrix} = 0.$$

What happens if the points p, q, r are collinear?

(6) The *Jacobian* of n differentiable functions $f_1, \ldots, f_n$ of n variables $x_1, \ldots, x_n$ is defined as the matrix whose (i, j)-entry is $\partial f_i/\partial x_j$:

$$\frac{\partial(f_1, \ldots, f_n)}{\partial(x_1, \ldots, x_n)} = \left(\frac{\partial f_i}{\partial x_j}\right).$$

If $x_1, \ldots, x_n$ are differentiable functions of $t_1, \ldots, t_n$ show that

$$\frac{\partial(f_1, \ldots, f_n)}{\partial(t_1, \ldots, t_n)} = \frac{\partial(f_1, \ldots, f_n)}{\partial(x_1, \ldots, x_n)} \cdot \frac{\partial(x_1, \ldots, x_n)}{\partial(t_1, \ldots, t_n)}.$$

Show that if there is a functional relation between $f_1, \ldots, f_n$, valid in a neighbourhood of a point p of $\mathbf{R}^n$:

$$\varphi(f_1, \ldots, f_n) = 0$$

then the Jacobian determinant vanishes throughout some neighbourhood of p.

(7)* The *Hessian* of a function of n variables is the Jacobian of its first derivatives:

$$H(f) = \left(\frac{\partial^2 f}{\partial x_i \partial x_j}\right).$$

If f is homogeneous of degree k in x and y over $\mathbf{C}$, show that $H(f) = 0$ iff $f = (ax+by)^k$. (Hint. Use Euler's theorem on homogeneous functions, Ex. (8), **6.8**, to show that fx and fy are linearly dependent over $\mathbf{C}$.)

(8) If the entries of a square matrix A are differentiable functions of a variable x, show that $(\det A)' = |a_1', a_2, \ldots, a_n| + |a_1, a_2', a_3, \ldots| + \cdots + |a_1, \ldots, a_{n-1}, a_n'|$.

(9) Let $f_1, \ldots, f_n$ be any differentiable functions of x, then the determinant formed from the successive derivatives of the fs:

$$W(f_1, \ldots, f_n) = |f_1, f_2', f_3'', \ldots, f_n^{(n-1)}|$$

(in Jacobi's notation) is called the *Wronskian* of $f_1, \ldots, f_n$. Show that

$$\frac{dW}{df} = |f_1, f_2', \ldots, f_{n-1}^{(n-2)}, f_n^{(n)}|.$$

(10)* Show that if n (sufficiently differentiable) functions are linearly dependent over $\mathbf{R}$, their Wronskian vanishes identically. If $f_1, \ldots, f_n$ are solutions of the linear differential equation

$$y^{(n)} + p_1 y^{(n-1)} + \cdots + p_n y = 0, \tag{1}$$

(with differentiable coefficients), it can be shown that the fs are linearly independent over $\mathbf{R}$ iff their Wronskian is different from 0. Show that the differential equation (1) satisfied by n linearly independent functions is unique up to a constant factor, and deduce that the Wronskian satisfies $W' = -p_1 W$.

7.4 The resultant

In Ch. **6** we saw that to find the zeros of a polynomial we need to know something of the field of coefficients, whereas to find whether two polynomials have a common factor, this is not necessary: the Euclidean algorithm provides a rational method of finding a highest common factor. One would therefore expect a criterion, in terms of the coefficients of f and g, for f and g to have a common factor. Such a criterion is provided by the resultant of f and g, to be described below. We begin with a lemma which expresses the existence of a common factor by the solubility of a system of linear equations.

LEMMA 1 *Let F be any field. Given* $f, g \in F[x]$, *not both zero, say*

$$f = a_0x^m+a_1x^{m-1}+\cdots+a_m, \tag{1}$$
$$g = b_0x^n+b_1x^{n-1}+\cdots+b_n, \tag{2}$$

then f, g *have a non-constant common factor or* $a_0 = b_0 = 0$ *if and only if there exist polynomials* f_1 *and* g_1 *not both zero, such that*

$$fg_1 = gf_1, \qquad \deg f_1 < m, \quad \deg g_1 < n. \tag{3}$$

Proof. Let $d = (f, g)$ have positive degree; then $f = df_1$, $g = dg_1$ and the equation in (3) holds. Further, $\deg f_1 < m$, $\deg g_1 < n$ and, if $a_0 = b_0 = 0$, this holds even if $\deg d = 0$.

Conversely, if (3) holds, suppose also that $(f, g) = 1$, then $f \mid f_1$, $g \mid g_1$, hence $\deg f < m$, $\deg g < n$, which is only possible if $a_0 = b_0 = 0$. ■

Strictly speaking, what we have found is not a criterion for f and g to have a common factor, but a test for f, g to have a common factor or vanishing highest terms (in the expressions (1), (2)). The significance of this condition is best seen by looking at the geometric interpretation: $f = 0$ represents a set of points on the line, in fact precisely m points if we operate in an algebraically closed field (where f splits into linear factors) and count multiple zeros appropriately. Further, if $a_0 = 0$, we must count a zero at infinity (this is seen by making the transformation $x' = 1/x$ and then putting $x' = 0$). Now Lemma 1 asserts the existence of f_1, g_1 to satisfy (3) iff f, g have a common zero, possibly infinite. However, we shall keep to the formulation given above, to save having to assume the field to be algebraically closed.

Let us return to (3). The degree of g_1 is at most $n-1$, so that fg_1 has degree at most $m+n-1$. Similarly gf_1 has degree at most $m+n-1$, and on equating the coefficients of the various powers of x in (3) we obtain a set of $m+n$ equations, homogeneous linear in the coefficients $u_0, u_1, \ldots, u_{m-1}$ of f_1 and $v_0, v_1, \ldots, v_{n-1}$ of g_1, viz.

$$\begin{array}{lll} a_0v_0 & -b_0u_0 & = 0 \\ a_1v_0+a_0v_1 & -b_1u_0-b_0u_1 & = 0 \\ \quad\cdots & \quad\cdots & \\ \qquad a_mv_{n-1} & \qquad -b_nu_{m-1} & = 0. \end{array}$$

This system has a non-trivial solution $(v_0, v_1, \ldots, v_{n-1}, u_0, u_1, \ldots, u_{m-1})$ iff the matrix of coefficients is singular, i.e. its determinant is zero. Thus if we put

$$R(f, g) = \begin{vmatrix} a_0 & & & & b_0 & & & \\ a_1 & a_0 & & & b_1 & b_0 & & \\ a_2 & a_1 & a_0 & & b_2 & b_1 & b_0 & \\ & \cdots & & & & \cdots & & \\ & & & a_m & & & & b_n \\ \multicolumn{4}{c}{\underbrace{\hspace{8em}}_{n \text{ columns}}} & \multicolumn{4}{c}{\underbrace{\hspace{8em}}_{m \text{ columns}}} \end{vmatrix} \tag{4}$$

then $R(f, g) = 0$ is necessary and sufficient for (3) to have a non-zero solution f_1, g_1. This polynomial R in the coefficients of f and g is called the *resultant* of f and g. Its properties are summed up in the next theorem. To state it we recall that the coefficients a_i of f are essentially the symmetric polynomials in the zeros of f: $a_i/a_0 = (-1)^i e_i$. Similarly for the b_j, and we may therefore speak of the weight of any expression in the a_i, b_j, this being the sum of the suffixes occurring. If this is the same in each term, the sum is said to be *isobaric* (cf. **6.9**).

THEOREM 2 *The resultant $R(f, g)$ of the polynomials f, g given by* (1), (2) *is a polynomial in the* as *and* bs *with integer coefficients, with the following properties*:

(i) *R is homogeneous of degree n in the* as *and m in the* bs, *and isobaric of weight mn*,

(ii) *$R = 0$ if and only if f, g have a non-constant common factor or* $a_0 = b_0 = 0$,

(iii) *there are polynomials F and G in x, of degrees less than m, n respectively, with coefficients which are polynomials in the* as *and* bs *such that*

$$R = fG + gF,$$

(iv) *if f and g can be written as product of linear factors,*

$$f = a_0 \prod_i (x - \alpha_i), \qquad g = b_0 \prod_j (x - \beta_j),$$

then

$$R(f, g) = a_0^n \prod_i g(\alpha_i) = (-1)^{mn} b_0^m \prod_j f(\beta_j) = a_0^m b_0^n \prod_{i,j} (\alpha_i - \beta_j). \tag{5}$$

Proof. (i) If in (4) f is replaced by tf, each a_i becomes ta_i and the first n columns of the determinant on the right are multiplied by t. Hence $R(tf, g) = t^n R(f, g)$ and this shows R to be homogeneous of degree n in the as. Similarly R is homogeneous of degree m in the bs. To compute the weight, multiply a_i by t^i and b_j by t^j, and then multiply each column on the right of (4) in order by $1, t, t^2, \ldots, t^{n-1}, 1, t, t^2, \ldots, t^{m-1}$. We can then take out the factors $1, t, t^2, \ldots, t^{m+n-1}$ from successive rows and get back R. Hence if R' is the resultant formed with $t^i a_i$ and $t^j b_j$, we find that $R' = t^N R$, where

$$N = 1+2+\cdots+(m+n-1)-(1+2+\cdots+(n-1))-(1+2+\cdots+(m-1))$$
$$= \tfrac{1}{2}[(m+n-1)(m+n)-n(n-1)-m(m-1)] = mn.$$

This shows R to be isobaric of weight mn.

(ii) follows directly from Lemma 1. To establish (iii) we multiply the ith row of R by x^{m+n-i} $(i = 1, \ldots, m+n-1)$ and add the result to the last row. This leaves all rows unchanged, except the last which becomes:

$$x^{n-1}f, x^{n-2}f, \ldots, xf, f, x^{m-1}g, x^{m-2}g, \ldots, xg, g.$$

If we now expand the determinant by the last row, we obtain

$$R = f(d_0 x^{n-1}+d_1 x^{n-2}+\cdots+d_{n-1})+g(c_0 x^{m-1}+\cdots+c_{m-1}),$$
$$= fG+gF,$$

where the coefficients of F, G are the cofactors of the last row of R, i.e. certain polynomials in the as and bs.

To prove (iv) denote by $R(y)$ the resultant of f and $g-y$ and put $g(\alpha_i) = \gamma_i$ for short, then $R(\gamma_i)$ is the resultant of $f(x)$ and $g(x)-\gamma_i$, which have the common zero $x = \alpha_i$, hence $R(\gamma_i) = 0$. By the remainder theorem, $y-\gamma_i \mid R(y)$ and this holds for all i. Now $R(y)$ has degree m in y and leading coefficient $(-1)^m a_0^n$, hence if $\gamma_1, \ldots, \gamma_m$ are distinct, we have $R(y) = a_0^n \prod (\gamma_i - y)$. This remains true over any extension of the field, which may therefore be taken infinite. By the irrelevance of algebraic inequalities this is true for all values of the γs, hence $R = R(0) = a_0^n \prod g(\alpha_i)$. Similarly we find $R = (-1)^{mn} b_0^m \prod f(\beta_j)$ and, expanding either of these expressions, we get the final formula of (iv). ■

COROLLARY *For any polynomials f, g, h,*

$$R(fg, h) = R(f, h)R(g, h). \tag{6}$$

To establish this formula we need the result, proved in Vol. 2, that any polynomial may be written as a product of linear factors over a suitable extension field of F. This will ensure that we can apply the formulae in (iv); alternatively the reader may limit himself to the case where F is the field of complex numbers, where this is certainly true. Now (6) follows immediately by applying the first expression given in (iv) of Th. 2. ■

An important special case is that where g is the derivative of f. In the determinant for $R(f, f')$ let us interchange neighbouring columns so as to map $\text{col}_i \to \text{col}_{2i} \pmod{2m-1}$; in detail: $\text{col}_m \to \text{col}_1$, $\text{col}_{m+1} \to \text{col}_3, \ldots$, $\text{col}_{2m-1} \to \text{col}_{2m-1}$. In all we have made $(m-1)+(m-2)+\cdots+1 = \binom{m}{2}$

transpositions. If we take out the factor a_0 from the first row we find an expression

$$D(f) = \begin{vmatrix} m & 1 & & & \\ (m-1)a_1 & a_1 & ma_0 & a_0 & \\ (m-2)a_2 & a_2 & (m-1)a_1 & a_1 & \\ \cdot & \cdot & \cdot & \cdot & \cdot \\ a_{m-1} & a_{m-1} & 2a_{m-2} & a_{m-2} & \cdots \\ & a_m & a_{m-1} & a_{m-1} & \cdots \\ & & & a_m & \cdots \\ & & & & \cdots \end{vmatrix}$$

This function is called the *discriminant* of f. It is related to the resultant of f and f' by the formula

$$D(f) = (-1)^{\frac{1}{2}m(m-1)}a_0^{-1}R(f, f'),$$

by what we have seen. From Th. 2 (i) we see that it is homogeneous of degree $2(m-1)$ in the as and isobaric of weight $m(m-1)$. Suppose that f splits into linear factors, say

$$f(x) = a_0 \prod_i (x-\alpha_i),$$

then $f'(\alpha_j) = a_0 \prod_{i \neq j} (\alpha_j - \alpha_i)$, hence

$$D(f) = a_0^{2(m-1)} \prod_{i>j} (\alpha_i - \alpha_j)^2 = [a_0^{m-1} \prod_{i>j} (\alpha_i - \alpha_j)]^2. \tag{7}$$

From Th. 2 (ii) we see that $D(f) = 0$ iff f has a repeated factor or $a_0 = a_1 = 0$ (corresponding to a repeated infinite zero).

Finally the multiplication formula for resultants leads to the following formula for the discriminant of a product:

$$D(fg) = D(f)D(g)R(f, g)^2.$$

Exercises

(1) Find the resultant of $a_0x^2+a_1x+a_2$ and $b_0x^2+b_1x+b_2$.

(2) Find the resultant of $a_0x^2+a_1x+a_2$ and x^3+px+q.

(3) Find the discriminants of the polynomials in Ex. (2).

(4) Verify that for a polynomial f and a linear polynomial $x-\alpha$, $R(x-\alpha, f) = f(\alpha)$.

(5) Show that $R(f, g)$ regarded as a polynomial in the coefficients of f and g is irreducible. (Hint. Use (5) and the fact that each factor of R will be symmetric in the zeros of f and in those of g.)

(6) Let f be a monic polynomial and denote by s_k the kth power sum of its zeros (i.e. $\sum \alpha_i^k$, where the α_i are the zeros of f). Use (5) to represent $D(f)$ as a product of

two Vandermonde determinants and hence show that

$$D(f) = \begin{vmatrix} s_0 & s_1 & s_2 & \dots & s_{n-1} \\ s_1 & s_2 & s_3 & \dots & s_n \\ \cdot & \cdot & \cdot & \cdot & \cdot \\ s_{n-1} & s_n & s_{n+1} & \dots & s_{2n-2} \end{vmatrix}$$

Further exercises on Chapter 7

(1) In the definition of a determinant, show that **D**.2 can be replaced by **D**.2′: $d(A)$ vanishes if two *neighbouring* columns of A are the same.

(2) Let X be a square matrix with entries x_{ij} which are indeterminates over a field. Show that det X as a polynomial in the n^2 elements x_{ij}, is irreducible. (Hint. Use the fact that det X is homogeneous of degree 1 in the entries of each row and column.)

(3) If A is a square matrix of odd order over a field of characteristic not 2, such that $A^{\mathrm{T}} = -A$, show that $\det A = 0$.

(4) Let A be an $n \times n$ matrix over a field F, partitioned as $A = \begin{pmatrix} P & Q \\ R & S \end{pmatrix}$ where $P \in F_r$ and Q, R, S are of appropriate sizes. If P is invertible, show that $\det A = \det P \,.\, \det(S - RP^{-1}Q)$.

(5) Let A, B be any $n \times n$ matrices over a field. Show that the matrix $P = \begin{pmatrix} I & B \\ -A & 0 \end{pmatrix}$ can be brought to the form $\begin{pmatrix} I & B \\ 0 & AB \end{pmatrix}$ by elementary operations on rows and hence show that $\det P = \det AB$. By suitably factorizing P show that its determinant is $\det A \,.\, \det B$ and hence obtain another proof of the multiplication theorem.

(6) Let A be a square matrix over a field. If the sum of the entries in each row and in each column is zero, show that all cofactors of A are equal.

(7) Show that for any matrices $A \in R_m$, $B \in R_n$, $C \in {}^mR^n$, where R is a commutative ring,

$$\det \begin{pmatrix} A & C \\ 0 & B \end{pmatrix} = \det A \,.\, \det B.$$

(Hint. By regarding this as an identity in the entries of the matrices, we may assume these entries to be indeterminates over the integers. Then A, B are non-singular and so may be inverted over the field of fractions. Thus we may take $C = APB$, and factorize the matrix on the left.)

(8) Suppose that $C = AB$, where $C \in F_r$, $A \in {}^rF^n$, $B \in {}^nF^r$. Show that if $n < r$, $\det C = 0$, while for $n \geqslant r$, $\det C$ is the sum of determinants $|A'| \,.\, |B'|$, where A' runs over all $r \times r$ matrices obtainable from A by selecting r columns, and B' represents the $r \times r$ matrix formed from the corresponding r rows of B. In symbols, if the columns of A, B^{T}, C are written as a_i, b_i', c_i respectively, then

$$d(c_1, \dots, c_r) = \sum_{i_1 < i_2 < \dots < i_r} d(a_{i_1}, \dots, a_{i_r}) d(b_{i_1}', \dots, b_{i_r}').$$

(9) Let F be a field and $A \in F_n$, $x \in {}^nF$, $y \in F^n$, $t \in F$. Obtain the following expansion for a bordered determinant:

$$\det \begin{pmatrix} A & x \\ y & t \end{pmatrix} = t \,.\, \det A - y(\operatorname{adj} A)x.$$

(10) Given $P \in F_n$, write $P = \begin{pmatrix} A & B \\ C & D \end{pmatrix}$ where for some $r < n$, $A \in F_r$ and B, C, D are of appropriate size. Let $Q = (q_{\lambda\mu})$ $(\lambda, \mu = r+1, \ldots, n)$ be the matrix of order $n-r$ whose (λ, μ)-entry is the determinant obtained by bordering A with the λth row and μth column of P. Use Ex. (9) to show that the equations $AX+B \,.\, \det A = 0$, $CX+D \,.\, \det A = Q$ have a solution $X \in {}^rF^{n-r}$. By expanding the product of P and $\begin{pmatrix} \operatorname{adj} A & X \\ 0 & \det A \,.\, I \end{pmatrix}$ show that $\det Q = \det P \,.\, (\det A)^{n-r-1}$. (Sylvester)

(11)* Given $a_1, \ldots, a_n, b_1, \ldots, b_n$ in a field F such that $1-a_ib_j \neq 0$, for all i, j, show that

$$\det\left(\frac{1}{1-a_ib_j}\right) = \frac{\Delta(a_1, \ldots, a_n)\Delta(b_1, \ldots, b_n)}{\prod_{i,j} (1-a_ib_j)} \qquad \text{(Cauchy)}$$

(Hint. Multiply up by the denominator and show that the left-hand side is a polynomial in the as and bs, alternating in each set.)

(12) Let g be a monic polynomial with n distinct zeros: $g = \prod(x-\beta_i)$, and let f be any polynomial of degree less than n. Show that

$$\frac{f}{g} = \frac{|1, \beta_2, \beta_3^2, \ldots, \beta_{n-1}^{n-2}, f(\beta_n)/(x-\beta_n)|}{|1, \beta_2, \beta_3^2, \ldots, \beta_{n-1}^{n-2}, \beta_n^{n-1}|},$$

and by expanding the determinant by its last row obtain a partial fraction expansion of the left-hand side (cf. Ch. **10**). Show how to modify this expression when several of the zeros of g coincide. (Hint. Differentiate with respect to β_1, if $\beta_2 = \beta_1$.)

(13) If two polynomials f, g have exactly one linear factor in common, $x-\alpha$, say, show that their resultant R satisfies $\partial R/\partial a_i = \partial R/\partial a_0 \,.\, \alpha^i$ $(i = 1, \ldots, n)$. What happens if f and g have a common factor of degree greater than 1?

(14)* Let $A \in F_n$ and $0 < r < n$. Show that $\det A = \sum \varepsilon(A') \det A' \,.\, \det A''$, where A' runs over all rth order matrices obtained from the first r rows of A and any choice of columns, A'' is the $(n-r)$th order matrix obtained by omitting these rows and columns and $\varepsilon(A') = (-1)^\nu$, where ν is the number of transpositions required to bring the columns of A' to the first r positions (Laplace). Illustrate this rule by evaluating

$$\begin{vmatrix} 1 & 2 & 1 & 3 \\ -1 & 0 & 0 & 5 \\ 4 & 0 & 0 & 3 \\ 2 & 3 & 1 & -2 \end{vmatrix}.$$

(15) Let R be any commutative ring and $n \geqslant 1$. If $A, B \in R_n$ and $AB = I$, show that $BA = I$.

8

Quadratic forms

8.1 Pairings and dual spaces

Let U and V be any vector spaces over a field F. Suppose we are given a bilinear form on $U \times V$ with values in F; writing its value at $x \in U$, $y \in V$ as $\langle x, y\rangle$, we have by definition,

$$\begin{aligned} \langle \lambda x+\lambda' x', y\rangle &= \lambda\langle x, y\rangle+\lambda'\langle x', y\rangle \qquad x, x' \in U,\ y, y' \in V, \\ \langle x, \lambda y+\lambda' y'\rangle &= \lambda\langle x, y\rangle+\lambda'\langle x, y'\rangle \qquad \lambda, \lambda' \in F. \end{aligned}$$

In the finite-dimensional case we can take bases $u_1, \ldots, u_m$ of U and $v_1, \ldots, v_n$ of V and write

$$a_{ij} = \langle u_i, v_j\rangle, \tag{1}$$

then the form is completely determined by the $m \times n$ matrix $A = (a_{ij})$. For, given $x = \sum \xi_i u_i$, $y = \sum \eta_j v_j$, we have

$$\langle x, y\rangle = \langle \sum \xi_i u_i, \sum \eta_j v_j\rangle = \sum \xi_i \eta_j \langle u_i, v_j\rangle = \sum \xi_i a_{ij} \eta_j.$$

In matrix notation this may be written

$$\langle x, y\rangle = \xi A \eta^{\mathrm{T}}, \tag{2}$$

where ξ, η are the rows representing the coordinates of x and y respectively. Conversely, any $m \times n$ matrix A gives rise to a bilinear form in this way and, if we denote the set of all bilinear forms on $U \times V$ by Bil (U, V), we have an obvious vector space structure on Bil (U, V) such that

$$\mathrm{Bil}\,(U, V) \cong {}^mF^n. \tag{3}$$

Of course the isomorphism (3) depends on the choice of bases in U and V. Let us see how the matrix of a bilinear form is affected by a change of bases in U and V. If the coordinates of x, y relative to new bases are ξ', η', where (as in **4.6**),

$$\xi = \xi' P, \qquad \eta = \eta' Q,$$

then

$$\langle x, y\rangle = \xi' P A(\eta' Q)^{\mathrm{T}} = \xi' P A Q^{\mathrm{T}} \eta'^{\mathrm{T}}.$$

Thus the matrix referred to the new bases is PAQ^{T}.

Here P, Q may be any invertible matrices; invoking Th. 1, **4.7**, we see that by a suitable choice of bases a bilinear form may always be expressed as

$$\langle x, y\rangle = \xi_1\eta_1 + \cdots + \xi_r\eta_r. \tag{4}$$

Here r, the *rank* of the bilinear form, is independent of the choice of coordinates and it can take any value between 0 and min $\{m, n\}$. A case of particular importance is that where $r = m = n$. We shall treat this in greater detail and give another proof of (4) in this case.

A bilinear form $\langle , \rangle$ on $U \times V$ is said to be *non-singular* or a *pairing* if

(i) $\langle x, b \rangle = 0$ for all $x \in U$ implies $b = 0$, and

(ii) $\langle a, y \rangle = 0$ for all $y \in V$ implies $a = 0$.

For example, let F^n again be the space of row vectors with n components and ${}^nF$ the corresponding space of column vectors, then there is a bilinear form on $F^n \times {}^nF$, given by

$$\langle x, y \rangle = xy;$$

here xy denotes the matrix product of the row x and the column y. Explicitly, if $x = (x_1, \ldots, x_n)$ and $y = (y_1, \ldots, y_n)^{\mathrm{T}}$, then $xy = \sum x_i y_i$. This is easily seen to be a pairing. In this example, both spaces have the same dimension; in fact this is a general property of pairings:

THEOREM 1 *Given two finite-dimensional vector spaces U and V over a field F, let $\langle , \rangle$ be a pairing on $U \times V$ into F. Then* dim U = dim $V = n$, *say, and to any basis $u_1, \ldots, u_n$ of U there corresponds just one family of elements $v_1, \ldots, v_n$ of V such that*

$$\langle u_i, v_j \rangle = \delta_{ij}; \tag{5}$$

moreover, the vs form a basis of V.

Any bases of U and V satisfying (5) are said to be *dual* to one another with respect to the given pairing.

Proof. Let $u_1, \ldots, u_n$ be any basis of U, then the rule

$$y \mapsto (\langle u_1, y \rangle, \ldots, \langle u_n, y \rangle) \tag{6}$$

defines a mapping $V \to F^n$ which is easily seen to be linear. The kernel consists of all $y \in V$ such that $\langle u_1, y \rangle = \cdots = \langle u_n, y \rangle = 0$. But when this holds, then for any $x \in U$, say $x = \sum \xi_i u_i$, we have

$$\langle x, y \rangle = \sum \xi_i \langle u_i, y \rangle = 0,$$

hence $y = 0$, because we have a pairing. Thus the mapping (6) is injective, whence dim $V \leqslant n$ = dim U. By symmetry we also have the opposite inequality, and combining the two we find dim U = dim V.

Now take the basis $u_1, \ldots, u_n$ of U and consider again the linear mapping (6). This is an injection from one n-dimensional space to another and hence is an isomorphism. Take the standard basis $e_1, \ldots, e_n$ in F^n and let v_i be the inverse image of e_i under the mapping (6), then $\langle u_i, v_j \rangle = \delta_{ij}$ as claimed. If $v_1', \ldots, v_n'$ is another family with this property, then each of $v_i - v_i'$ lies in the kernel of the mapping (6) and so must be zero. Thus there is only one such family. Now the fact that the vs form a basis follows because they are the inverse image, under an isomorphism, of a basis of F^n. ■

Given any space V over F, we can define another space V^* and a bilinear form on $V \times V^*$, as follows. We take $V^* = \text{Hom}(V, F)$, thus the elements of V^* are the linear mappings of V into F, also called *linear forms* on V. Given $\alpha \in V^*$, we denote its value on $x \in V$ by $\langle x, \alpha \rangle$. The linearity of α is expressed by the equation

$$\langle \lambda x + \lambda' x', \alpha \rangle = \lambda \langle x, \alpha \rangle + \lambda' \langle x', \alpha \rangle \qquad x, x' \in V, \lambda, \lambda' \in F. \tag{7}$$

Further we know that Hom (V, F) is itself a vector space over F; the operations are given by

$$\langle x, \lambda\alpha + \lambda'\alpha' \rangle = \lambda \langle x, \alpha \rangle + \lambda' \langle x, \alpha' \rangle \qquad \alpha, \alpha' \in V^*, \lambda, \lambda' \in F. \tag{8}$$

Together (7) and (8) show $\langle , \rangle$ to be bilinear on $V \times V^*$; we claim that when dim V is finite, $\langle , \rangle$ is a pairing. For, given $\alpha \in V^*$, if $\langle x, \alpha \rangle = 0$ for all $x \in V$, then $\alpha = 0$, by definition of V^*; to prove the other half of the condition, take a basis $v_1, \ldots, v_n$ of V and observe that any $x \in V$ can be written

$$x = \sum \xi_i v_i, \tag{9}$$

with uniquely determined coefficients $\xi_i \in F$. It is clear that for each $i = 1, \ldots, n$, the mapping of V into F given by

$$\alpha_i : x \mapsto \xi_i \tag{10}$$

is linear, hence $\alpha_i \in V^*$. In fact we can now rewrite (9) as

$$x = \sum \langle x, \alpha_i \rangle v_i. \tag{11}$$

This equation shows immediately that if $\langle x, \alpha \rangle = 0$ for all $\alpha \in V^*$, then $x = 0$, hence $\langle , \rangle$ is a pairing. In fact the elements (10) of V^* just form the basis of V^* dual to the vs. The space $V^* = \text{Hom}(V, F)$ is called the *dual* of V.

The passage from V to V^* is a functor; in fact this is just the contravariant hom functor h_F which as we saw in **4.8** exists in every category. It is here applied to the category of vector spaces over F and linear mappings; in this case the values of h_F are not merely sets, but vector spaces over F. We go on to describe explicitly the effect of this functor on linear mappings.

Given any vector spaces U and V, and a linear mapping $f: U \to V$, the corresponding mapping $f^*: V^* \to U^*$ is defined as follows. With each $\beta \in V^*$ we associate the element $f^*\beta$ of U^* given by

$$\langle x, f^*\beta \rangle = \langle xf, \beta \rangle \qquad x \in U. \tag{12}$$

The mapping f^* so defined is linear, since $\langle x, f^*(\lambda\beta + \lambda'\beta') \rangle = \langle xf, \lambda\beta + \lambda'\beta' \rangle = \lambda \langle xf, \beta \rangle + \lambda' \langle xf, \beta' \rangle = \lambda \langle x, f^*\beta \rangle + \lambda' \langle x, f^*\beta' \rangle = \langle x, \lambda f^*\beta + \lambda' f^*\beta' \rangle$. It is called the *transpose* or *adjoint* of f. Note in particular that whereas f goes from U to V, f^* goes from V^* to U^*. That * is a functor follows from the verification in **4.8**.

In the finite-dimensional case it remains to describe the matrix for f^* in terms of that of f. Let us take dual bases, (u_i), (α_j) say in U and U^* and dual bases (v_r), (β_s) in V and V^*, and assume that f has the matrix (a_{ir}) relative to the given bases of U and V, i.e.

$$u_i f = \sum a_{ir} v_r.$$

Let f^* have the matrix (b_{js}):

$$f^*\beta_s = \sum \alpha_j b_{js}.$$

The the definition (12) shows that

$$b_{ir} = \langle u_i, f^*\beta_r\rangle = \langle u_i f, \beta_r\rangle = a_{ir}.$$

Hence, relative to the dual bases in V^* and U^*, f^* has the same matrix as f.

The reader who expected f^* to have the transpose (rather than the same matrix) should observe that in passing from a space to its dual we have switched the side on which f operates (cf. (12)). If the coordinate vectors of U and V are written as columns, those of U^* and V^* are naturally written as rows. This change makes transposition unnecessary; it is only if we continue to write operators on the same side that transposition is needed. The above convention has the advantage that it does not rely on the commutativity of the ground field; this will be useful when we wish to consider vector spaces over skew fields and modules over general rings.

To return to our earlier illustration, an $m \times n$ matrix A may be thought of as defining a linear mapping from m-dimensional to n-dimensional row vectors:

$$A: F^m \to F^n.$$

The transpose acts on the dual spaces:

$$A^*: {}^nF \to {}^mF.$$

If we iterate the above construction by forming the dual of V^*, we obtain V^{**}, the second dual or *bidual* of V. Let us denote elements of V^{**} by the letters $f, g, \ldots$ and write (f, α) for the value of f at $\alpha \in V^*$. We notice that each $x \in V$ can be used to define an element $\hat{x}$ say of V^{**}; its value on $\alpha \in V^*$ is given by the equation

$$(\hat{x}, \alpha) = \langle x, \alpha\rangle. \tag{13}$$

In words, to evaluate $\hat{x}$ at α, evaluate α at x. The correspondence $x \mapsto \hat{x}$ is a mapping of V into V^{**}; clearly it is linear and if $\hat{x} = 0$, then $\langle x, \alpha\rangle = 0$ for all $\alpha \in V^*$ and it follows that $x = 0$. Thus we have an injection $V \to V^{**}$ and, since $\dim V = \dim V^* = \dim V^{**}$, it must be an isomorphism.

Of course the assertion that V^{**} is isomorphic to V only tells us that $\dim V^{**} = \dim V$. In fact the isomorphism $\wedge$ is *natural* in the following sense, Given any linear mapping $f: U \to V$, let $f^*: V^* \to U^*$ be the adjoint of f and $f^{**}: U^{**} \to V^{**}$ the adjoint of f^*. Then the diagram

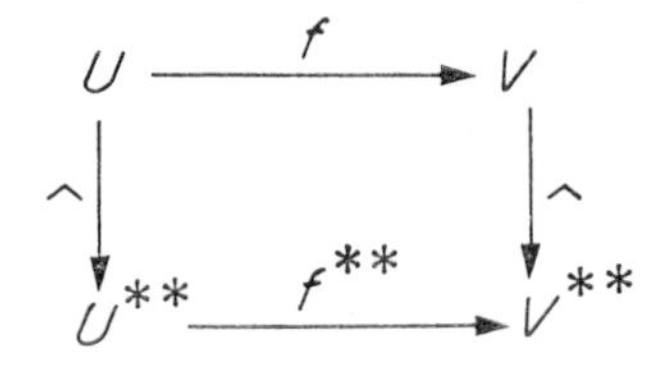

commutes. For, given $x \in U$, we have, for all $\alpha \in V^*$, $(\hat{x}f^{**}, \alpha) = (\hat{x}, f^*\alpha) = \langle x, f^*\alpha\rangle = \langle xf, \alpha\rangle = (\widehat{xf}, \alpha)$, hence

$$\hat{x}f^{**} = \widehat{xf},$$

which proves the assertion. We state the result as

THEOREM 2 *Let V be a finite-dimensional vector space over a field F, then its bidual V^{**} is isomorphic to V under a natural isomorphism.* ∎

To give a general definition of naturality, let F and G be two functors from a category $\mathscr{A}$ to another category $\mathscr{B}$. Then a *natural transformation* from F to G is given by a family of $\mathscr{B}$-morphisms $t_A: AF \to AG$ for the $\mathscr{A}$-objects, such that for any $\mathscr{A}$-morphism $\alpha: A_1 \to A_2$ the diagram

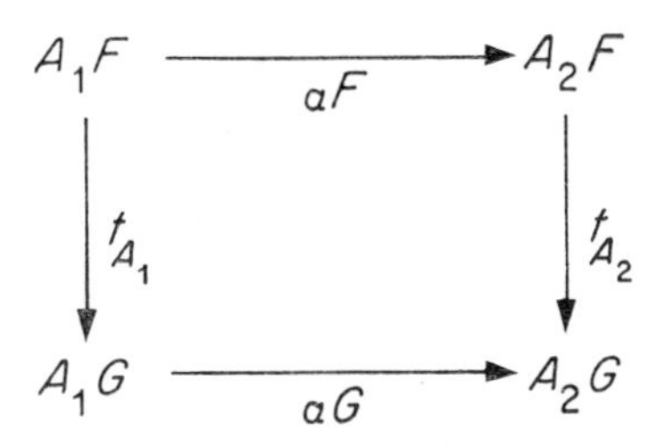

commutes. If t_A is an isomorphism for each $\mathscr{A}$-object A, t is called a *natural isomorphism*. It is clear how this definition generalizes the terminology used in Th. 2: there the mapping $x \mapsto \hat{x}$ is a natural isomorphism from a finite-dimensional vector space to its bidual.

Exercises

(1) Show that the form $\langle x, y\rangle = \sum x_i y_i$ on $F^n \times {}^nF$ defines a pairing.

(2) For any field F, verify that $e_1, e_1+e_2, e_1+e_2+e_3$ form a basis of ${}^3F$ and find the dual basis of F^3.

(3) If F is a field of characteristic 0, show that $e_2-e_1, e_3-e_1, \ldots, e_n-e_1, \sum e_i$ form a basis of ${}^nF$ and find the dual basis of F^n.

(4) Let $\langle,\rangle$ be a bilinear form defined on $U \times V$, and for any sets $X \subseteq U$, $Y \subseteq V$, put $X^\perp = \{y \in V \mid \langle x, y\rangle = 0 \text{ for all } x \in X\}$, ${}^\perp Y = \{x \in U \mid \langle x, y\rangle = 0 \text{ for all } y \in Y\}$. Verify that $X^\perp$, ${}^\perp Y$ are subspaces of V, U respectively and that ${}^\perp(X^\perp) \supseteq X$, $({}^\perp Y)^\perp \supseteq Y$.

(5) Let U, V be finite-dimensional vector spaces over F, and $\langle,\rangle$ a bilinear form on $U \times V$. Find a relation between $\dim U$, $\dim V$, $\dim U^\perp$, $\dim {}^\perp V$.

(6) Let U, V be finite-dimensional spaces with a pairing $\langle,\rangle$. If X is a subspace of U, show that $\dim X + \dim X^\perp = \dim U$. Deduce that ${}^\perp(X^\perp) = X$.

(7) Show that in Th. 1 it is enough to assume that *one* of U, V is finite-dimensional.

(8) Is the commutativity of the ground field needed in Th. 1?

(9) If V is the space of infinite sequences $(x_1, x_2, \ldots)$ with finitely many non-zero components, show that $V^* = F^N$.

(10) Show that the mapping $V \to V^{**}$ described in the text is injective even for an infinite-dimensional space V with a basis.

(11) If U and V are finite-dimensional, show that $(U \oplus V)^* \cong U^* \oplus V^*$.

(12) If U, V are finite-dimensional spaces, show that a linear mapping $f: U \to V$ is injective (surjective) iff f^* is surjective (injective). Deduce that f is an isomorphism iff f^* is.

8.2 Inner products; quadratic and hermitian forms

Throughout this section we shall restrict the field to be of characteristic not two. For readers only interested in the real and complex cases this will be no restriction; those who want to know what happens in the forbidden case should turn to the exercises. Moreover, all vector spaces are assumed finite-dimensional unless the contrary is stated.

Let V be a vector space over a field F and consider a bilinear form $\langle,\rangle$ on V that is also *symmetric*:

$$\langle y, x\rangle = \langle x, y\rangle. \tag{1}$$

A symmetric bilinear form is often called an *inner product* and a space carrying such a form is called an *inner product space*. We observe that when (1) holds, it is enough to assume that $\langle,\rangle$ is linear in the first argument; linearity in the second argument follows then by symmetry. Likewise, $\langle,\rangle$ will be non-singular as soon as

$$\langle a, x\rangle = 0 \text{ for all } x \in V \text{ implies } a = 0.$$

For example, the usual scalar product of vectors in 3-space is a symmetric bilinear form. More generally, in any space V, if the coordinates of x, y relative to a fixed basis are given by rows ξ, η respectively, then the equation

$$\langle x, y\rangle = \xi\eta^{\mathrm{T}}$$

defines an inner product.

Returning to the general case, we observe that the values of $\langle,\rangle$ are completely determined by the values of $\langle x, x\rangle$. For we can write

$$2\langle x, y\rangle = \langle x+y, x+y\rangle - \langle x, x\rangle - \langle y, y\rangle, \tag{2}$$

and this determines $\langle x, y\rangle$ because the characteristic is not 2, by hypothesis. This process of expressing the inner product in terms of the function $f(x) = \langle x, x\rangle$ is known as *polarization*; f itself is called the *quadratic form* associated with $\langle,\rangle$.

In **8.1** we saw that bilinear forms correspond to rectangular matrices. In the same way an inner product on a space V with the basis $v_1, \ldots, v_n$ is completely described by the square matrix whose entries are the products of the basis vectors:

$$a_{ij} = \langle v_i, v_j \rangle, \tag{3}$$

but now the matrix $A = (a_{ij})$ is no longer arbitrary. From the symmetry (1) of our form we see that $a_{ij} = a_{ji}$, i.e. the matrix equals its transpose:

$$A^{\mathrm{T}} = A. \tag{4}$$

A square matrix satisfying (4) is said to be *symmetric*. Clearly any symmetric matrix can be used to define an inner product on V, by the rule (3). The value on $x = \sum \xi_i v_i$, $y = \sum \eta_i v_i$ is then

$$\langle x, y \rangle = \sum a_{ij} \xi_i \eta_j,$$

or in matrix form

$$\langle x, y \rangle = \xi A \eta^{\mathrm{T}}. \tag{5}$$

If we change coordinates in V, from ξ to ξ' say, where $\xi = \xi' P$, then

$$\langle x, y \rangle = \xi' P A (\eta' P)^{\mathrm{T}} = \xi' P A P^{\mathrm{T}} \eta'^{\mathrm{T}}.$$

Hence in the new coordinates the matrix of the form is PAP^{T}. Two matrices A, B are said to be *congruent* if $B = PAP^{\mathrm{T}}$ for some invertible matrix P. It is easily seen that any matrix congruent to a symmetric matrix is itself symmetric. Our findings may be expressed as follows.

THEOREM 1 *Under a change of basis the matrix of an inner product undergoes a congruence transformation.* ■

A natural question to ask at this point is: How far can the expression for an inner product be simplified by a suitable choice of coordinates? We shall find that the matrix of an inner product can always be brought to diagonal form. This is essentially the familiar process of completing the square.

THEOREM 2 *Let $\langle , \rangle$ be an inner product on a finite-dimensional vector space V over a field of characteristic not two. Then V has a basis $v_1, \ldots, v_n$ such that $\langle v_i, v_j \rangle = 0$ for $i \neq j$.*

Proof. We use induction on $n = \dim V$. For $n = 1$ there is nothing to prove, so let $n > 1$. If $\langle x, y \rangle = 0$ for all $x, y \in V$, the result again holds, so we may assume that the form does not vanish identically. In fact we can find a vector $v_1 \in V$ such that $\langle v_1, v_1 \rangle \neq 0$, for if $\langle x, x \rangle = 0$ for all $x \in V$, then by polarization $\langle , \rangle$ vanishes identically, against the hypothesis.

With the vector v_1 we have chosen, let us define

$$V_1 = \{x \in V \mid \langle x, v_1 \rangle = 0\}. \tag{6}$$

Clearly V_1 is a subspace of V and, since $v_1 \notin V_1$, $\dim V_1 < n$, so by induction V_1 has a basis $v_2, \ldots, v_m$ such that $\langle v_i, v_j \rangle = 0$ for $i \neq j$ and $i, j > 1$. By the definition (6) and the symmetry of the form, this holds even if i or j is 1 and the theorem will follow if we show that $v_1, \ldots, v_m$ is a basis of V. Given

$x \in V$, we can verify that the vector

$$x_1 = x - \frac{\langle x, v_1 \rangle}{\langle v_1, v_1 \rangle} v_1$$

lies in V_1, hence x_1 is linearly dependent on $v_2, \ldots, v_m$ and so x is linearly dependent on $v_1, \ldots, v_m$. Thus the vs span V and satisfy the conclusion. We can therefore pick out a subset which forms a basis and still satisfies the conclusion (in fact the $v_1, \ldots, v_m$ already form a basis, as the reader may verify). ■

In terms of matrices the conclusion may be stated as

COROLLARY *Every symmetric matrix (over a field of characteristic not two) is congruent to a diagonal matrix.* ■

The hypothesis on the field is necessary, as we see by taking the matrix $\begin{pmatrix} 0 & 1 \\ 1 & 0 \end{pmatrix}$ over a field of characteristic 2, or equivalently the form defined by $\xi_1\eta_2 + \xi_2\eta_1$. This cannot be transformed to diagonal form (in characteristic 2) and it does not arise from a quadratic form by polarization. By contrast, in characteristic not two, as we have seen, the inner product $\langle x, y \rangle$ is completely determined by the corresponding quadratic form

$$\langle x, x \rangle = \xi A \xi^{\mathrm{T}}.$$

For this reason we often use the terms 'inner product' and 'quadratic form' interchangeably.

Although Th. 2 enormously simplifies the study of inner products and of quadratic forms, it does not provide a complete classification, since we have so far no means of deciding when two diagonal matrices are congruent. In fact, for most fields, this classification leads to difficult questions that are far from completely solved. But for the real and complex fields we can easily give a complete answer.

THEOREM 3 *Let $\langle , \rangle$ be a quadratic form on a vector space V over the complex field* $\mathbf{C}$. *Then for a suitable coordinate system in V, the form can be expressed as*

$$\langle x, x \rangle = \xi_1^2 + \cdots + \xi_r^2,$$

where r, the rank of the form, is independent of the choice of coordinates.

We note that $r \leqslant \dim V$, but equality need not hold, e.g. $\langle , \rangle$ might be identically zero.

Proof. By Th. 2 we can find a basis $v_1', \ldots, v_n'$ of V, such that

$$\langle x, x \rangle = a_1 \xi_1'^2 + \cdots + a_n \xi_n'^2 \qquad \text{where } a_i \in \mathbf{C}.$$

We may take the basis to be ordered so that $a_1, \ldots, a_r \neq 0$, $a_{r+1} = \cdots = a_n = 0$. Then r is the rank of the matrix of $\langle , \rangle$, clearly unaffected by congruence transformations and so independent of the choice of basis. Now in **C** every element has a square root, so we can find $b_i \in \mathbf{C}$ to satisfy $a_i = b_i^2$ $(i = 1, \ldots, r)$. Put $v_i = b_i^{-1}v_i'$ $(i \leqslant r)$, $v_j = v_j'$ $(j \geqslant r+1)$, then $\xi_i = b_i\xi_i'$ $(i \leqslant r)$ and in the new coordinates,

$$\langle x, x\rangle = \xi_1^2 + \cdots + \xi_r^2. \blacksquare$$

The content of Th. 3 can also be expressed by saying that the rank forms a complete set of invariants for classifying quadratic forms over **C**. Of course the rank of a quadratic form is an invariant whatever the field, but it is usually not enough to distinguish incongruent forms. In such cases we may be able to find other invariants which together with the rank give us a complete classification. An example is the real field **R**, where just one other invariant, the signature, is needed to classify quadratic forms completely.

THEOREM 4 *Let $\langle , \rangle$ be a quadratic form on a vector space V over the real field* **R**. *Then for a suitable coordinate system in V, the form can be expressed as*

$$\langle x, x\rangle = \xi_1^2 + \cdots + \xi_p^2 - \xi_{p+1}^2 - \cdots - \xi_r^2,$$

where p and r depend only on the form and not on the choice of basis.

This means that p, r form a complete set of invariants. The number $2p-r$, representing the excess of the positive over the negative terms, is called the *signature* of the form and the assertion that this is an invariant of the form is known as *Sylvester's law of inertia.*

To prove the result, we can again write

$$\langle x, x\rangle = a_1\xi_1'^2 + \cdots + a_r\xi_r'^2 \qquad \text{where } a_i \in \mathbf{R},$$

where r is the rank of the form and so $a_i \neq 0$ for $i = 1, \ldots, r$. We may take the basis of V ordered so that $a_1, \ldots, a_p > 0$ and $a_{p+1}, \ldots, a_r < 0$. Since every positive number has a square root in **R**, on writing $|a_i| = b_i^2$, $v_i = b_i^{-1}v_i'$ $(i \leqslant r)$, $v_j = v_j'$ $(j > r)$, $\xi_i = b_i\xi_i'$ $(i \leqslant r)$, we obtain

$$\langle x, x\rangle = \xi_1^2 + \cdots + \xi_p^2 - \xi_{p+1}^2 - \cdots - \xi_r^2.$$

We already know r to be an invariant. If in another coordinate system we have

$$\langle x, x\rangle = \eta_1^2 + \cdots + \eta_q^2 - \eta_{q+1}^2 - \cdots - \eta_r^2,$$

where $q \neq p$, say $q > p$, consider the subspace of V given by the equations

$$\xi_1 = \cdots = \xi_p = \eta_{q+1} = \cdots = \eta_n = 0 \quad \text{where } n = \dim V. \tag{7}$$

These are $n-(q-p) < n$ linear equations, in n unknowns (the coordinates in some system); they have a non-trivial solution and so the subspace of V they define is non-zero. If x is a vector in it, then

$$\langle x, x\rangle = \eta_1^2 + \cdots + \eta_q^2 = -\xi_{p+1}^2 - \cdots - \xi_r^2.$$

This equation, in real coordinates, is possible only when $\eta_1 = \cdots = \eta_q = \xi_{p+1} = \cdots = \xi_r = 0$, hence all ηs vanish and so $x = 0$, which contradicts the fact that the space defined by (7) was non-zero. Therefore $q = p$, as claimed. ■

A quadratic form on a real vector space is said to be *positive-definite* if

$$\langle x, x\rangle > 0 \qquad \text{for all } x \neq 0;$$

if $\langle x, x\rangle \geqslant 0$ for all vectors x, the form is said to be *positive-semidefinite*. If $-\langle,\rangle$ is positive-(semi)definite, we call $\langle,\rangle$ *negative-(semi)definite*.

We observe that a positive-definite form on an n-dimensional space has rank and signature equal to n and may be written as a sum of n squares in a suitable coordinate system. We state this as a

COROLLARY *A positive-definite quadratic form on a real n-dimensional space can in a suitable coordinate system be written as*

$$\langle x, x\rangle = \xi_1^2 + \cdots + \xi_n^2. \ ■$$

In particular this shows that any positive-definite quadratic form can be transformed into any other (on the same space).

As we shall see in **8.3**, positive-definite forms play a rather special role: they can be used to define a norm or length function which has all the properties usually required of such a function. On a complex vector space there can be no positive-definite quadratic forms, for if $\langle x, x\rangle > 0$, then $\langle ix, ix\rangle = -\langle x, x\rangle < 0$. Nevertheless there is a trick which enables one to construct a very similar theory, using instead of quadratic forms a variant introduced by C. Hermite and named after him:

DEFINITION A *hermitian form* on a complex vector space V is a function h on $V \times V$ with complex values, which is linear in the first argument:

$$h(\lambda x + \lambda' x', y) = \lambda h(x, y) + \lambda' h(x', y),$$

but (in place of symmetry) possesses *hermitian symmetry:*

$$h(y, x) = \overline{h(x, y)}, \tag{8}$$

where the bar denotes the complex conjugate.

As a result h is not linear but *antilinear* in the second argument:

$$h(x, \lambda y + \lambda' y') = \bar{\lambda} h(x, y) + \bar{\lambda}' h(x, y').$$

A hermitian form is again determined by the values of $h(x, x)$, by polarization, which now takes the form

$$4h(x, y) = f(x+y) - f(x-y) + if(x+iy) - if(x-iy),$$

where $f(x) = h(x, x)$. In terms of a basis $v_1, \ldots, v_n$ of V, h is given by the matrix $A = (a_{ij})$, where

$$a_{ij} = h(v_i, v_j).$$

By (8) this matrix satisfies the condition $a_{ji} = \overline{a_{ij}}$, i.e.

$$\bar{A}^{\mathrm{T}} = A. \tag{9}$$

We write A^{H} for $\bar{A}^{\mathrm{T}}$ and call it the *hermitian* conjugate of A. A square matrix equal to its hermitian conjugate, i.e. satisfying (9), is said to be *hermitian.* Thus a hermitian form is described by a hermitian matrix and clearly any hermitian matrix defines, relative to a given basis, a hermitian form. The value of the form on $x = \sum \xi_i v_i$, $y = \sum \eta_i v_i$ is given by

$$h(x, y) = \xi A \eta^{\mathrm{H}}.$$

Under change of coordinates, from ξ to ξ', where $\xi = \xi' P$, the form becomes

$$h(x, y) = \xi' PA(\eta' P)^{\mathrm{H}} = \xi' PAP^{\mathrm{H}} \eta'^{\mathrm{H}}.$$

Thus in the new coordinates the matrix of the form is PAP^{H}. We have a reduction theorem for hermitian forms entirely analogous to Th. 4.

THEOREM 5 *Let h be a hermitian form on a complex vector space V. Then for a suitable coordinate system in V, the form can be expressed as*

$$h(x, x) = |\xi_1|^2 + \cdots + |\xi_p|^2 - |\xi_{p+1}|^2 - \cdots - |\xi_r|^2,$$

where p and r depend only on the form and not on the choice of basis.

The proof is exactly as for Th. 4; this time we have to use the fact that for any $\lambda \in \mathbf{C}$, $|\lambda|^2 = \lambda\bar{\lambda} \geqslant 0$ and a sum of squares of absolute values can vanish only when each summand is zero. ∎

As before we can define the *signature* of a hermitian form, and Th. 5 shows that rank and signature form a complete set of invariants for hermitian forms. The forms of rank and signature equal to the dimension of the space are again called *positive-definite* and the terms *positive-semidefinite* and *negative (semi) definite* can be defined as before.

From the definition it is clear that a hermitian form is positive-definite iff it can be transformed so as to have the unit-matrix as its matrix. But there is a criterion which makes it unnecessary to reduce the form. For any square matrix A let us define the *leading minor* of order r as the determinant formed from the first r rows and r columns; this definition makes sense for any r not exceeding the order of A.

THEOREM 6 *A hermitian matrix is positive-definite if and only if all its leading minors are positive.*

Proof. Let A be a positive-definite hermitian matrix; then there is an invertible matrix P such that $PAP^{\mathrm{H}} = I$, hence $\det A = |\det P|^{-2} > 0$. Now denote by A_r the $r \times r$ matrix consisting of the first r rows and columns of A. This is still positive-definite, since it is the matrix of the form in r variables obtained by putting $\xi_{r+1} = \cdots = \xi_n = 0$. Hence $\det A_r > 0$ for $r = 1, 2, \ldots, n$ and so all leading minors are positive.

Conversely, assume that all leading minors of A are positive, then in particular $a_{11} > 0$. Consider the following transformation of coordinates:

$$\eta_1 = 1/a_{11}(a_{11}\xi_1 + a_{21}\xi_2 + \cdots + a_{n1}\xi_n), \qquad \eta_i = \xi_i \; (i = 2, \ldots, n).$$

In terms of these coordinates,

$$\sum a_{ij}\xi_i\bar{\xi}_j = a_{11}\eta_1\bar{\eta}_1 + \sum_2^n b_{ij}\eta_i\bar{\eta}_j, \tag{10}$$

where

$$b_{ij} = \overline{b_{ji}} = a_{ij} - a_{i1}a_{1j}/a_{11}.$$

Now for $r > 1$ we have, by subtracting a_{1j}/a_{11} times the first column from the jth,

$$\det A_r = \begin{vmatrix} a_{11} & a_{12} & \ldots & a_{1r} \\ a_{21} & a_{22} & \ldots & a_{2r} \\ \cdot & \cdot & \cdot & \cdot \\ a_{r1} & a_{r2} & \ldots & a_{rr} \end{vmatrix} = \begin{vmatrix} a_{11} & 0 & 0 & \ldots & 0 \\ a_{21} & b_{22} & b_{23} & \ldots & b_{2r} \\ \cdot & \cdot & \cdot & \cdot & \cdot \\ a_{r1} & b_{r2} & b_{r3} & \ldots & b_{rr} \end{vmatrix}$$

It follows that the form in $n-1$ variables, $\sum_2^n b_{ij}\eta_i\bar{\eta}_j$, has all its leading minors positive, hence by induction it is positive-definite. Now (10) shows that the original form is positive-definite. ■

We observe that the same criterion applies for real symmetric matrices; because for a real matrix 'symmetric' and 'hermitian' mean the same thing.

Exercises

(1) Let V be a space with an inner product and define

$$N = \{x \in V \mid \langle x, y\rangle = 0 \text{ for all } y \in V\}.$$

Verify that N is a subspace of V (N is called the *radical* of the form). If N' is any complementary subspace of N in V, show that the product induced on N' is non-singular.

(2) Show that the relation of congruence between matrices is an equivalence relation.

(3) In the proof of Th. 2, if $\dim V = n$ and V_1 is as in (6), show that $\dim V_1 = n-1$. Deduce that any basis of V_1 together with v_1 forms a basis of V.

(4) Let $\langle,\rangle$ be an inner product on a space V over any field. Show that the function $f(x) = \langle x, x\rangle$ satisfies

Q.1 $f(\lambda x) = \lambda^2 f(x)$.

Q.2 $b(x, y) = f(x+y) - f(x) - f(y)$ is bilinear.

Any function on a vector space V with values in the ground field, satisfying **Q**. 1–2 is called a *quadratic form* on V. If char $F \neq 2$, show that any quadratic form f can be used to define an inner product $\langle,\rangle$ by the rule $2\langle x, y\rangle = b(x, y)$, where b is given by **Q**. 2.

(5) Let V be a vector space over a field of characteristic 2. Show that any quadratic form f defines a symmetric bilinear form b (by **Q**. 2) and, moreover, $b(x, x) = 0$

for all $x \in V$. Given a quadratic form f on V, by introducing a basis of V, show that a bilinear form b_0 on V can always be found such that $b_0(x, x) = f(x)$, but that b_0 cannot in general be chosen to be symmetric.

(6) A vector x in an inner product space is said to be *isotropic* if it is orthogonal to itself: $\langle x, x \rangle = 0$. A vector space with an inner product is said to be *anisotropic* if it contains no non-zero isotropic vector. Show that a real inner product space is anisotropic iff the product is (positive or negative-)definite. Give an example of a vector space over **Q** with an inner product, which is anisotropic but indefinite.

(7) A 2-dimensional space over a field of characteristic $\neq 2$ with a non-singular inner product which is not anisotropic is called a *hyperbolic plane*. Show that every hyperbolic plane has a basis u, v such that $\langle u, u \rangle = \langle v, v \rangle = 0$, $\langle u, v \rangle = 1$, and that every space of this form is a hyperbolic plane.

(8) Show that every non-singular inner product space can be written as a direct sum of pairwise orthogonal hyperbolic planes and an anisotropic space.

(9) If $\langle , \rangle$ is an inner product on a space V, and $\langle , \rangle$ is non-singular on a subspace U of V, show that V is a direct sum: $V = U \oplus U^{\perp}$, where $U^{\perp}$ is defined as in Ex. (4) **8.1**. Deduce that the inner product is completely determined by its restrictions to U and $U^{\perp}$.

(10) Let V be an inner product space. Given a basis $u_1, \ldots, u_n$ of V, define the *discriminant* of $\langle , \rangle$ relative to this basis as $D(u_1, \ldots, u_n) = \det(\langle u_i, u_j \rangle)$. Show that the inner product is non-singular iff the discriminant is non-zero. Show that the ratio of the discriminants relative to any two bases is a square of an element of F, and that every square from F occurs in this way.

(11) Let $f(x_1, \ldots, x_n)$ be a real-valued function of n real variables which has a power series expansion near the origin $x = 0$. Then a necessary condition for a maximum or minimum at $x = 0$ is that $\partial f/\partial x_i = 0$ $(i = 1, \ldots, n)$ (then f is said to be *stationary* at $x = 0$ and $x = 0$ is called a *critical point* of f). Verify that this is equivalent to the condition that in the first approximation, $f - f(0)$ be a quadratic form: $f - f(0) = \Sigma a_{ij} x_i x_j + \cdots$ $(a_{ij} = a_{ji})$. Show that if in the above expression for f, (a_{ij}) is positive-(negative-) definite, then f has a minimum (maximum) at $x = 0$ (though not conversely). Investigate the following cases for a maximum or minimum: (i) $x_1^2 + x_2^2 + 3x_1x_2$, (ii) $x_1^2 + x_2^2 + x_3^2 - 2x_2x_3$, (iii) $x_1^2 + ax_2^4$ for varying a, (iv) $x_1^3 - 3x_1x_2^2$, (v) $x_1^2 + x_2^2 - 2x_1^2x_2 + x_1^4$.

(12) Find the rank and signature of the following forms: (i) $x_1^2 + x_2^2 + x_3^2 - 2x_1x_3 - 2x_2x_3$, (ii) $x_1x_2 + x_2x_3 + x_3x_1$, (iii) $x_1^2 + 2x_2^2 - 2x_3^2 - 4x_1x_2 - 4x_2x_3$.

8.3 Euclidean and unitary spaces

The analysis of inner products in the last section will now be used to define the notion of length. To justify the definition given below we recall that in elementary geometry the length of a vector $x = (\xi_1, \xi_2)$ is (on the basis of Pythagoras' theorem) defined as

$$|x| = (\xi_1^2 + \xi_2^2)^{\frac{1}{2}}. \tag{1}$$

By polarization of $|x|^2$ one obtains the familiar scalar product of two vectors:

$$\langle x, y\rangle = \xi_1\eta_1+\xi_2\eta_2. \tag{2}$$

Expressing the vectors in polar coordinates, we have $x = (r\cos\alpha, r\sin\alpha)$, $y = (r'\cos\alpha', r'\sin\alpha')$, hence

$$\langle x, y\rangle = rr'(\cos\alpha\cos\alpha'+\sin\alpha\sin\alpha') = rr'\cos(\alpha-\alpha').$$

Thus $\langle x, y\rangle$ represents the product of the lengths of x and y by the cosine of the angle between them.

Similarly, the scalar product of two vectors in $\mathbf{R}^3$ is defined by

$$\langle x, y\rangle = \xi_1\eta_1+\xi_2\eta_2+\xi_3\eta_3. \tag{3}$$

The definition given below is at first sight more general, but in fact reduces precisely to the above form in 2 and 3 dimensions.

DEFINITION A *Euclidean space* is a finite-dimensional real vector space equipped with a positive-definite quadratic form.

We shall denote the associated bilinear form by $\langle x, y\rangle$ and call this the *inner product* of x and y.

For example, $\mathbf{R}^n$ becomes a Euclidean space if we define

$$\langle x, y\rangle = \xi_1\eta_1+\cdots+\xi_n\eta_n. \tag{4}$$

For $n = 2, 3$ this reduces to the cases (2), (3) mentioned above; in fact, Th. 4, **8.2**, Cor. shows that in the general n-dimensional Euclidean space the inner product can always be taken to be of this form. We shall return to this point later, but for the moment merely note that Euclidean spaces may nevertheless arise in circumstances that are at first sight quite unrelated to (4).

Thus let V be the set of all continuous real functions defined for $0 \leqslant x \leqslant 1$. This set forms a real vector space in an obvious way (cf. Ex. (7), **4.1**). If we put

$$\langle f, g\rangle = \int_0^1 f(x)g(x)\,dx, \tag{5}$$

it is easily verified that this defines an inner product on V whose associated quadratic form is positive-definite. In fact, V only fails to be a Euclidean space in the sense defined earlier because it is not finite-dimensional. But the functions allowed in (5) may be restricted in some way, e.g. to be solutions of a given differential equation, so as to form a finite-dimensional space. Then we have a Euclidean space and all the results proved below are applicable.

Let V be any Euclidean space and take $x\in V$. Then $\langle x, x\rangle \geqslant 0$, with equality iff $x = 0$. We take the positive square root of $\langle x, x\rangle$ and call it the *norm* or *length* of x, written as $\|x\|$:

$$\|x\| = \langle x, x\rangle^{\frac{1}{2}}.$$

If x is any non-zero vector, we can always find a vector proportional to it, of unit length. We need $\lambda \in \mathbf{R}$ such that $\|\lambda x\| = 1$. Now $\|\lambda x\|^2 = \lambda^2 \|x\|^2$, so we need only put $\lambda = 1/\|x\|$ to achieve our objective. A vector of length 1 is said to be *normalized* and the process just described, of replacing a non-zero vector by a normalized vector proportional to it is called *normalization.*

Two vectors x, y are said to be *orthogonal* if

$$\langle x, y \rangle = 0.$$

By the symmetry of the inner product this is equivalent to the condition $\langle y, x \rangle = 0$. Given any subset X of V, we define

$$X^{\perp} = \{y \in V \mid \langle x, y \rangle = 0 \text{ for all } x \in X\}.$$

It is easily seen that $X^{\perp}$ is a subspace of V. This notion is of particular importance when X is itself a subspace; then $X^{\perp}$ is called the *orthogonal complement* of X.

Our first task is to establish some well known inequalities for Euclidean space.

THEOREM 1 (Schwarz's inequality) *In a Euclidean space V, any $x, y \in V$ satisfy*

$$|\langle x, y \rangle| \leqslant \|x\| \, . \, \|y\|, \tag{6}$$

with equality if and only if x and y are linearly dependent.

Proof. For $y = 0$ this is clear. Otherwise we have, for any $\lambda \in \mathbf{R}$,

$$\langle x - \lambda y, x - \lambda y \rangle \geqslant 0, \tag{7}$$

with equality precisely when x and y are linearly dependent. The left-hand side becomes, on expansion,

$$\|x\|^2 - 2\lambda \langle x, y \rangle + \lambda^2 \|y\|^2 \geqslant 0.$$

Multiplying by $\|y\|^2$ and putting $\lambda = \langle x, y \rangle / \langle y, y \rangle$, we find that

$$\|x\|^2 \, . \, \|y\|^2 - 2\langle x, y \rangle^2 + \langle x, y \rangle^2 \geqslant 0,$$

which simplifies to (6). Equality means that we have equality in (7), i.e. that x and y are linearly dependent. ■

COROLLARY *For any x, y in a Euclidean space,*

$$\|x + y\| \leqslant \|x\| + \|y\|. \tag{8}$$

For $\|x+y\|^2 = \|x\|^2 + 2\langle x, y \rangle + \|y\|^2 \leqslant \|x\|^2 + 2\|x\| \, . \, \|y\| + \|y\|^2 \leqslant (\|x\| + \|y\|)^2$. Now (8) follows on taking square roots and observing that both sides of (8) are positive. ■

In any Euclidean space we can define the *distance* between two vectors x, y as

$$d(x, y) = \|x - y\|. \tag{9}$$

In the examples of $\mathbf{R}^2$ and $\mathbf{R}^3$ given earlier, this corresponds to the distance between the end-points of x and y, e.g. in $\mathbf{R}^2$

$$\|x-y\| = [(\xi_1-\eta_1)^2+(\xi_2-\eta_2)^2]^{\frac{1}{2}}.$$

In a general Euclidean space, the distance defined by (9) satisfies the three axioms usually required of a metric and is compatible with the addition:

M. 1 $d(x, y) \geqslant 0$, *with equality iff* $x = y$.

M. 2 $d(x, y) = d(y, x)$ (*symmetry*).

M. 3 $d(x, y)+d(y, z) \geqslant d(x, z)$ (*triangle inequality*).

M. 4 $d(x+a, y+a) = d(x, y)$ (*translation invariance*).

M. 1, 2 and 4 are obvious from the definition; **M.** 3, which corresponds to the fact that in a triangle the sum of any two sides is at least equal to the third, follows from (8) on replacing x, y in (8) by $x-y, y-z$. On account of this interpretation (8) itself is also called the *triangle inequality*.

For any non-zero x, y we have, by Schwarz's inequality,

$$-1 \leqslant \frac{\langle x, y\rangle}{\|x\| \,.\, \|y\|} \leqslant 1, \tag{10}$$

hence there is a unique angle α between 0° and 180° such that $\cos \alpha = \langle x, y\rangle/\|x\| \,.\, \|y\|$. We call α the *angle* between x and y. Thus we have, by definition,

$$\langle x, y\rangle = \|x\| \,.\, \|y\| \,.\, \cos \alpha. \tag{11}$$

The following assertions are left for the reader to verify:

(i) Two vectors are linearly dependent iff one of them is 0 or the angle between them is 0° or 180°,

(ii) the angle between x and y is 90° iff x, y are orthogonal as defined earlier,

(iii) if the angle between x and y is α, then the angle between x and $-y$ (or between $-x$ and y) is $180°-\alpha$.

If V is an n-dimensional Euclidean space, then we know from **8.2** that for a suitable choice of basis $v_1, \ldots, v_n$ in V, the inner product takes the form

$$\langle x, y\rangle = \xi_1\eta_1+ \cdots +\xi_n\eta_n, \tag{12}$$

because the given quadratic form was positive-definite. This means that the products of the basis vectors are given by

$$\langle v_i, v_j\rangle = \delta_{ij}. \tag{13}$$

Conversely, if the basis is chosen to satisfy (13), the inner product will take the form (12). By the *standard norm* on V we shall understand a norm defined by an inner product of the form (12); the corresponding basis, to satisfy (13) is called an *orthonormal basis* of V. Generally, by an *orthonormal set* in a Euclidean space we understand a set of normalized vectors that are mutually orthogonal. What has been said shows that every Euclidean space has an orthonormal basis and, relative to any such basis, the space has the standard norm. In practice one is often faced with the task of constructing such an orthonormal basis explicitly. This is accomplished by the *Gram–Schmidt orthogonalization process*, which we now explain.

Let us assume that we have a Euclidean space V with a basis $u_1, \ldots, u_n$. By normalizing u_1 we obtain a unit vector v_1 linearly dependent on u_1. Next we take u_2 and form $u_2' = u_2 - \langle u_2, v_1 \rangle v_1$ (in geometrical terms, u_2' is the result of projecting u_2 on the orthogonal complement of v_1). Then $\langle u_2', v_1 \rangle = 0$ but $u_2' \neq 0$, for $u_2' = 0$ would mean that u_2 is linearly dependent on v_1, hence on u_1, which contradicts the fact that the us form a basis. If we normalize u_2' we obtain a vector v_2 which is still orthogonal to v_1 and linearly dependent on u_1 and u_2. If we have already found an orthonormal set $v_1, \ldots, v_{r-1}$ linearly dependent on $u_1, \ldots, u_{r-1}$, we replace u_r by $u_r' = u_r - \langle u_r, v_1 \rangle v_1 - \cdots - \langle u_r, v_{r-1} \rangle v_{r-1}$, then $u_r' \neq 0$ (because u_r is not linearly dependent on $u_1, \ldots, u_{r-1}$), so we can normalize u_r' and obtain a vector v_r forming an orthonormal set with $v_1, \ldots, v_{r-1}$. By induction on r, we find an orthonormal set $v_1, \ldots, v_n$ which spans the same space as $u_1, \ldots, u_n$, i.e. it spans V. We see that it forms a basis of V by counting, or by using the following

LEMMA 2 *Any orthonormal set on a Euclidean space is linearly independent.*

For if $v_1, \ldots, v_n$ is an orthonormal set and we have a relation $\sum \alpha_i v_i = 0$, then on multiplying by v_j we find that $0 = \langle \sum \alpha_i v_i, v_j \rangle = \alpha_j$, for $j = 1, \ldots, n$, hence every linear relation between the vs is trivial. ∎

This then shows that the orthonormal set we have constructed is a basis. More generally, given any orthonormal set $v_1, \ldots, v_r$ (not necessarily a basis) we can apply the Gram–Schmidt process, starting from $v_1, \ldots, v_r$ and using any basis of V. After $\dim V - r$ steps we get an orthonormal set spanning V, i.e. a basis. The result may be stated as follows.

THEOREM 3 *Let V be a Euclidean space; then any orthonormal set $v_1, \ldots, v_r$ may be completed to an orthonormal basis of V.* ∎

There is an entirely analogous development of complex spaces with a positive-definite hermitian form; we shall fairly rapidly go through the corresponding definitions and results, pausing only where they differ significantly from the real case.

A *unitary space* is defined as a finite-dimensional complex vector space V equipped with a positive-definite hermitian form, again denoted by $\langle , \rangle$.

The *length* or *norm* of a vector x is again defined as the positive square root of $\langle x, x \rangle$ and is denoted by $\|x\|$. Any non-zero vector can be normalized. Orthogonality of two vectors is defined as before; although $\langle , \rangle$ is no longer symmetric, it is still true that $\langle x, y \rangle = 0$ iff $\langle y, x \rangle = 0$.

Schwarz's inequality still holds; when we expand (7) we now have

$$\|x\|^2 - \lambda\langle y, x\rangle - \bar{\lambda}\langle x, y\rangle + |\lambda|^2\|y\|^2 \geqslant 0.$$

When we multiply by $\|y\|^2$ and put $\lambda = \langle x, y\rangle/\langle y, y\rangle$, we find

$$\|y\|^2\|x\|^2 - 2\langle x, y\rangle\langle y, x\rangle + |\langle x, y\rangle|^2 \geqslant 0.$$

Remembering that $\langle y, x\rangle = \overline{\langle x, y\rangle}$, we can again reduce this to (6). Now the corollary follows in essentially the same way as before; the reader should write this out for himself, noting the differences.

We can now define the distance between two vectors and show that we have a metric on V. In defining the angle between two vectors we no longer have (10), but we can still use (11) to define α; the only difference is that α need no longer be real.

Next we define orthonormal sets and as before show that they are linearly independent. The Gram–Schmidt process can again be used to construct orthonormal bases; we leave the reader to verify that Th. 3 still holds for unitary spaces.

Exercises

(1) Apply the Gram–Schmidt process to the following vectors in $\mathbf{R}^3$ (with the standard norm) to get an orthonormal basis: (1, 2, 0), (1, 0, 1), (2, 3, 1).

(2) Find an orthonormal basis for the subspace of $\mathbf{R}^4$ (with the standard norm) spanned by (1, 2, 0, −1) and (2, 1, 3, 0).

(3) Show that the functions $\exp 2\pi inx$ ($n \in \mathbf{Z}$, $x \in \mathbf{R}$) form an orthonormal set relative to the inner product

$$\langle f, g\rangle = \int_0^1 f(x)\overline{g(x)}\,\mathrm{d}x.$$

Show that the functions $\sin 2\pi nx$, $\cos 2\pi nx$, ($n \in \mathbf{Z}$) are mutually orthogonal and construct an orthonormal set from them.

(4) Find an orthonormal basis for the space of real polynomials of degree at most 3, relative to the inner product

$$\langle f, g\rangle = \int_{-1}^1 f(x)g(x)\,\mathrm{d}x,$$

by applying the Gram–Schmidt process to the basis $1, x, x^2, x^3$. (The resulting polynomials are the *Legendre polynomials,* except for normalization.)

(5) Do the same relative to the same space with the inner product

$$\langle f, g\rangle = \int_{-1}^1 \frac{f(x)g(x)}{\sqrt{1-x^2}}\,\mathrm{d}x.$$

(These are the *Čebyšev polynomials.*)

(6) Let V be a Euclidean or unitary space. Verify that the function $\|x\| = \langle x, x\rangle^{\frac{1}{2}}$ satisfies the following properties of a norm:

N. 1 $\|x\| \geqslant 0$ *with equality iff* $x = 0$,

N. 2 $\|\alpha x\| = |\alpha| \,.\, \|x\|$,

N. 3 $\|x+y\| \leqslant \|x\|+\|y\|$.

Moreover, it also satisfies the *parallelogram law*:

N. 4 $\|x+y\|^2+\|x-y\|^2 = 2(\|x\|^2+\|y\|^2)$.

Interpret this equation geometrically.

(7) Let $u_1, \ldots, u_n$ be an orthonormal set in a unitary space U. Given $x \in U$, determine scalars $\alpha_1, \ldots, \alpha_n$ such that $x - \Sigma\alpha_i u_i$ is orthogonal to x. Deduce that

$$\Sigma|\langle x, u_i\rangle|^2 \leqslant \|x\|^2,$$

with equality iff x is linearly dependent on the us. (This is known as *Bessel's inequality*.)

(8) Show that an orthonormal set $u_1, \ldots, u_n$ in a unitary space U forms a basis iff for any $x, y \in U$,

$$\langle x, y\rangle = \sum\langle x, u_i\rangle\langle u_i, y\rangle.$$

(This is known as *Parseval's identity*.)

(9) Prove that for any n vectors $x_1, \ldots, x_n$ in a unitary space

$$\det(\langle x_i, x_j\rangle) \geqslant 0,$$

with equality iff the xs are linearly dependent. (The determinant is known as *Gram's determinant*; it may be interpreted as the volume of the n-cell determined by the xs, cf. the cases $n = 2, 3$.)

8.4 Orthogonal and unitary matrices

In the study of vector spaces we found that the notion of a linear mapping was of fundamental importance. When our spaces are Euclidean or unitary, we shall naturally want to confine our attention to those linear mappings that preserve the norm. Thus we make the

DEFINITION A linear mapping $f: U \to V$ of Euclidean spaces is said to be an *orthogonal mapping* or an *orthogonal transformation* if it preserves the norm:

$$\|xf\| = \|x\| \qquad \text{for all } x \in U. \tag{1}$$

Similarly, a norm preserving transformation of unitary spaces is called *unitary*.

Thus when we wish to consider Euclidean spaces as a category, the morphisms will usually be orthogonal mappings and correspondingly for unitary spaces. As an example of an orthogonal transformation in $\mathbf{R}^2$ we may take a rotation; for an example of a unitary transformation in $\mathbf{C}^n$ consider multiplication by $e^{i\lambda}$ ($\lambda \in \mathbf{R}$).

To avoid repetition we shall in what follows concentrate on the unitary case and only briefly point out differences that arise in the real case. Formally we may regard Euclidean space and orthogonal transformations as the special case of the unitary situation where the ground field is restricted to be real, for here the operation of taking complex conjugates reduces to the identity.

Any unitary transformation $f\colon U \to V$ has zero kernel and hence is injective; moreover, it preserves the inner product, as we see by squaring and then polarizing both sides of the equation (1). Hence f transforms any orthonormal basis of U into an orthonormal set in V. A unitary isomorphism between two spaces is called an *isometry*, and two unitary spaces are said to be *isometric* if there is an isometry between them. What we have said shows that any unitary transformation between spaces of the same dimension must be an isometry; in particular, every unitary mapping of a space into itself is an isometry. Now the fact that every unitary space has an orthonormal basis may be expressed by saying that any n-dimensional unitary space is isometric to $\mathbf{C}^n$ and that any n-dimensional Euclidean space is isometric to $\mathbf{R}^n$, each taken with the standard norm.

The unitary mappings of a space into itself clearly form a group, called the *unitary group* of the space. If the space is n-dimensional, and hence isometric to $\mathbf{C}^n$, then its group is isomorphic to the unitary group of $\mathbf{C}^n$, which is usually written $\mathbf{U}_n(\mathbf{C})$. Similarly the orthogonal mappings of a Euclidean space into itself form a group, called the *orthogonal group*, and for an n-dimensional space this is isomorphic to $\mathbf{O}_n(\mathbf{R})$, the orthogonal group of $\mathbf{R}^n$.

What is the condition satisfied by the matrix of a unitary transformation? Let f be a unitary transformation of V into itself, and let its matrix, relative to an orthonormal basis of V, be $A = (a_{ij})$, then

$$v_i f = \sum a_{ij} v_j.$$

The unitarity condition states that

$$\begin{aligned}\delta_{ij} = \langle v_i f, v_j f\rangle &= \langle \textstyle\sum a_{ir} v_r, \sum a_{js} v_s\rangle \\ &= \textstyle\sum a_{ir} \bar{a}_{js} \langle v_r, v_s\rangle,\end{aligned}$$

hence

$$\sum a_{ir} \bar{a}_{jr} = \delta_{ij}. \tag{2}$$

In matrix language (3) can be restated as

$$AA^{\mathrm{H}} = I, \tag{3}$$

where $A^{\mathrm{H}} = \bar{A}^{\mathrm{T}}$ is the hermitian conjugate, as in **8.2**. Any matrix A satisfying (3) is said to be *unitary* and what we have shown can be expressed by saying that a linear mapping of a unitary space into itself is unitary iff its matrix, relative to some orthonormal basis, is unitary. Looking at (2), we can also express this condition by saying that the rows of A form an orthonormal set, in the standard norm. From (3) we see that A is invertible and A^{H} is its inverse.

Hence

$$A^{\mathrm{H}}A = I, \tag{4}$$

or, equivalently,

$$A^{\mathrm{T}}\bar{A} = I.$$

This states that the columns of A form an orthonormal set.

A real unitary matrix is one satisfying

$$AA^{\mathrm{T}} = I. \tag{5}$$

Such a matrix is called *orthogonal*, so that a linear mapping of a Euclidean space into itself is orthogonal iff its matrix, relative to some orthonormal basis, is orthogonal.

Let us return to (3) and take determinants; we obtain

$$\det A \,.\, \det \bar{A} = 1,$$

hence any unitary matrix has a determinant of absolute value 1; in particular, any orthogonal matrix has determinant ± 1.

We now look at the problem of classifying hermitian forms on unitary spaces (or quadratic forms on Euclidean spaces). In **8.2** we saw that hermitian forms on a complex vector space are completely determined by their rank and signature. Since the possibilities of changing the basis are more restricted in a unitary space, the rank and signature will still be invariant, but we expect to find other invariants as well. We shall find that every hermitian form on a unitary space can, by a unitary transformation, be diagonalized (i.e. changed to a form with diagonal matrix) and that the diagonal coefficients are invariants of the given form.

This problem has an important geometrical interpretation, at least in the real case. A quadratic form represents a conic (ellipse, hyperbola or parabola) in the plane or a quadric (ellipsoid, hyperboloid etc.) in 3-space and the fact that this form can be diagonalized means geometrically that every quadric in Euclidean space can be referred to a set of mutually orthogonal axes associated with the quadric, its *principal axes*, while the diagonal coefficients are related to the lengths of these axes. We shall not make use of this interpretation here (except to motivate the terminology).

Given a hermitian form h on a unitary space V, we seek an orthonormal basis of V for which h is reduced to diagonal form. If we fix an orthonormal basis in V, the matrix A of h will be hermitian and we therefore need a unitary matrix P such that PAP^{H} is a diagonal matrix, Λ say. Multiplying by P on the right and recalling that P satisfies (4), we find

$$PA = \Lambda P. \tag{6}$$

If the rows of P are $p_1, \ldots, p_n$ and the diagonal elements of Λ are $\lambda_1, \ldots, \lambda_n$, this equation may be rewritten as

$$p_i A = \lambda_i p_i. \tag{7}$$

In terms of the linear mapping f defined by A this states that $u_i f = \lambda_i u_i$, where $u_i \in V$ is the vector with coordinates p_i. Since P is unitary, the u_i form an orthonormal set.

Thus, given a hermitian matrix A, we must find a normalized row vector x such that for some $\lambda \in \mathbf{C}$,

$$xA = \lambda x. \tag{8}$$

For a given value of λ this is a homogeneous system of equations for x and we can obtain a normalized solution from any non-zero solution by normalization. The system has a non-zero solution iff the matrix of the system is singular, i.e.

$$\det(\lambda I - A) = 0. \tag{9}$$

This is a monic equation of degree n in λ; clearly we may consider it for any square matrix, not necessarily hermitian. It is called the *characteristic equation* of the matrix A and its left-hand side is the *characteristic polynomial.* The roots of (9) are called the *eigenvalues* of A and if λ is such a root, any solution x of (8) is called an *eigenvector* corresponding to λ. Thus the eigenvalues are found by solving (9) and for each eigenvalue we can obtain a non-zero eigenvector by solving (8). In order to find a unitary matrix P satisfying (6) we therefore need n eigenvectors of A forming an orthonormal system. We now show that for a hermitian matrix such a system always exists.

LEMMA 1 *The eigenvalues of a hermitian matrix A are all real; more precisely, the characteristic polynomial of an $n \times n$ hermitian matrix splits into n real linear factors.*

In the proof we shall use the fact that $\mathbf{C}$ is algebraically closed and so every polynomial of degree n in x splits into n linear factors $x - \lambda_i$, where the λ_i are the zeros of the given polynomial. Thus the second assertion of the lemma follows from the first, and we need only prove the latter.

Let c be an eigenvalue of A (possibly complex) and u a non-zero eigenvector, then $uA = cu$ and, multiplying by $u^{\mathrm{H}} = \bar{u}^{\mathrm{T}}$, we obtain

$$uAu^{\mathrm{H}} = cuu^{\mathrm{H}}.$$

Now $uu^{\mathrm{H}} = \sum |u_i|^2 > 0$, because $u \neq 0$, and $\overline{uAu^{\mathrm{H}}} = (uAu^{\mathrm{H}})^{\mathrm{H}} = uA^{\mathrm{H}}u^{\mathrm{H}} = uAu^{\mathrm{H}}$, because $A^{\mathrm{H}} = A$. This shows that uAu^{H} is real, and on dividing by uu^{H} we find that c is real. ■

We observe that this lemma shows in particular that a real symmetric (= real hermitian) matrix has only real eigenvalues.

LEMMA 2 *Two eigenvectors of a hermitian matrix corresponding to different eigenvalues are orthogonal.*

Proof. Let $uA = cu$, $vA = dv$, then, on transposing the second equation, we find $Av^{\mathrm{H}} = dv^{\mathrm{H}}$. Multiplying this equation by u on the left we get $uAv^{\mathrm{H}} =$

duv^{H}; if the first equation is multiplied by v^{H} on the right we get $uAv^{\mathrm{H}} = cuv^{\mathrm{H}}$; a comparison shows $(c-d)uv^{\mathrm{H}} = 0$. Hence, whenever $c \neq d$, we have $uv^{\mathrm{H}} = 0$. ■

LEMMA 3 *The eigenvectors corresponding to a fixed eigenvalue c of a hermitian matrix A form a space whose dimension equals the multiplicity of c as root of the characteristic equation.*

Proof. Put

$$V_c = \{x \in \mathbf{C}^n \mid xA = cx\},$$

then it is easily verified that V_c is a subspace of V; it is called the *eigenspace* for c. If $xA = cx$, then $(xA)A = c(xA)$, hence V_c is mapped into itself by A. Let U be the orthogonal complement of V_c: $U = V_c^{\perp}$, then

$$\mathbf{C}^n = U \oplus V_c; \tag{10}$$

we claim that A also maps U into itself: given $y \in U$, we have $xy^{\mathrm{H}} = 0$ for all $x \in V_c$, hence $x(yA)^{\mathrm{H}} = xA^{\mathrm{H}}y^{\mathrm{H}} = cxy^{\mathrm{H}} = 0$ and so $yA \in U$. Since U is orthogonal to V_c we can introduce an orthonormal basis in $\mathbf{C}^n$ adapted to the decomposition (10); relative to this basis, A takes on the form

$$A = \begin{pmatrix} A_1 & 0 \\ 0 & A_2 \end{pmatrix},$$

and it follows that $\det(xI - A) = \det(xI - A_1) \,.\, \det(xI - A_2)$. Now $A_2 = cI$, by the definition of V_c, so if $\dim V_c = m$, $\det(xI - A_2) = (x-c)^m$. On the other hand A_1 does not have c as eigenvalue, by the definition of U, hence m is the multiplicity of c as claimed. ■

It is now an easy matter to perform the diagonalization. Denote the different eigenvalues of A by $c_1, \ldots, c_r$ and write $\ker(c_iI - A) = V_i$, $\dim V_i = m_i$, so that V_i is the eigenspace and m_i the multiplicity for c_i. Take an orthonormal basis in each V_i, then the union of these bases is again orthonormal (by Lemma 2), with n elements and hence an orthonormal basis of V. If P is the matrix with these vectors as rows, then $PP^{\mathrm{H}} = I$ and $PA = \Lambda P$, where Λ is the diagonal matrix with the eigenvalues of A (according to their proper multiplicity) along the main diagonal. Thus we have proved

THEOREM 4 (Transformation to principal axes) *Let A be any hermitian matrix; then there is a unitary matrix P such that*

$$PAP^{\mathrm{H}} = \Lambda, \tag{11}$$

where Λ is a diagonal matrix with the eigenvalues of A (according to their multiplicity) along the diagonal. ■

In the real case A is symmetric, and its eigenvalues are again real, so P can be taken real, and is therefore orthogonal; now the equation (11) reads

$$PAP^{\mathrm{T}} = \Lambda. \tag{12}$$

It is clear that the eigenvalues of a hermitian matrix are invariant under unitary transformation, because $P(xI-A)P^{\mathrm{H}} = xI-PAP^{\mathrm{H}}$, hence we see that the eigenvalues of a hermitian matrix form a complete set of invariants under unitary transformations. Of course the rank and signature of a hermitian matrix can be expressed in terms of the eigenvalues.

From Th. 4 it is easy to deduce a seemingly more general result on the simultaneous reduction of two hermitian forms:

COROLLARY *Let A and B be two hermitian matrices, of which one, say A, is positive-definite. Then there is an invertible matrix Q such that*

$$QAQ^{\mathrm{H}} = I, \qquad QBQ^{\mathrm{H}} = \Lambda, \tag{13}$$

where Λ is a diagonal matrix whose diagonal elements are the roots of the equation

$$\det(xA-B) = 0. \tag{14}$$

This result follows by applying Th. 4 to the unitary space whose inner product is defined by A. In detail, by Th. 5, **8.2**, A is congruent to I, say

$$UAU^{\mathrm{H}} = I. \tag{15}$$

Write $UBU^{\mathrm{H}} = B'$ and apply Th. 4 to find a unitary matrix P such that $PB'P^{\mathrm{H}} = \Lambda$, where Λ is diagonal with the roots of $\det(xI-B') = 0$ along the diagonal. Since $xI-B' = xUAU^{\mathrm{H}}-UBU^{\mathrm{H}} = U(xA-B)U^{\mathrm{H}}$, this equation is equivalent to (14). If we now put $Q = PU$, then $QAQ^{\mathrm{H}} = PUAU^{\mathrm{H}}P^{\mathrm{H}} = PP^{\mathrm{H}} = I$ and $QBQ^{\mathrm{H}} = PB'P^{\mathrm{H}} = \Lambda$, i.e. (13). ∎

The equation (14) (and sometimes also the characteristic equation, which is the special case $A = I$) is sometimes called the *secular equation*, because it arises in the calculation of secular perturbations of planetary orbits. The result has many applications in physics and elsewhere, a typical one being in mechanics. A mechanical system is expressed in terms of general coordinates $q_1, \ldots, q_n$ and their velocities $\dot{q}_1, \ldots, \dot{q}_n$ (e.g. if the system consists of k particles, the qs might be their coordinates in 3-space; here $n = 3k$). The motion of the system is described by the kinetic energy T which is a positive-definite quadratic form in the $\dot{q}_i$, and the potential energy V. If $q_i = 0$ corresponds to an equilibrium position, then for small values of the q_i we can expand V in powers and products of the q_i. There are no linear terms because $\partial V/\partial q_i = 0$ at equilibrium, hence $V = \sum b_{ij}q_iq_j + \cdots$ (where dots indicate higher terms); thus in the first approximation we may treat V also as a quadratic form and the Cor. shows that in suitable coordinates T takes the form $T = \sum \dot{q}_i^2$ and $V = \sum c_iq_i^2 + \cdots$. Now the equations of motion (which express the fact that the total energy $T+V$ is constant) take on a particularly simple form, viz.

$$\ddot{q}_i + c_iq_i = 0 \qquad (i = 1, \ldots, n).$$

The resulting coordinates are called *normal coordinates*.

For another interpretation of Th. 4 let us rewrite equation (11) as

$$PAP^{-1} = \Lambda, \tag{16}$$

and interpret A as a linear mapping of V into itself (relative to a given orthonormal basis). We saw that this mapping preserves norms iff A is unitary and we now ask: what mappings have a hermitian matrix? Let

$$v_i f = \sum a_{ij} v_j,$$

then $\langle v_i f, v_k \rangle = \sum a_{ij} \langle v_j, v_k \rangle = a_{ik}$, $\langle v_i, v_k f \rangle = \sum \bar{a}_{kj} \langle v_i, v_j \rangle = \bar{a}_{ki}$, hence A is hermitian precisely when $\langle v_i f, v_k \rangle = \langle v_i, v_k f \rangle$, i.e. by linearity, iff

$$\langle xf, y \rangle = \langle x, yf \rangle \qquad \text{for all } x, y \in V. \tag{17}$$

A linear mapping f of a unitary space into itself satisfying (17) is said to be *hermitian* or *self-adjoint*. What we have shown can then be expressed by saying that a linear mapping of a unitary space is self-adjoint iff its matrix (relative to some orthonormal basis) is hermitian. If we identify V with its dual by letting $v \in V$ correspond to the linear form $x \mapsto \langle x, v \rangle$ then (17) just expresses the fact that f coincides with its adjoint mapping defined in **8.1**. Now Th. 4 may be restated as

THEOREM 4′ *Any self-adjoint mapping of a unitary space, $f\colon V \to V$ can, in a suitably chosen orthonormal basis, $u_1, \ldots, u_n$ of V, be written as*

$$u_i f = \lambda_i u_i,$$

where the λ_i are real and are the eigenvalues of f. ■

The set of eigenvalues of a linear mapping is frequently called its *spectrum* and the following version of Th. 4 is called the *spectral theorem*:

THEOREM 4″ *Let f be a self-adjoint mapping on a unitary space V, whose different eigenvalues are $\lambda_1, \ldots, \lambda_r$, with corresponding eigenspaces $V_1, \ldots, V_r$. Then*

$$V = V_1 \oplus \cdots \oplus V_r,$$

vectors in different V_is are orthogonal and dim V_i *is the multiplicity of λ_i.* ■

In this form the theorem is easily adaptable to infinite-dimensional generalizations.

There is an interesting application of Th. 4″ to functions of matrices. We recall (from Ex. (31), Further Exercises Ch. **6**) that for any distinct $\lambda_1, \ldots, \lambda_r$ and any $\mu_1, \ldots, \mu_r$ there exists a polynomial f of degree less than r such that $f(\lambda_i) = \mu_i$, namely

$$f(x) = \sum \mu_i \frac{p_i(x)}{p_i(\lambda_i)} \qquad \text{where } p_i = \prod_{j \neq i} (x - \lambda_j).$$

Of course there will be many such polynomials once we lift the restriction on the degree.

THEOREM 5 *Let A be a hermitian matrix whose different eigenvalues are $\lambda_1, \ldots, \lambda_r$. Given any real numbers $\mu_1, \ldots, \mu_r$ and any real polynomial f such that $f(\lambda_i) = \mu_i$, the matrix $f(A)$ is hermitian with eigenvalues $\mu_1, \ldots, \mu_r$, where μ_i has the same multiplicity as λ_i.*

Proof. Choose an orthonormal basis of eigenvectors for A. If u is any eigenvector belonging to the eigenvalue λ_1 say, then $Au = \lambda_1 u$, hence $f(A)u = f(\lambda_1)u = \mu_1 u$. This shows that relative to the given basis, $f(A)$ is diagonal with elements $\mu_1, \ldots, \mu_r$ along the diagonal, with the appropriate multiplicities. Hence $f(A)$ is hermitian, with the stated eigenvalues. ∎

This result can of course be applied to find the eigenvalues of polynomial functions of matrices, but there are other applications. E.g., let A be a hermitian matrix which is positive-semidefinite, thus its eigenvalues $\lambda_1, \ldots, \lambda_r$ are non-negative. Put $\mu_i = \lambda_i^{\frac{1}{2}}$, and let f be a polynomial such that $f(\lambda_i) = \mu_i$. By Th. 5, $B = f(A)$ is a hermitian matrix with eigenvalues μ_i. More precisely, if u is an eigenvector of A for λ_1, thus $Au = \lambda_1 u$, then $Bu = \mu_1 u$, and hence $B^2 u = \mu_1^2 u = \lambda_1 u = Au$. This shows that B^2 and A agree on all eigenvectors of A and so are equal:

$$A = B^2.$$

We have therefore expressed a square root of A as a polynomial in A. Now the μ_i were chosen non-negative and B is therefore positive-semidefinite. We claim that A has exactly one positive-semidefinite square root; for we have seen that B is such a square root and, if C is another, let us take a basis of eigenvectors of C. Since $B = f(A) = f(C^2)$, they are also eigenvectors of B and, if $Cu = \mu_1 u$ say, then $Au = \lambda_1 u$, hence $Bu = \mu_1 u = Cu$. This holds for all eigenvectors of C, therefore $B = C$. We sum up these results as

THEOREM 6 (Square root lemma) *Let A be a positive-semidefinite hermitian matrix. Then there exists exactly one positive-semidefinite hermitian matrix B such that $B^2 = A$. Moreover, B can be expressed as a polynomial in A.* ∎

Exercises

(1) Show that the eigenvalues of a unitary matrix have absolute value 1.

(2) In any category a morphism f is called a *monomorphism* or *monic* if $\alpha f = \beta f$ implies $\alpha = \beta$. Verify that in the category of unitary spaces and unitary mappings every morphism is monic.

(3) An orthogonal transformation is said to be *proper* if its determinant is 1. Show that the set $\mathbf{O}_n^+(\mathbf{R})$ of proper orthogonal transformations is a subgroup of index 2 in the orthogonal group.

(4) Show that every proper orthogonal transformation in $\mathbf{R}^3$ has 1 as eigenvalue. Deduce that every orthogonal transformation in $\mathbf{R}^3$ leaves a line pointwise fixed.

Show also that every proper real orthogonal 2×2 matrix has the form

$$\begin{pmatrix} \cos\alpha & -\sin\alpha \\ \sin\alpha & \cos\alpha \end{pmatrix}.$$

(5) Show that in Euclidean 2-space every conic of the form $\sum a_{ij}x_i x_j = 1$, where the left-hand side is positive-definite, can be transformed to the form $x_1^2/a_1^2 + x_2^2/a_2^2 = 1$. Interpret a_1, a_2.

(6) For any matrix A, if $Ax = \lambda x$ and f is any polynomial, then $f(A)x = f(\lambda)x$. Deduce that every hermitian matrix satisfies its characteristic equation (this is the Cayley–Hamilton theorem, and is in fact true for all square matrices, cf. **11.2**).

(7) Express the rank and signature of a hermitian matrix in terms of its eigenvalues.

(8) Show that the hermitian matrices of order n form a real subspace of $\mathbf{C}_n$. If A, B are hermitian matrices, show that $AB+BA$ and $i(AB-BA)$ are again hermitian; give an example to show that AB need not be hermitian.

(9) Find a coordinate system in which $2x_1^2+3x_2^2+2x_1x_3+2x_3^2$ and $2x_2^2+x_3^2-4x_1x_2-2x_1x_3-6x_2x_3$ are simultaneously transformed to diagonal form.

(10) In the matrix $\begin{pmatrix} \frac{1}{3} & \frac{2}{3} & * \\ * & \frac{1}{3} & * \\ * & * & \frac{1}{3} \end{pmatrix}$ find all ways of replacing the asterisks by numbers so as to obtain an orthogonal matrix. Find the determinant.

(11) Find a simultaneous reduction to diagonal form for $5x_1^2+x_2^2+5x_3^2+8x_1x_3-2x_1x_2$ and $6x_1^2+2x_2^2+3x_3^2+8x_1x_3-4x_1x_2$.

(12) Show that $x_1^2+3x_2^2+6x_3^2+4x_1x_2+2x_2x_3$ is not positive-definite and find values of x_1, x_2, x_3 for which the form takes negative values.

(13) By considering the secular equation for $x_1^2-x_2^2$ and x_1x_2 show that these two forms cannot be simultaneously transformed to diagonal form.

(14) Find the characteristic equation for the following matrices and hence a basis of eigenvectors:

$$\text{(i)}\ \begin{pmatrix} 3 & -2 & 1 \\ -2 & 6 & -2 \\ 1 & -2 & 3 \end{pmatrix} \qquad \text{(ii)}\ \begin{pmatrix} 0 & 1 & 1 \\ 1 & 0 & 1 \\ 1 & 1 & 0 \end{pmatrix}$$

(15) If $A \in \mathbf{C}_n$ is non-singular, show that $A^{\mathrm{H}}A$ is hermitian positive-definite. By writing $A^{\mathrm{H}}A = P^2$, where P is positive-definite, show that every non-singular matrix A can be written in the form $A = UP$, where U is unitary and P positive-definite hermitian (this is the *polar decomposition* of A, corresponding to the expression $c = e^{i\lambda}r$ for complex numbers).

(16)* If $A \in \mathbf{C}_n$ is non-singular, find unitary matrices U, V such that $UAV = D$, where D is diagonal, with diagonal elements the positive square roots of the eigenvalues of AA^{H}. (Hint. Find U from the equation $UAA^{\mathrm{H}}U^{\mathrm{H}} = D^2$.)

8.5 Alternating forms

A bilinear form g on a vector space V is said to be *alternating* if

$$g(x, x) = 0 \qquad \text{for all } x \in V. \tag{1}$$

By polarization we find that

$$g(x, y)+g(y, x) = g(x+y, x+y)-g(x, x)-g(y, y) = 0,$$

hence every alternating bilinear form is also *antisymmetric*, i.e.

$$g(x, y) = -g(y, x). \tag{2}$$

Conversely, if g is an antisymmetric form, we have on putting $y = x$ in (2), $2g(x, x) = 0$. If the ground field has characteristic not 2, we conclude that g must be alternating; thus in characteristic not 2 there is no need to distinguish between alternating and antisymmetric forms. But in characteristic 2 this reasoning does not apply and, since we do not wish to restrict the field, we shall make the stronger assumption (1).

Relative to a basis $v_1, \ldots, v_n$ of V, an alternating form g is specified by its matrix $A = (a_{ij})$, where

$$a_{ij} = g(v_i, v_j).$$

From (1) and (2) we see that $a_{ii} = 0$, $a_{ji} = -a_{ij}$; a matrix satisfying these conditions is called *alternating* or *skew-symmetric*. We note that in characteristic not 2, a matrix A is alternating iff $A^{\mathrm{T}} = -A$.

The reduction of alternating forms is particularly simple: the rank is a complete system of invariants.

THEOREM 1 *Given an alternating form g on a space V (over any field), the form has even rank $2r$, say, and in a suitably chosen coordinate system its matrix is*

$$\left(\begin{array}{cc|c} 0 & I & \\ -I & 0 & 0 \\ \hline \multicolumn{2}{c|}{0} & 0 \end{array}\right). \tag{3}$$

Proof. We shall show that V has a basis $u_1, \ldots, u_r, v_1, \ldots, v_r, w_1, \ldots, w_s$, where $\dim V = 2r+s$, rank $g = 2r$ and

$$g(u_i, v_i) = -g(v_i, u_i) = 1,$$

while all other values are 0. Clearly this will prove the theorem.

If $g = 0$, then $r = 0$ and there is nothing to prove. Otherwise take $x, y \in V$ such that $g(x, y) \neq 0$, then on dividing x or y by $g(x, y)$ we obtain vectors u_1, v_1 such that $g(u_1, v_1) = 1$. We observe that u_1 and v_1 must be linearly independent, for if $v_1 = \lambda u_1$ say, then $g(u_1, v_1) = \lambda g(u_1, u_1) = 0$, a contradiction. Let V_1 be the space spanned by u_1 and v_1 and put

$$U = \{z \in V \mid g(z, u_1) = g(z, v_1) = 0\}.$$

Clearly U is a subspace of V and any $x \in V$ can be written as

$$x = g(x, v_1)u_1 + g(u_1, x)v_1 + x',$$

for a unique vector x'. It is easily verified that $x' \in U$, from which it follows that $V = V_1 \oplus U$; moreover, any vector in U is orthogonal to any vector in V_1. Thus $\dim U = \dim V - 2$, and now the result follows by induction on $\dim V$. ∎

COROLLARY *If g is a non-singular alternating form on a finite-dimensional space V, then V is even-dimensional, with basis $u_1, \ldots, u_r, v_1, \ldots, v_r$ such that $g(u_i, v_i) = -g(v_i, u_i) = 1$, while all other products are zero.* ∎

In terms of matrices we can interpret the result by saying that every alternating matrix is congruent to one of the form (3). In particular, if A is a non-singular alternating $n \times n$ matrix, then n is even and there is a matrix P such that

$$A = PJP^{\mathrm{T}}, \qquad \text{where } J = \begin{pmatrix} 0 & I \\ -I & 0 \end{pmatrix} \tag{4}$$

Taking determinants and observing that $\det J = 1$, we find that

$$\det A = (\det P)^2. \tag{5}$$

In particular we may for A take the matrix with $a_{ij} = t_{ij}$ $(i < j)$ where the t_{ij} are $\binom{n}{2}$ independent indeterminates over the rational field and $a_{ii} = 0$, $a_{ji} = -a_{ij}$ $(i < j)$. We then have equation (5) with $F = \mathbf{Q}(t_{12}, t_{13}, \ldots, t_{n-1n})$ as our ground field. Hence in this case, $\det P = f/g$, where f, g are polynomials in the ts with rational coefficients, and we may assume that f and g are coprime. By (5), $\det A = f^2/g^2$, i.e. $f^2 = g^2 \,.\, \det A$, and by unique factorization in a polynomial ring over a field (cf. Ex. (7), **6.7**) $g \mid f$, so on replacing f by f/g, we may write $\det A = f^2$, where f is a polynomial in the ts. This polynomial is determined up to sign, which we may fix so that f reduces to 1 when $A = J$. The polynomial determined in this way is called the *Pfaffian* of order $n = 2r$, and denoted by $Pf\, A$, or Pf_r. E.g., for $r = 1, 2$ we have $Pf_1 = t_{12}$, $Pf_2 = -(t_{12}t_{34} + t_{13}t_{42} + t_{14}t_{23})$.

By a *symplectic space* we understand a finite-dimensional vector space with a non-singular alternating form defined on it. By the Cor. to Th. 1 every symplectic space is even-dimensional; a basis of the form described in the Cor. is said to be *symplectic*. A linear mapping between symplectic spaces which preserves the value of the alternating form is called a *symplectic mapping*; when it is also an isomorphism we speak of an *isometry*. From the Cor. it is clear that any two symplectic spaces of the same dimension are isometric.

The isometries of a symplectic space of dimension $2m$ over a field F form a group, called the *symplectic group* and denoted by $\mathbf{Sp}_{2m}(F)$. Thus $\mathbf{Sp}_{2m}(F)$ may be taken to consist of all matrices P transforming a given symplectic basis into another such basis. They correspond to matrices P such that

$$PJP^{\mathrm{T}} = J. \tag{6}$$

Any matrix P satisfying this condition is said to be *symplectic*.

Let A be an alternating matrix, so that $\det A = (Pf\,A)^2$, and let P be any matrix with indeterminate coefficients p_{ij}, then

$$\det(PAP^{\mathrm{T}}) = (\det P)^2 \det A,$$

hence by taking square roots on both sides, we find

$$Pf(PAP^{\mathrm{T}}) = \delta \,.\, \det P \,.\, Pf\,A,$$

where $\delta = \pm 1$. This is a polynomial identity in the p_{ij}; taking $P = I$, we see that $\delta = 1$. Hence we obtain the identity

$$Pf(PAP^{\mathrm{T}}) = \det P \,.\, Pf(A), \tag{7}$$

valid for any alternating matrix A and any P.

In particular, let P be a symplectic matrix and take $A = J$ in (7), then the equation reduces to $\det P = 1$. Hence we have proved

THEOREM 2 *Every symplectic matrix has determinant* 1. ■

A geometric interpretation of alternating forms may be given as follows. Let V be an n-dimensional space and V^* its dual (cf. **8.1**). Then V^* is again n-dimensional and, relative to a given pair of dual bases $u_1, \ldots, u_n$ of V and $\alpha_1, \ldots, \alpha_n$ of V^*, an $n \times n$ matrix A may be taken to define a linear mapping $f: V \to V^*$:

$$u_i f = \sum a_{ij}\alpha_j. \tag{8}$$

The adjoint f^* of f is a mapping from V^{**} to V^*, i.e. if we identify V^{**} and V according to the natural isomorphism, f^* is again from V to V^*. It is given by the equation

$$\langle u_i, u_j f\rangle = \langle u_j, u_i f^*\rangle,$$

so its matrix is the transpose of A:

$$u_i f^* = \sum a_{ji}\alpha_j. \tag{9}$$

Projective geometry is concerned with 1-dimensional subspaces of V and V^* (points and primes, i.e. hyperplanes, respectively). A mapping of the form (8) or (9) is called a *correlation*. Now (8) and (9) represent the same correlation (i.e. the same mapping of 1-dimensional subspaces) iff

$$A^{\mathrm{T}} = \lambda A \qquad \text{for some scalar } \lambda. \tag{10}$$

The transpose of this equation is $A = \lambda A^{\mathrm{T}}$ and, combining this with (10) itself, we obtain $\lambda^2 A = \lambda A^{\mathrm{T}} = A$; hence if we exclude the trivial case $A = 0$, we have $\lambda^2 = 1$, i.e. $\lambda = \pm 1$.

Thus (in characteristic not 2) there are two cases. If $\lambda = 1$, A is symmetric; it then represents a quadric (i.e. hypersurface defined by an equation of degree 2) and (8) is the mapping which associates with each point a prime, its *polar* relative to the quadric defined by A. In this case A is called a *polarity*.

Secondly, if $\lambda = -1$, f is called a *null-system*: each point lies on its transform under f, a fact expressed by the equation

$$\langle x, xf\rangle = 0,$$

where $\langle , \rangle$ is the pairing between V and V^*. The results found earlier show that a non-singular null-system exists only in odd dimensions, because n-dimensional projective space consists of the 1-dimensional subspaces of F^{n+1}. For example, every null-system in the plane is singular, but there is a non-singular null-system in 3-space.

Exercises

(1) Let V be a vector space and $\alpha, \beta \in V^*$. Show that

$$f(x, y) = \langle x, \alpha \rangle \langle y, \beta \rangle - \langle x, \beta \rangle \langle y, \alpha \rangle$$

is an alternating form on V and find the possible values of the rank.

(2) Let A be a hermitian positive-definite matrix of even order, and J as in (4). Show that $\det(xJ - A) = 0$ has no real roots.

(3) Show that J is proper orthogonal; deduce that J is also symplectic.

(4) If A is any square matrix, show that

$$Pf\begin{pmatrix} 0 & A \\ -A^{\mathrm{T}} & 0 \end{pmatrix} = \det A.$$

(5) Show that every matrix is congruent to a solution of $X^{\mathrm{T}} = X + H$, where H is of the form (3). Express the rank of H in terms of the given matrix.

(6) Find all triangular symplectic 4×4 matrices. (Hint. Find the diagonal ones first.)

(7) Show that every symplectic space of dimension $2m$ has an m-dimensional subspace on which the alternating form is identically zero. Given an n-dimensional space V with an alternating form of rank $2r$, show how to construct a symplectic space U containing V as subspace such that the given form on V is induced by that of U. What is the least dimension of U?

(8) Show that $\mathbf{Sp}_{2m}(F)$ admits the operation $P \mapsto (P^{-1})^{\mathrm{T}}$. Verify that this is an automorphism.

Further exercises on Chapter 8

(1) Let U, V be finite-dimensional vector spaces. If $\langle , \rangle$ is a bilinear form on $U \times V$ and U', V' are subspaces of U, V respectively, show that $\langle , \rangle$ defines a bilinear form on $U' \times V'$ by restriction. If U' is chosen to be a complement of ${}^{\perp}V$ in U, and V' a complement of $U^{\perp}$ in V, show that U', V' are maximal subspaces of U, V such that the restriction of $\langle , \rangle$ to $U' \times V'$ is a pairing. Show also that $\dim U' = \dim V'$ and the common value is the rank of $\langle , \rangle$.

(2) For two finite-dimensional spaces U, V verify that $\mathrm{Bil}(U, V) \cong \mathrm{Hom}(U, V)$, but that the isomorphism depends on the choice of bases. (Hint. Compare the transformation laws for the representing matrices.)

(3) Given finite-dimensional spaces U, V, show that $\mathrm{Hom}(U, V) \cong \mathrm{Bil}(U, V^*)$ and prove that this isomorphism is natural in U and V.

(4) If V is an infinite-dimensional space with basis $v_1, v_2, \ldots$ define a bilinear form on $V \times V$ by $\langle v_i, v_j \rangle = 1$ if $j = i+1$ and 0 otherwise. Show that $V^\perp \neq {}^\perp V$.

(5) If $\langle , \rangle$ is any bilinear form on V, show that for any $x, y, z \in V$, x and $\langle x, y\rangle z - \langle x, z\rangle y$ are orthogonal.

(6) Let $\langle , \rangle$ be a non-singular bilinear form on V such that for all $x, y \in V$, $\langle x, y\rangle = 0 \Rightarrow \langle y, x\rangle = 0$. Show that $\langle x, y\rangle\langle z, x\rangle = \langle y, x\rangle\langle x, z\rangle$, and deduce that for any $x, y \in V$, either $\langle x, y\rangle = \langle y, x\rangle$ or $\langle x, x\rangle = \langle y, y\rangle = 0$. If $\langle x, x\rangle \neq 0$ for some $x \in V$, show that $\langle , \rangle$ is symmetric, hence $\langle , \rangle$ is either symmetric or alternating.

(7) Let $\langle , \rangle$ be the pairing between an n-dimensional space V and its dual. Given $a \in V$ and $\alpha \in V^*$ such that $\langle a, \alpha\rangle = 0$, but $a, \alpha \neq 0$, show that the mapping

$$\tau: x \mapsto x + \langle x, \alpha\rangle a$$

is linear and leaves an $(n-1)$-dimensional subspace H of V elementwise fixed (τ is called a *transvection* in the direction a in the hyperplane determined by α). Show that the elementary matrix operations of type γ defined in **5.2** correspond to transvections and that in suitable coordinates every transvection can be represented by such a matrix.

(8) For any 2×2 matrix A show that $a_{12} - a_{21}$ is an invariant under congruence transformations of determinant 1.

(9) Let $A = \begin{pmatrix} a & h \\ h & b \end{pmatrix}$ be a matrix over $\mathbf{Z}$, positive-definite over $\mathbf{Q}$; if $-a < 2h \leqslant a \leqslant b$, A is said to be *reduced*. Show that every positive-definite matrix of order 2 is congruent to exactly one reduced matrix. (Hint. Transform A so as to minimize the (1, 2)-entry.)

(10) If $A = ((r, s))$, $1 \leqslant r, s \leqslant n$, where (r, s) is the HCF of r and s, show that A is congruent to $\mathrm{diag}(\varphi(1), \varphi(2), \ldots, \varphi(n))$, where φ is the Euler function. (H. J. S. Smith. Hint. Use $P = (p_{ij})$, where $p_{ij} = 1$ if $j \mid i$ and 0 otherwise, and recall Ex. (16), **2.3**.)

(11) Show that a real skew-symmetric matrix does not have -1 as eigenvalue. If S is any skew-symmetric matrix such that $I+S$ is non-singular, show that $U = (I-S)(I+S)^{-1}$ is proper orthogonal. Verify that every proper orthogonal matrix which does not have -1 as eigenvalue can be expressed in this way. (A. Cayley)

(12) Show that every orthogonal matrix A of odd order has $\det A$ as eigenvalue.

(13) If P, Q are proper orthogonal matrices of order 3, show that $P+Q$ cannot be of rank 2.

(14) If A is hermitian positive-definite, show that $A = PP^H$, where P is triangular. (O. Toeplitz)

(15) Show that any non-singular matrix over **C** can be written as UP, where U is unitary and P is triangular. (Hint. Use Ex. (14) and Ex. (15), **8.4**.)

(16) Let A be a square matrix over **C**, with columns $a_1, \ldots, a_n$. By applying Ex. (15), show that $|\det A|^2 \leqslant \prod a_i^{\mathrm{H}} a_i$. (J. Hadamard. Geometrically this means that a parallelepiped with edges of given lengths has maximum volume when the edges meeting at a point are perpendicular.)

(17)* A symmetric matrix of rank r is said to be *regularly arranged* if no two consecutive leading minors of rank less than r are **0**. Show that any symmetric matrix can be regularly arranged by permuting the rows (and permuting the columns correspondingly); give an example to show that this does not remain true if 'two consecutive' is omitted in the above definition. In a real symmetric regularly arranged matrix, if one of the leading minors of rank less than r is zero, show that the adjacent leading minors have opposite signs.

(18)* Show that the number of negative eigenvalues of a real symmetric regularly arranged matrix is equal to the number of sign changes in the sequence of leading minors. (Frobenius)

(19) If A is a real symmetric matrix, show that there exists an integer m such that $mI+A$ is positive-definite. Find the least m when $A = \begin{pmatrix} -10 & 5 & 2 \\ 5 & 0 & 3 \\ 2 & 3 & 6 \end{pmatrix}$.

(20) Find an orthogonal transformation of $x_1^2+x_2^2+x_3^2-x_1x_2-x_1x_3-x_2x_3$ to diagonal form.

(21) Let A, B be square matrices satisfying $AB+BA = B$. Show that $AB^2 = B^2A$. If B is non-singular, show that, for each eigenvalue λ of A, $1-\lambda$ is also an eigenvalue and that, if the eigenvalues of A are all distinct, B can be chosen so that $B^2 = I$. Find such a matrix B when

$$A = \begin{pmatrix} 0 & 0 & 0 \\ 0 & \frac{1}{2} & 0 \\ 0 & 0 & 1 \end{pmatrix}.$$

(22) Let A be a square matrix. Show that A is skew-symmetric iff $xAx^{\mathrm{T}} = 0$ for all rows x.

(23) Let h be a hermitian form on a complex vector space V and define $V_0 = \{x \in V \mid h(x, x) = 0\}$. Show that V_0 is a subspace of V. Show further that $h(x, y) = 0$ for all $x \in V$, $y \in V_0$.

(24) Show that $\mathbf{Sp}_2(F)$ consists of all matrices with determinant 1. If $A \in \mathbf{Sp}_4(F)$ and if in 2×2 block form $A = \begin{pmatrix} P & PQ^2 \\ R & S \end{pmatrix}$, where Q is non-singular, show how to reduce A by right multiplication in $\mathbf{Sp}_4(F)$ to the form where the (1, 2)-block is 0 and hence describe all matrices A of the given form.

9

Further group theory

9.1 The isomorphism theorems

Let G and H be any groups and let

$$f: G \to H \tag{1}$$

be a homomorphism between them. By definition this means that f is a mapping satisfying $(xy)f = xf \,.\, yf$ and it follows that $1f = 1$ and $(xf)^{-1} = (x^{-1})f$.

It is clear that the image of (1), $\operatorname{im} f$ or Gf, is a subgroup of H. For, given $x', y' \in \operatorname{im} f$, say $x' = xf$, $y' = yf$ (where $x, y \in G$), then $x'y' = (xy)f \in \operatorname{im} f$, hence $\operatorname{im} f$ is closed under multiplication; the other properties are proved similarly.

Consider the kernel of f; by definition this is the inverse image of 1_H under f:

$$\ker f = \{x \in G \mid xf = 1\}.$$

This is a subgroup of G, for $1f = 1$, and if $xf = yf = 1$, then $(xy^{-1})f = (xf)(yf)^{-1} = 1$. The kernel has a further property, but to state it we need a definition.

Given a subset X of a group G, an element $c \in G$ is said to *normalize* X if

$$c^{-1}Xc = X. \tag{2}$$

Of course this does not mean that $c^{-1}xc = x$ for all $x \in X$, but only that $c^{-1}xc \in X$ and $x \in c^{-1}Xc$ for each $x \in X$. A set C is said to normalize X if each $c \in C$ normalizes X. The set of all elements of G which normalize X is called the *normalizer* of X, and if the whole group G normalizes X we also say: X is *normal* in G.

We claim that the kernel of a homomorphism is always normal in G. For if $a \in \ker f$ and $c \in G$, then $af = 1$ and so $(c^{-1}ac)f = (cf)^{-1}cf = 1$. This then shows that the kernel of any homomorphism is a normal subgroup of G.

We shall find that every normal subgroup occurs as the kernel of some homomorphism. Let N be a normal subgroup of G; this is often abbreviated by writing $N \lhd G$. We have $c^{-1}Nc = N$ for all $c \in G$, hence

$$Nc = cN \qquad \text{for all } c \in G, \tag{3}$$

i.e. every left coset of N is a right coset and vice versa. This condition (3) is sufficient as well as necessary for N to be normal in G. Let us write G/N for the collection of all cosets of N and define a multiplication of cosets as follows: every coset has the form aN ($a \in G$) and $aN = a'N$ iff $aa'^{-1} \in N$. We put

$$aN \,.\, bN = abN. \tag{4}$$

At first sight the coset on the right appears to depend on a and b. But if we replace b by another element in its coset, say b', then $bN = b'N$ and hence $abN = ab'N$. Similarly, if we replace a by a', where $aN = a'N$, then $Nab = Na'b$, i.e. $abN = a'bN$. This shows that abN in fact depends only on the cosets aN, bN and not on the choice of a and b within these cosets. We have therefore defined a multiplication of cosets. This multiplication is associative:

$$(aN \,.\, bN)cN = abN \,.\, cN = abcN = aN \,.\, bcN = aN(bN \,.\, cN),$$

there is a neutral element, namely N, and each coset aN has the inverse $a^{-1}N$, as is easily verified.

Thus the set G/N of cosets is a group under the multiplication (4). Moreover, as (4) shows, the mapping

$$a \mapsto aN \tag{5}$$

from G to G/N is a homomorphism, called the *natural* homomorphism and sometimes abbreviated nat. Its kernel, not surprisingly, is N. What we have shown above can be summed up as

THEOREM 1 *Given a homomorphism of groups, $f\colon G \to H$, its image is a subgroup of H and its kernel is a normal subgroup of G. Conversely, given any normal subgroup N of G, a group structure can be defined on the set of cosets G/N in such a way that the mapping* (5) *from G to G/N is a homomorphism with kernel N.* ■

The group G/N defined in the proof is called the *quotient group* of G by the normal subgroup N; its elements may be represented by a transversal of N in G. We note that in an abelian group, *every* subgroup is normal. To give an example, for any $n \in \mathbf{Z}$, $n\mathbf{Z}$ is a normal subgroup of $\mathbf{Z}$ and if $n > 1$, $\mathbf{Z}/n\mathbf{Z}$ is the cyclic group of order n: $\mathbf{Z}/n\mathbf{Z} \cong \mathbf{Z}/n$ (as additive groups). We note that the above construction of G/N corresponds precisely to the way in which the integers mod n were obtained in Ch. **2**.

A useful consequence of Th. 1 is the *factor theorem*:

THEOREM 2 *Given a homomorphism of groups $f\colon G \to H$ and a normal subgroup N of G such that $N \subseteq \ker f$, there is a unique mapping $f'\colon G/N \to H$ such that the triangle*

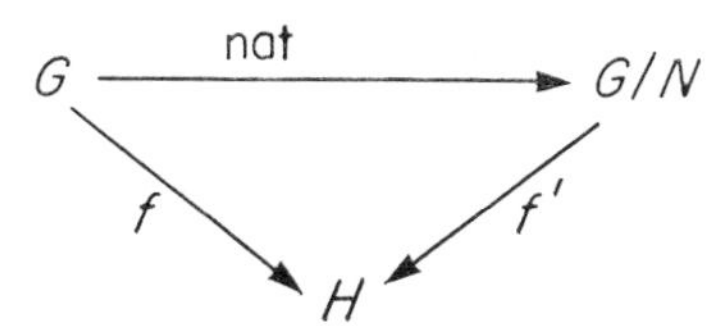

commutes. Moreover, f' is a homomorphism, which is injective if and only if $N = \ker f$.

Proof. If f' exists at all, it must satisfy

$$(aN)f' = af, \tag{6}$$

so that there can be at most one such mapping. In fact, a mapping f' satisfying (6) exists, for af is independent of the choice of a in its coset: if $aN = a'N$, then $aa'^{-1} \in N \subseteq \ker f$ (by hypothesis) and so $af = a'f$. Thus the mapping f' given by (6) is well-defined. It is a homomorphism because $(abN)f' = (ab)f = af \,.\, bf = (aN)f' \,.\, (bN)f'$. The cosets mapped to 1 by f are precisely the cosets of N in $\ker f$, hence f' is injective iff $\ker f = N$. ■

This result may be expressed by saying that G/N with the natural homomorphism is universal for homomorphisms from G which map N to **1**. As the solution to a universal problem G/N with its natural mapping is therefore determined up to isomorphism by the property of Th. 2.

If we apply Th. 2 to an arbitrary homomorphism f, with $N = \ker f$, we obtain an injective homomorphism $G/\ker f \to H$. Thus we have an isomorphism between $G/\ker f$ and $\operatorname{im} f$ and this provides an analysis of group homomorphisms:

THEOREM 3 (First isomorphism theorem) *Given any homomorphism $f\colon G \to H$, there is a factorization $f = \alpha f_1 \beta$,*

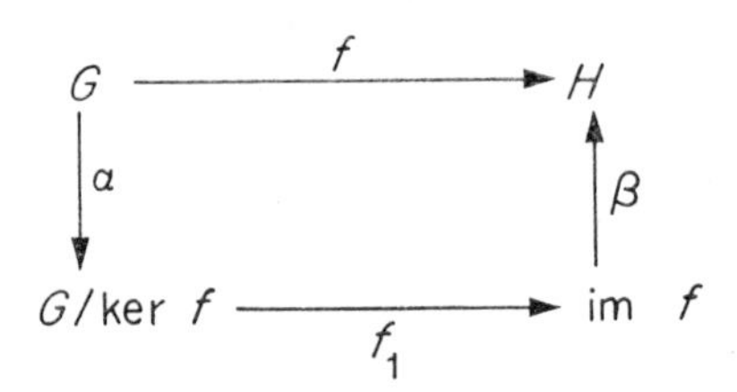

where $\alpha\colon G \to G/\ker f$ is the natural homomorphism, $\beta\colon \operatorname{im} f \to H$ the inclusion mapping, and $f_1\colon G/\ker f \to \operatorname{im} f$ an isomorphism. ■

As an illustration take the mapping $\sigma \mapsto \operatorname{sgn} \sigma$, which assigns to each permutation its sign. This is a homomorphism $\mathrm{Sym}_n \to \mathbf{C}_2$ with kernel Alt_n.

Every group G has itself and **1** as normal subgroups, hence by Th. 1, it has the trivial group and itself as quotients. If there are no other quotients and G is non-trivial, G is said to be *simple*. E.g., every group G of prime order p must be simple, for any subgroup must have order dividing p (by Lagrange's theorem), i.e. p or 1, so there are no subgroups other than G or **1**. As is easily seen, any abelian simple group must be of this form. However, there are many non-abelian simple groups, e.g. Alt_n is simple for $n > 4$, as we shall see in **9.5**.

We recall from Ch. **3** that if in a group G we take a subset X, then $gp\{X\}$, the subgroup generated by X, consists of all products

$$a_1a_2 \ldots a_n \qquad \text{where } a_i \in X \cup X^{-1}, \quad \text{and } n \geqslant 0.$$

If $X = H \cup K$, where H and K are subgroups of G, then $X^{-1} = X$ and so in this case $gp\{H \cup K\}$, also written $gp\{H, K\}$ and called the *join* of H and K, consists of all products $a_1 \ldots a_n$ with $a_i \in H \cup K$. In particular,

$$HK \subseteq gp\{H, K\}; \tag{7}$$

here equality need not hold, e.g. Sym_4 is generated by (1 2) and (1 2 3 4). The subgroups A, B generated by these cycles respectively have orders 2 and 4, hence AB has at most 8 elements, whereas $gp\{A, B\} = \text{Sym}_4$ has 24 elements.

Since HK always contains H and K, equality holds in (7) iff HK is a subgroup, and this is so precisely when

$$HK = KH. \tag{8}$$

For when (8) holds, then $HK \,.\, HK = HHKK = HK$ and $(HK)^{-1} = KH = HK$, hence HK is a subgroup. Conversely, if HK is a subgroup, then $HK = (HK)^{-1} = KH$, i.e. (8).

Condition (8) and with it equality in (7) holds in particular if one of H, K is normal in G. In this case there is an important relation between quotient groups, described in

THEOREM 4 (Second isomorphism theorem or parallelogram rule) *Let G be a group and H, K subgroups of G, where $K \lhd G$. Then $H \cap K \lhd H$ and there is an isomorphism*

$$H/(H \cap K) \cong HK/K.$$

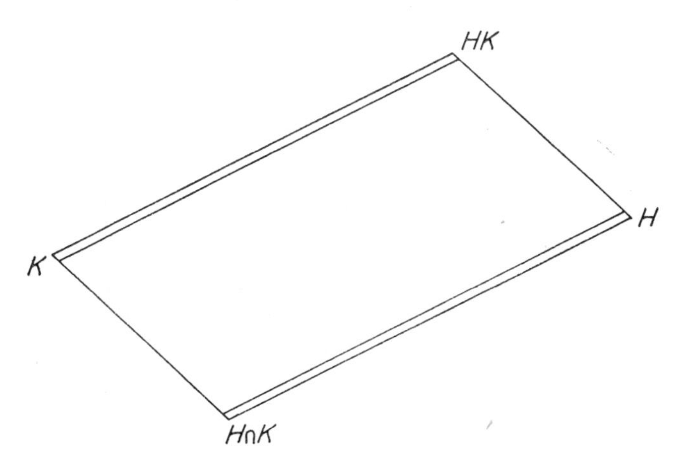

Figure 7

The result may be illustrated by the type of diagram used for vector spaces in **4.4**. In the parallelogram shown in Fig. 7 each vertex represents a subgroup and each side rises from one subgroup to another containing it. Moreover, HK is the least subgroup containing H and K, represented by the vertices below it, and $H \cap K$ is the greatest subgroup contained in H and K, represented by the vertices above it. Finally, the double lines represent the quotient groups whose isomorphism is asserted in the theorem.

Proof. Consider the natural homomorphism $f\colon G \to G/K$ and let $f_1 = f \mid H$ be its restriction to H. The kernel of f_1 consists of the elements of H mapped to 1, i.e. $\ker f_1 = H \cap K$, while the image consists of the union of the cosets of K meeting H, i.e. HK. Hence by Th. 3, $H/(H \cap K) \cong HK/K$, as asserted. ∎

For example, let H be a subgroup of Sym_n. Either $H \subseteq \mathrm{Alt}_n$, then H consists entirely of even permutations, or H contains an odd permutation, then $H . \mathrm{Alt}_n = \mathrm{Sym}_n$ and

$$H/(H \cap \mathrm{Alt}_n) \cong \mathrm{Sym}_n/\mathrm{Alt}_n \cong \mathbf{C}_2.$$

Thus if H contains any odd permutations at all, then the even permutations in H form a normal subgroup of index 2 in H.

Given $N \lhd G$, the natural homomorphism $G \to G/N$ maps each subgroup H of G to a subgroup of G/N, namely HN/N. If we restrict ourselves to subgroups H containing N, we obtain a bijection between the set of subgroups between G and N and the set of subgroups of G/N; moreover, as is easily checked, $H \lhd G$ iff $H/N \lhd G/N$. Writing nat for the natural homomorphism in each case, we have the following commutative diagram:

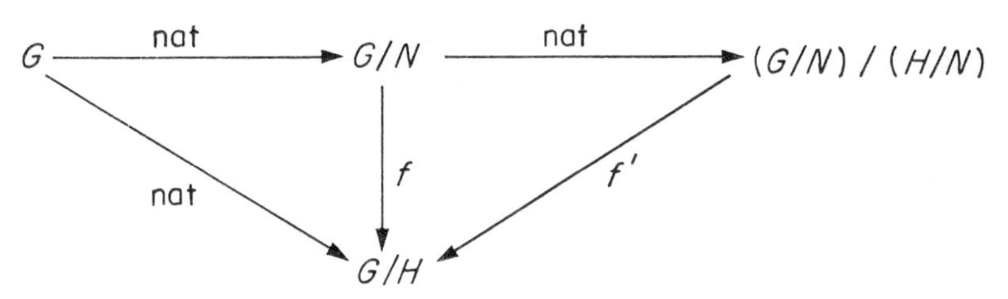

Here f is the mapping which exists by Th. 2, because $N \lhd H$, and f' is the mapping obtained, because $H/N = \ker f$. Since this is an equality, f' is injective and hence is an isomorphism. Thus we have proved

THEOREM 5 (Third isomorphism theorem) *Given a group G and $N \lhd G$, there is a natural bijection between the subgroups of G containing N and the subgroups of G/N: $H \leftrightarrow H/N$, and if $H \lhd G$, there is an isomorphism*

$$(G/N)/(H/N) \cong G/H. \blacksquare$$

Exercises

(1) Show that every subgroup of index 2 in a group is normal.

(2) Let H and K be subgroups of a group G. Show that HK is a union of right cosets of K (and a union of left cosets of H). Show also that the number of these cosets is $(H: H \cap K)$.

(3) Let G be a group and H a subgroup. Show that the least normal subgroup containing H is the subgroup generated by all the conjugates of H in G.

(4) If every subgroup of G is normal, show that any two elements of coprime orders commute.

(5) Verify that the mapping $x \mapsto \exp 2\pi i x$ is a homomorphism of the additive group of $\mathbf{R}$ into the multiplicative group of non-zero complex numbers. Find the image and kernel; do the same for the restriction of this mapping to $\mathbf{Q}$.

(6) Give an example to show that a normal subgroup of a normal subgroup of G need not be normal in G. (Hint. Try Alt_4.)

(7) If G is finite, $N \lhd G$ and the order of N is prime to its index, show that every element of order dividing $|N|$ is contained in N.

(8) Let G, G' be groups with normal subgroups N, N' and natural mappings ν, ν' onto the quotients G/N, G'/N' respectively. If $f: G \to G'$ is a homomorphism such that $Nf \subseteq N'$, show that there is a unique homomorphism $f_1: G/N \to G'/N'$ such that $f\nu' = \nu f_1$.

(9) Let H_1, H_2 be distinct conjugates of a subgroup of G; show that $H_1 H_2 \neq G$ and deduce that $(H_1: H_1 \cap H_2) < (G: H_2)$.

(10) Let G be a simple group with a subgroup of index $n > 1$. Show that G can be represented as a permutation group on n letters. Deduce that $|G| \leqslant n!$, in particular, G, must be finite.

(11) Show that every group with a subgroup H of finite index also has a normal subgroup contained in H and of finite index.

(12) Let G be a group; show that a subset is normal in G iff it is a union of complete conjugacy classes. If C is a normal subset of G, show that the centralizer of C in G is a normal subgroup of G. In particular, the centralizer of any normal subgroup of G is again normal in G.

9.2 The Jordan–Hölder theorem

Our first use of the isomorphism theorems will be to prove some general results on the structure of groups. Let us define a *factor* of a group G as a quotient H/K, where H and K are subgroups of G such that $K \lhd H$. Now the idea is to take in a given group G a chain of subgroups

$$G = G_0 \supseteq G_1 \supseteq \ldots \supseteq G_r = 1. \tag{1}$$

If $G_i \lhd G_{i-1}$ for $i = 1, \ldots, r$, the chain is said to be *normal*. In that case we can form the quotients G_{i-1}/G_i and we can sometimes obtain information about G from a knowledge of the factors G_{i-1}/G_i. Of course it should be borne in mind that G_i need not be normal in G (but only in G_{i-1}).

Any chain obtained from (1) by inserting further terms is called a *refinement* of (1); we allow (1) as a refinement of itself, besides the *proper* refinements, where new subgroups are actually inserted. Assume now that (1) is a normal chain; a second normal chain in G,

$$G = H_0 \supseteq H_1 \supseteq \ldots \supseteq H_s = \mathbf{1} \tag{2}$$

is said to be *isomorphic* to (1) if $s = r$ and there is a permutation $i \mapsto i'$ such that

$$G_{i-1}/G_i \cong H_{i'-1}/H_{i'}.$$

Our first task is to find a means of comparing different chains in G. We begin with a lemma comparing two factors in G, which is based on the parallelogram rule.

LEMMA 1 (Zassenhaus lemma) *Given a group G and subgroups H, H', K, K', where $H' \lhd H$, $K' \lhd K$, then $K'(H' \cap K) \lhd K'(H \cap K)$, $H'(H \cap K') \lhd H'(H \cap K)$ and there are isomorphisms*

$$\frac{K'(H \cap K)}{K'(H' \cap K)} \cong \frac{H \cap K}{(H \cap K')(H' \cap K)} \cong \frac{H'(H \cap K)}{H'(H \cap K')}. \tag{3}$$

These relations are again indicated in Fig. 8 where sloping lines run down to subgroups, two lines running down to a point represent the intersection of groups and two lines running up to a point represent the join. The reader is advised to keep in mind the figure and the parallelogram illustrating Th. 4, **9.1**.

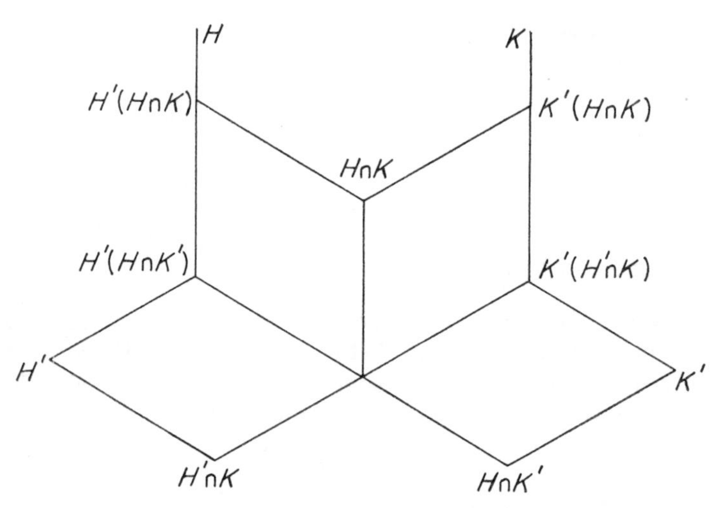

Figure 8

To prove the lemma we first note the following general relation, sometimes called the *modular law*:

Given three subgroups A, B, C of a group G, if $A \subseteq C$, then

$$A(B \cap C) = AB \cap C. \tag{4}$$

Any member of the left-hand side has the form ax, where $a \in A$, $x \in B \cap C$; hence $ax \in AB$ and $ax \in C$ because $A \subseteq C$; it follows that $ax \in AB \cap C$. Conversely, take $c \in AB \cap C$, then $c \in C$ and $c = ab$ where $a \in A$, $b \in B$. Hence $b = a^{-1}c \in C$, so that $b \in B \cap C$ and, therefore, $ab \in A(B \cap C)$. This proves (4).

We now come to the proof of the lemma. We apply the parallelogram rule to the subgroups K' and $H \cap K$ of K. Since $K' \lhd K$ and $K' \cap H \cap K = H \cap K'$, we find

$$K'(H \cap K)/K' \cong (H \cap K)/(H \cap K'), \tag{5}$$

and it follows that $H \cap K' \lhd H \cap K$. By symmetry, $H' \cap K \lhd H \cap K$ and hence

$$(H \cap K')(H' \cap K) \lhd H \cap K.$$

Now in the isomorphism (5), the subgroup $(H \cap K')(H' \cap K)$ corresponds to $K'(H \cap K')(H' \cap K) = K'(H' \cap K)$, hence $K'(H' \cap K) \lhd K'(H \cap K)$ and by Th. 5, **9.1**,

$$K'(H \cap K)/K'(H' \cap K) \cong (H \cap K)/(H' \cap K)(H \cap K').$$

This proves one half of (3); now the other half follows by symmetry. ■

Given two factors H/H', K/K' of a group G, the factor $K'(H \cap K)/K'(H' \cap K)$ is sometimes referred to as the *projection* of H/H' on K/K'. In this terminology Lemma 1 just asserts that the projection of one factor on a second is isomorphic to the projection of the second factor on the first. Let us now take two normal chains in G and project their factors on each other:

The normal chains are

$$G = G_0 \supseteq G_1 \supseteq \ldots \supseteq G_r = \mathbf{1}, \tag{6}$$

$$G = H_0 \supseteq H_1 \supseteq \ldots \supseteq H_s = \mathbf{1}. \tag{7}$$

If we put

$$G_{ij} = G_i(H_j \cap G_{i-1}) \; i = 1, \ldots, r, j = 1, \ldots, s,$$

then

$$G_{i-1} = G_{i0} \supseteq G_{i1} \supseteq \ldots \supseteq G_{is} = G_i$$

is a normal chain from G_{i-1} to G_i; putting these pieces together, we get a refinement of the chain (6):

$$G = G_{00} \supseteq G_{01} \supseteq \ldots \supseteq G_{0s} = G_{10} \supseteq \ldots \supseteq G_{r-1s} = G_{r0} \supseteq \ldots \supseteq G_{rs} = \mathbf{1}. \tag{8}$$

Similarly the groups

$$H_{ji} = H_j(G_i \cap H_{j-i}) \; i = 1, \ldots, r, j = 1, \ldots, s,$$

provide a refinement of (7):

$$G = H_{00} \supseteq H_{01} \supseteq \ldots \supseteq H_{0r} = H_{10} \supseteq \ldots \supseteq H_{s-1\,r} = H_{s0} \supseteq \ldots \supseteq H_{sr} = \mathbf{1}. \tag{9}$$

By the Zassenhaus lemma,

$$G_{ij-1}/G_{ij} \cong H_{ji-1}/H_{ji},$$

which shows the chains (8) and (9) to be isomorphic. Hence we have proved

THEOREM 2 (Schreier refinement theorem) *Any two normal chains in a group G have isomorphic refinements.* ∎

If we omit all repetitions from the chains (8), (9) (corresponding to trivial factors), they remain isomorphic. Thus we may take the isomorphic chains in Th. 2 to be without repetitions.

A normal chain in G which has no proper refinements is called a *composition series* of G and its factors (necessarily simple groups) are called the *composition factors* of G. E.g., in a finite group any normal chain without repetitions can be refined to a composition series: we just keep inserting terms as long as possible; the process must stop because the group is finite. But, as we shall see later, composition series also exist in many infinite groups.

Let G be any group with a composition series; any refinement of this series reduces to the series itself once we omit the repetitions. Now take any normal chain in G and construct the isomorphic refinements of this chain and the given composition series, which exist by Th. 2. They must be composition series (possibly with repetitions), hence we obtain

THEOREM 3 (Jordan–Hölder theorem) *If a group has a composition series, then any normal chain without repetitions can be refined to a composition series and any two composition series are isomorphic.* ∎

As an illustration consider $\mathbf{Z}/m$, the additive group of integers mod m. This is a finite abelian group, hence it has a composition series. Every factorization $m = a_1 a_2 \ldots a_r$ corresponds to a chain $\mathbf{Z}/m \supseteq a_1\mathbf{Z}/m \supseteq a_1 a_2 \mathbf{Z}/m \supseteq \ldots \supseteq 0$ with factors isomorphic to $\mathbf{Z}/a_i$. Now Th. 3 tells us that every factorization of m can be refined to a factorization into primes and that any two such factorizations are essentially the same, except for the order of the factors. Thus we have another proof that $\mathbf{Z}$ is a UFD (the fundamental theorem of arithmetic).

Exercises

(1) Show that a cyclic group of prime power order has only one composition series.

(2) Find all composition series of Sym_4.

(3) Let G be a group, H a maximal proper subgroup and N any normal subgroup of G distinct from H. Show that $G/N \cong H/(H \cap N)$; show that this holds for any

subgroup H not contained in N, provided that either H is also normal and G/N is simple, or G/N has prime order.

(4) Let G be a group with two composition series $H_1, \ldots, H_r$ and $K_1, \ldots, K_s$. Show that either $H_1 = K_1$ or $H_1K_1 = G$. Use induction on $\min\{r, s\}$ to prove that any two composition series of G are isomorphic.

(5) Let G be a group with a composition series. Show that for any normal subgroup N of G there is a composition series through N.

(6) Let G be a finite group and H a subgroup of G. Show that every composition factor of H is a factor of some composition factor of G.

(7) Let G be a finite group and H a subgroup occurring in a normal chain of G. Show that the composition factors of H are also composition factors of G. Given the existence of non-abelian finite simple groups, show that this does not generally hold for all subgroups.

9.3 Groups with operators

Many of the results on groups established in this chapter hold for rings and as well for modules (cf. Ch. **10**). In order to take account of this fact we shall treat a slight extension of the group concept by allowing operators to act on the group. To keep to the simplest form we have confined ourselves to unary operators, i.e. operators depending on a single argument. Although strictly speaking, the case of rings is not covered thereby, it is not worth describing the more general multiple operators needed for this single case.

Let G be a group and Ω a set, then Ω is said to act on G by endomorphisms if each $\omega \in \Omega$ defines an endomorphism of G, i.e. there is a mapping $x \mapsto x\omega$ satisfying

$$(xy)\omega = (x\omega)(y\omega) \qquad \text{for all } x, y \in G. \tag{1}$$

The pair consisting of the group G and the operator domain Ω is called an Ω-*group*, the elements of Ω *operators*. Each operator defines an endomorphism of G, but, of course, different operators may well define the same endomorphism.

By an Ω-*subgroup* of an Ω-group G one understands a subgroup of G which is mapped into itself by all the operators in Ω. E.g. if Ω consists of all conjugations of G, an Ω-subgroup is just a normal subgroup of G.

If we have two Ω-groups G, H, i.e. two groups with the same operator domain Ω, we can speak of Ω-homomorphisms: they are group homomorphisms $f\colon G \to H$ such that

$$(xf)\omega = (x\omega)f \qquad \text{for all } \omega \in \Omega,\ x \in G. \tag{2}$$

We express (2) by saying that f is *compatible* with the operators. We note that (2) may also be described by saying that the following diagram commutes:

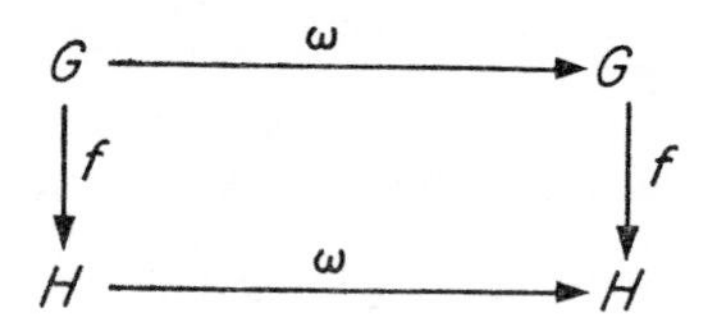

Given an Ω-group G and an Ω-subgroup N which is moreover normal in G, we can form the quotient group G/N; now an Ω-action may be defined on G/N by the rule

$$(xN)\omega = (x\omega)N. \tag{3}$$

This is well-defined because N is mapped into itself by each $\omega \in \Omega$. Thus each quotient group of G by a normal Ω-subgroup is again an Ω-group and, from the definition (3), it is easily seen that the natural mapping $G \to G/N$ is compatible with the action of Ω.

Examples of Ω*-actions.* (i) $\Omega = \varnothing$. In this case the notions Ω-group, Ω-subgroup, Ω-homomorphism etc. reduce to the usual notions of group, subgroup, homomorphism.

(ii) Given any group G, let Ω consist of all conjugations $x \mapsto a^{-1}xa$. The Ω-subgroups of G are just the normal subgroups and every homomorphism is an Ω-homomorphism.

(iii) Let F be a field. The vector spaces over F are abelian groups on which there is an endomorphism defined for each $\alpha \in F$ (satisfying certain conditions). Thus we have in effect F-groups; this is an important special case of Ω-groups, which will be generalized later, by replacing F by an arbitrary ring. This means that many of our results will apply to vector spaces. For example, the quotient group construction yields the notion of a quotient space of a vector space. We were able to dispense with this notion in Ch. **4**, because every subspace of a vector space has a complement, unique up to isomorphism. However, for groups (and for the modules to be considered later) it is no longer true that each subgroup has a complement and from this circumstance the notion of quotient group derives its importance.

In the last two sections we purposely described the isomorphism and refinement theorems in the abstract setting, so as not to burden the reader with unnecessary detail. In fact all these results can be taken over for arbitrary Ω-groups; the reader is urged to go through **9.1** and **9.2** and satisfy himself that all the results stated there still hold for groups with operators.

For example, an Ω-group is called *simple* if it has no normal Ω-subgroup other than G and **1**. The bigger Ω is, the fewer Ω-subgroups there are, so a group which is far from simple as an abstract group may be simple as an Ω-group, for suitably defined Ω, and this greatly increases the scope of the results proved.

As a first application, let G be an abstract group and take Ω to be the action of G on itself by conjugation. The Ω-subgroups of G are just the normal subgroups and a chain of Ω-subgroups is a chain

$$G = G_0 \supseteq G_1 \supseteq \ldots \supseteq G_r = \mathbf{1},$$

such that $G_i \lhd G$ (and not merely $G_i \lhd G_{i-1}$). Such a chain is said to be *invariant* and the corresponding composition series is called a *chief series*, its factors being the *chief factors* of G. Now we have the following analogue of the Jordan–Hölder theorem:

THEOREM 1 *If a group G has a chief series, then any invariant series without repetitions can be refined to a chief series and any two chief series of G are isomorphic.* ■

Let G and A be any groups, written additively (but not necessarily abelian). Given any two homomorphisms α, β from G to A, we can define their sum as a mapping from G to A by the equation

$$x(\alpha+\beta) = x\alpha+x\beta \qquad (x \in G). \tag{4}$$

This mapping $\alpha+\beta$ need not be a homomorphism; the condition for it to be one is that $(x+y)(\alpha+\beta) = x(\alpha+\beta)+y(\alpha+\beta)$, i.e.

$$(x+y)\alpha+(x+y)\beta = x\alpha+x\beta+y\alpha+y\beta.$$

Expanding the left-hand side, and cancelling the terms $x\alpha$ and $y\beta$ common to both sides, we obtain

$$y\alpha+x\beta = x\beta+y\alpha \qquad (x, y \in G). \tag{5}$$

This then is the condition for $\alpha+\beta$ as defined by (4) to be a homomorphism. In particular, when A is abelian, this condition is always satisfied. In that case we have, on writing Hom (G, A) for the set of all homomorphisms from G to A:

PROPOSITION 2 *Let G be any group and A an abelian group, written additively, then* Hom (G, A) *is an abelian group under the operation defined by* (4).

Proof. It only remains to verify the group laws: Clearly $(\alpha+\beta)+\gamma$ and $\alpha+(\beta+\gamma)$, applied to $x \in G$, both give $x\alpha+x\beta+x\gamma$. Further, the operation $x \mapsto 0$ is the neutral element and $-\alpha$, defined by $x(-\alpha) = -x\alpha$, is the inverse of α. The details may be left to the reader. ■

When G and A are both written multiplicatively (but the group operation on homomorphisms is still written as addition), the homomorphisms are often written as exponents, then (4) takes the form:

$$x^{\alpha+\beta} = x^{\alpha} \,.\, x^{\beta}.$$

Returning to the case where A is abelian and additive, let us now take $G = A$; the set Hom (A, A) is then the set of all endomorphisms of A, usually denoted by End (A). Since the product of two endomorphisms (i.e. the mapping obtained by performing them in succession) is still an endomorphism, we have a multiplication defined on End (A), which is clearly associative and has the identity mapping as neutral element. Moreover, the defining property of endomorphisms shows that $(\alpha+\beta)\gamma = \alpha\gamma+\beta\gamma$, while the definition of addition in End (A) ensures that $\alpha(\beta+\gamma) = \alpha\beta+\alpha\gamma$. This proves that End (A) is a ring, called the *endomorphism ring* of A.

Since we have a natural ring structure on End (A), it is reasonable, in considering operator domains for an abelian group A, to put a ring structure on the operator domain. This leads to the following

DEFINITION Let R be a ring. An abelian group A with R as operator domain is called an *R-module* if

$$\begin{aligned} x(\alpha+\beta) &= x\alpha+x\beta, & x &\in A, \\ x(\alpha\beta) &= (x\alpha)\beta, & \alpha, \beta &\in R, \\ x \, . \, 1 &= x. \end{aligned} \tag{6}$$

The rules (6) ensure that the ring operations in R correspond to the way endomorphisms of A are combined. From the definition of an operator, we also have in addition to the rules (6),

$$(x+y)\alpha = x\alpha+y\alpha \qquad x, y \in A, \ \alpha \in R.$$

The above definition may also be expressed by saying that an R-module structure on A is given by a ring homomorphism $R \to$ End (A). For each $\alpha \in R$ defines an endomorphism of A, i.e. an element of End (A), and the equations (6) ensure that this correspondence is a ring homomorphism.

A variant of the above definition is obtained if we replace the second equation in (6) by

$$x(\alpha\beta) = (x\beta)\alpha. \tag{7}$$

In this case it is usual to write the operators on the left, i.e. to write αx instead of $x\alpha$. Then (7) reads

$$(\alpha\beta)x = \alpha(\beta x). \tag{8}$$

Such modules are called *left R-modules*, in contrast to the sort defined earlier, which are also called *right R-modules*. When R is commutative, (7) reduces to (6) and the difference between left and right R-modules is merely one of notation. But for general rings there is an actual difference; we shall return to this question in Ch. **10**.

Exercises

(1) Show that for any Ω-homomorphism the image and kernel are Ω-subgroups.

(2) Let C be a cyclic group. If C is defined as an Ω-group in some way, show that all its subgroups are necessarily Ω-subgroups. Give an example of an Ω-group structure

on $\mathbf{Z}^2$, the additive group of pairs (a, b) $(a, b \in \mathbf{Z})$ and a subgroup which is not an Ω-subgroup.

(3) Prove Prop. 2 in detail.

(4) Prove the Schreier refinement theorem for groups with operators. By including conjugations among the operators show that in the isomorphism constructed between refinements of invariant series, the isomorphisms of corresponding factors are compatible with the action of G by conjugation.

(5) Let G be a finite group and H a normal subgroup of G of order n prime to its index. Show that every subgroup of order dividing n is contained in H. If a normal chain consists of subgroups each of which has order prime to its index in G, show that it is an invariant chain.

(6) Let G be any group and α an endomorphism of G. Show that α commutes with every conjugation in G iff $x^\alpha x^{-1}$ lies in the centralizer of im α for all $x \in G$. An endomorphism with this property is called *normal.* Show that for any normal automorphism α, the mapping $x \mapsto x^\alpha x^{-1}$ is an endomorphism.

(7) Let G be an Ω-group. Show that the image of a normal Ω-endomorphism is a normal Ω-subgroup. If G is Ω-simple, deduce that every normal Ω-endomorphism is either trivial (i.e. it maps G to the trivial subgroup) or an Ω-automorphism.

(8)* Let A be any group (not necessarily abelian) written additively. Show that Map (A), the set of all mappings from A to itself, with pointwise addition (as in (4)) and composition as multiplication, is an additive group (not necessarily abelian), is a multiplicative monoid and that it satisfies one distributive law: $\alpha(\beta+\gamma) = \alpha\beta+\alpha\gamma$. Such a structure is called a *near-ring*. What goes wrong if we try to use these rules to define End (A) as a near-ring?

9.4 Automorphisms

In the brief discussion of symmetries in Ch. **3** we saw that the symmetries of a geometrical figure form a group. This applies much more generally: for any mathematical structure we can define its *symmetries* as the bijections with itself that preserve all the structure. Then it is easily seen that products and inverses of symmetries are again symmetries, i.e. all the symmetries form a group. In the case of a group G, a bijection preserving the group structure was called an automorphism of G; thus we see that the set of all automorphisms of G is a group. It is called the *automorphism group* of G and written Aut (G). E.g., $\mathbf{C}_3$ has just one automorphism, apart from the trivial † automorphism which leaves everything fixed. In additive notation it is $i \mapsto 2i$. Thus Aut $(\mathbf{C}_3) \cong \mathbf{C}_2$. Similarly $\mathbf{C}_6$ has precisely two automorphisms, while $\mathbf{C}_5$ has 4.

In any abelian group (written additively) the mapping

$$x \mapsto nx \qquad \text{for a given } n \in \mathbf{Z} \tag{1}$$

† Note that whereas an endomorphism is trivial when it maps everything to the neutral element, a trivial automorphism is one which leaves everything fixed.

is an endomorphism and, if it happens to be a bijection, it is an automorphism. E.g., this always holds for $n = \pm 1$ and, when A is finite, it holds whenever n is prime to $|A|$.

In a non-abelian group the power mapping $x \mapsto x^n$ need not be an endomorphism, but in this case we have another important class of automorphisms. Let $a \in G$, then the conjugation by a:

$$\alpha_a : x \mapsto a^{-1}xa \tag{2}$$

is an automorphism of G. It is a homomorphism, because $a^{-1}xa \,.\, a^{-1}ya = a^{-1}xya$, and it is bijective, since it has the inverse $\alpha_{a^{-1}}$. This is called the *inner automorphism* of G defined by a; it is trivial iff $ax = xa$ for all $x \in G$, i.e. precisely when a lies in the centre of G. Thus a non-abelian group always has non-trivial inner automorphisms. An automorphism which is not inner is said to be *outer*.

Let us return to the inner automorphism (2); we have already seen in Ch. **3** that (2) defines a G-action, i.e. $\alpha_1 = 1$ and $\alpha_{ab} = \alpha_a\alpha_b$. This shows that we have a homomorphism

$$G \to \text{Aut}\,(G), \qquad \text{given by } a \mapsto \alpha_a. \tag{3}$$

The image, by definition, is the group of inner automorphisms of G, denoted by Inn (G). The kernel consists of all elements lying in the centre of G, Z say. Hence by the first isomorphism theorem,

$$\text{Inn}\,(G) \cong G/Z.$$

PROPOSITION 1 *For any group G, the inner automorphisms form a normal subgroup of* Aut G. *More precisely, if $a \in G$ and $\sigma \in$* Aut G, *then*

$$\sigma^{-1}\alpha_a\sigma = \alpha_{a\sigma}. \tag{4}$$

Proof. The normality follows once we have established (4) and this is a simple verification. We take $x \in G$ and apply both sides of (4) to $x\sigma$: $x\sigma \,.\, \sigma^{-1}\alpha_a\sigma = x\alpha_a\sigma = (a^{-1}xa)\,\sigma = (a\sigma)^{-1}.\, x\sigma \,.\, a\sigma = x\sigma \,.\, \alpha_{a\sigma}$. ∎

The quotient Aut (G)/Inn (G) is called the *automorphism class group.*

From the definition, $A =$ Aut G can be regarded as a permutation group of G, i.e. a subgroup of $\Sigma(G)$. Now $\Sigma(G)$ contains two other subgroups of interest, the groups $R = R(G)$ of right multiplications and $L = L(G)$ of left multiplications. Consider the former; since

$$\sigma^{-1}\rho_a\sigma = \rho_{a\sigma} \qquad \text{for all } \sigma \in A, \tag{5}$$

we see that $RA = AR$ and this is therefore the subgroup of $\Sigma(G)$ generated by R and A. This subgroup of G is called the *holomorph* of G and is denoted by Hol (G) or, more briefly, by H:

$$H = AR = RA.$$

Since L consists of the permutations $\lambda_a = \rho_a\alpha_a^{-1}$, H also contains L. The connexion between these subgroups is given in

THEOREM 2 *Let G be any group, $R = R(G)$ its group of right multiplications, $L = L(G)$ its group of left multiplications and $A = \text{Aut } G$, then*

(i) *each of R, L is the centralizer of the other in $\Sigma(G)$,*

(ii) *$H = AR$ is the normalizer of R in $\Sigma(G)$,*

(iii) *A is the stabilizer of $1 \in G$ in the G-action of H.*

Proof. (i) Let $\lambda \in \Sigma(G)$ commute with each element of R, then $\rho_a\lambda = \lambda\rho_a$, i.e. $(xa)\lambda = (x\lambda)a$. Put $x = 1$ and write $b = 1\lambda$, then $a\lambda = ba$, hence $\lambda = \lambda_b$. Conversely, each λ_b clearly centralizes R, hence L is the precise centralizer of R in $\Sigma(G)$. By symmetry R is the centralizer of L in $\Sigma(G)$.

We shall prove (ii) and (iii) together. Let N be the normalizer of R in $\Sigma(G)$, then by (5) $H \subseteq N$; we shall prove (iii) by showing that the stabilizer of $1 \in G$ in the group action of N on G is A. Let $\sigma \in N$, then for each $a \in G$ there exists $a' \in G$ such that

$$\sigma^{-1}\rho_a\sigma = \rho_{a'}, \tag{6}$$

and if σ fixes 1, then by applying (6) to 1 we find that $a' = a\sigma$. Hence $(xa)\sigma = x\rho_a\sigma = x\sigma\rho_{a\sigma} = x\sigma \,.\, a\sigma$, i.e. $\sigma \in A$.

To prove (ii), let $\sigma \in N$ and write $1\sigma = a$, then $\sigma\rho_a^{-1}$ lies in N and fixes 1, hence $\sigma\rho_a^{-1} \in A$, and so $\sigma \in AR = H$. ■

Let G be any group. A subgroup H of G is said to be *characteristic* in G if it is mapped into itself by all automorphisms of G. Any characteristic subgroup H of G in particular admits all inner automorphisms and hence is normal in G. Moreover, if K is a characteristic subgroup of H, any automorphism of G induces an automorphism of H and hence maps K into itself, so that K is characteristic in G. If we merely know that H is normal in G, this still applies to all inner automorphisms of G. Thus we have

PROPOSITION 3 *Let G be any group. Any characteristic subgroup of a characteristic* (resp. *normal*) *subgroup of G is characteristic* (resp. *normal*) in G. ■

Any group G may be regarded as an Ω-group, where $\Omega = \text{Aut } G$ is the set of all automorphisms of G. The characteristic subgroups are then just the Ω-subgroups; as we have seen they are normal in G and, if H is an Ω-subgroup, G/H again has a natural Ω-group structure as defined in **9.3**, i.e. Aut G again acts on G/H.

By a *characteristic series* of G we understand a chain

$$G = G_0 \supset G_1 \supset \cdots \supset G_r = \mathbf{1}$$

of characteristic subgroups of G which has no proper refinements; its factors are called the *characteristic factors* of G. Now the theory of Ω-groups yields again an analogue of the Jordan–Hölder theorem:

THEOREM 4 *If a group G has a characteristic series* (e.g. *when G is finite*), *then any chain of characteristic subgroups of G without repetitions can* *be refined to a characteristic series of G, and any two characteristic series of G are isomorphic.* ■

Exercises

(1) If G is cyclic, show that Aut G is abelian.

(2) Show that the centre of a group is a characteristic subgroup.

(3) Show that every group with more than 2 elements has a non-trivial automorphism.

(4) If Inn (G) is cyclic, show that G is abelian; deduce that Inn (G) is then trivial.

(5) Show that every group with exactly 2 automorphisms must be abelian. Show that there are 4 cyclic groups with exactly 2 automorphisms.

(6) For any automorphism α, show that $(x^{-1})\alpha = (x\alpha)^{-1}$.

(7) Let V be the 4-group. Show that Aut $V \cong \mathrm{Sym}_3$, Hol $V \cong \mathrm{Sym}_4$.

(8) Let G be a group, H a subgroup, C its centralizer and N its normalizer in G. Show that $C \lhd N$ and that N/C is isomorphic to a subgroup of Aut H.

(9)* Let G be a group and P the subset of $\Sigma(G)$ consisting of all permutations $x \mapsto x'$ such that $(xy^{-1}z)' = x'y'^{-1}z'$. Verify that P is a subgroup of $\Sigma(G)$. Show that (i) $P \supseteq H$ and (ii) P normalizes $R(G)$ in $\Sigma(G)$. Deduce that $P = H$.

(10) Show that the homomorphism (3) from G to Aut G is an isomorphism iff G has trivial centre, and no outer automorphisms. Such a group is called *complete.*

(11)* Show that Sym_3 and Sym_4 are complete.

(12)* Let G be a simple non-abelian group. Show that G has a trivial centre and hence can be embedded in Aut G. Prove that Aut G is complete. (Hint. Show that the image of G under the embedding has trivial centralizer in Aut G, and use Ex. (7), **9.5.**)

9.5 The derived group; soluble groups and simple groups

Let G be a group and $x, y \in G$, then the expression

$$(x, y) = x^{-1}y^{-1}xy \tag{1}$$

is called the *commutator* of x and y. Clearly $(x, y) = 1$ iff x and y commute; more precisely, by writing (1) as

$$xy = yx(x, y), \tag{2}$$

we see that the commutator of x and y is the correction term needed to shift y past x. A most important property of commutators is contained in the following almost obvious remark:

LEMMA 1 *Any homomorphism $f: G \to H$ maps the commutator (x, y) to (xf, yf).*

The proof is a simple verification, which may be left to the reader. ■

In particular, if, in the lemma, H is abelian, then $(x, y) \in \ker f$. Consider the subgroup G' generated by all commutators; it is called the *commutator subgroup* of G or also the *derived group* of G. We have seen that for any $c \in G$, the mapping $x \mapsto c^{-1}xc$ is an automorphism of G, hence the conjugate of a commutator is again a commutator. It follows that G' consists of complete conjugacy classes and is therefore normal in G. The quotient G/G' is sometimes written G^{ab}; clearly it is abelian, as we see by applying the natural homomorphism $G \to G^{ab}$ to (2). Conversely, given any homomorphism into an abelian group $f: G \to A$, we have $(x, y) \in \ker f$ for all $x, y \in G$, hence $G' \subseteq \ker f$. By the factor theorem we can factorize f uniquely as $G \overset{ab}{\to} G^{ab} \to A$. This group G^{ab} is often called the group G *abelianized*, or *made abelian*, and the property found above is just the universal mapping property for G^{ab}. Thus we have

THEOREM 2 *To any group G there corresponds an abelian quotient group G^{ab} with a homomorphism $\nu_G: G \to G^{ab}$, which is universal for homomorphisms of G into abelian groups. Thus for each homomorphism f from G to an abelian group A there exists a unique homomorphism $f': G^{ab} \to A$ such that the diagram shown commutes.* ■

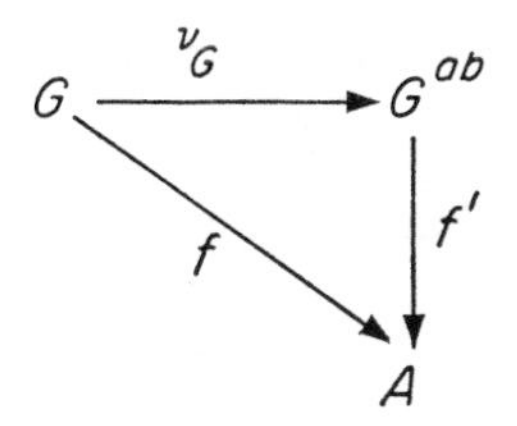

Here the mapping ν_G is just the natural mapping from G to its quotient group G^{ab}. We now show that the correspondence $G \mapsto G^{ab}$ is actually a functor from groups to abelian groups, while ν is a natural transformation from the identity functor to *ab*. Our first task is to define the functor *ab* on morphisms. Thus let $f: G \to H$ be a homomorphism of groups; by combining this with the natural mapping ν_H we obtain a homomorphism from G to H^{ab} and, by Th. 2, there is a unique homomorphism, f^{ab}, say, from G^{ab} to

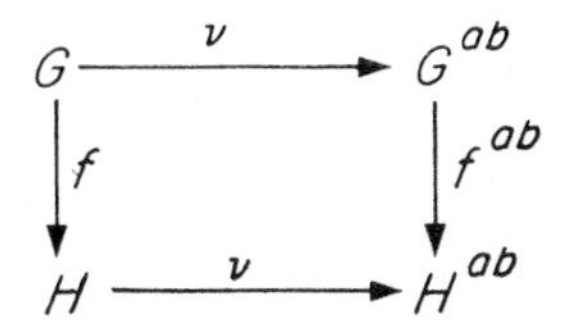

H^{ab} to make the square shown in the diagram commutative. If $g: H \to K$ is a homomorphism from H, it is easily shown, from the uniqueness of f^{ab}, that

$$(fg)^{ab} = f^{ab} \cdot g^{ab},$$

and clearly $1^{ab} = 1$, hence we have indeed a functor. Moreover, the commutativity of the square shown expresses the fact that ν is a natural transformation.

With each group G we can associate a chain of subgroups, its *derived series*:

$$G \supseteq G' \supseteq G'' \supseteq \ldots \supseteq G^{(i)} \supseteq \ldots,$$

in which each term is the derived group of the preceding term. From Lemma 1 it follows that G' is characteristic in G; by induction we see that each term of the derived series is characteristic in G, and hence is normal in G. If after a finite number of terms we reach 1, the group G is said to be *soluble* or *solvable*. If H is a subgroup of G, then clearly $H' \subseteq G'$; it follows that every subgroup of a soluble group is again soluble. If $K = G/N$ is a quotient group of G, then the natural homomorphism $G \to K$ maps G' onto K' (by Lemma 1) and, more generally, maps $G^{(i)}$ onto $K^{(i)}$. This shows that every quotient (and hence every factor) of a soluble group is again soluble.

For finite groups an alternative description of soluble groups is provided by

THEOREM 3 *A finite group G is soluble if and only if the factors in a composition series of G are all of prime order.*

Proof. Assume that G is soluble and let its derived series be

$$G \supset G' \supset G'' \supset \cdots \supset G^{(k)} = \mathbf{1}.$$

Each factor in this chain is finite abelian and, if we refine it to a composition series of G, each factor is then simple abelian, and hence of prime order. Conversely, if G has a composition series with factors of prime order:

$$G = G_0 \supset G_1 \supset \cdots \supset G_r = \mathbf{1}, \tag{3}$$

then $G' \subseteq G_1$ because G_0/G_1 is abelian; similarly $G'' \subseteq G_1' \subseteq G_2$ and generally we find that $G^{(i)} \subseteq G_i$, by induction on i, hence $G^{(r)} \subseteq G_r = \mathbf{1}$, i.e. G is soluble, as claimed. ■

In the proof of the sufficiency we only used the fact that the factors in (3) were abelian, hence we obtain the

COROLLARY 1 *A group G is soluble if and only if it has a finite normal chain with abelian factors.* ■

Let G be soluble, then as we saw, for any $N \lhd G$, both N and G/N are soluble. Conversely, assume that G is a group with a normal subgroup N such that N and G/N are soluble. Then the normal chain $G \supseteq N \supseteq \mathbf{1}$ for G can be refined to one with abelian factors; for the section from N to $\mathbf{1}$ this follows by taking the derived series for N; between G and N we take the groups corresponding to the derived series of G/N. Thus we have a normal chain with abelian factors for G, and by Cor. 1, G must be soluble. Hence we have

COROLLARY 2 *A group G with a normal subgroup N is soluble if and only if both N and G/N are soluble.* ■

As we know, any finite group has a composition series (3). By the third isomorphism theorem, a composition series can be characterized as a normal chain with simple factors, and we have just seen that soluble groups are characterized by the fact that their composition factors are abelian (of prime order). This makes it natural to ask what the non-abelian simple groups look like.

Several infinite families of non-abelian simple groups have long been known, as well as the five simple groups discovered by Mathieu in 1861 and 1873. Infinite families of new finite simple groups were found by Chevalley in 1955 and, since then, a number of simple groups not fitting into any known family (as well as more families) have been discovered. All these simple groups are of even order, confirming the conjecture stated by Burnside in 1902 (but going back still further) that no non-abelian simple group of odd order exists or, equivalently, that every group of odd order is soluble. This conjecture was proved by Feit and Thompson in 1962.

We shall confine ourselves here to exhibiting one family of simple groups, the alternating groups.

THEOREM 4 Alt_n *is simple for* $n > 4$.

Proof. Let N be a normal subgroup of $A = \mathrm{Alt}_n$; we assume that $N \neq \mathbf{1}$ and shall show that $N = A$.

To prove this we pick a permutation $\alpha \in N$ which is not the identity but fixes as many symbols as possible. Clearly α cannot move just one symbol, nor (since it is even) just two. We claim that α moves exactly three symbols and fixes the rest. If this is not so, α must move at least four symbols. We list two cases:

(i) α contains a cycle with at least three symbols, so in cycle notation it has the form

$$\alpha = (1\ 2\ 3\ \ldots)\ldots$$

By hypothesis α moves at least four symbols; if it moved exactly four, it would have to be odd: $\alpha = (1\ 2\ 3\ 4)$, so in fact α moves at least five symbols in this case, say $1, 2, \ldots, 5$.

(ii) α consists entirely of 2-cycles:

$$\alpha = (1\ 2)(3\ 4)\ldots,$$

We transform α by $\beta = (3\ 4\ 5)$, then in case (i),

$$\alpha_1 = \beta^{-1}\alpha\beta = (1\ 2\ 4\ldots)\ldots,$$

and in case (ii),

$$\alpha_1 = \beta^{-1}\alpha\beta = (1\ 2)(4\ 5)\ldots$$

In both cases, $\alpha_1 \neq \alpha$, i.e. $(\alpha, \beta) = \alpha^{-1}\alpha_1 \neq 1$. Clearly $(\alpha, \beta) \in N$, and if α moves 5, then (α, β) leaves fixed any symbol fixed by α and also fixes 2, so it moves fewer symbols than α. If α leaves 5 fixed, then $\alpha = (1\ 2)(3\ 4)$ and $(\alpha, \beta) = (3\ 5\ 4)$, which again moves fewer symbols than α.

It follows that α moves exactly three symbols and fixes the rest, and it is therefore a 3-cycle; thus N contains a 3-cycle, say $(1\ 2\ 3)$. If we transform by $\gamma = (1\ 2)(3\ i)$ for $i > 3$, we get

$$\gamma^{-1}(1\ 2\ 3)\gamma = (1\ 2\ i),$$

hence N contains all 3-cycles $(1\ 2\ i)$ and, since the latter generate Alt_n, we find that $N = \mathrm{Alt}_n$, as claimed. ■

We note that the symbols 1, 2, . . ., 5 enter explicitly, so the proof does not apply for $n < 5$. For $n = 1, 2, 3$ Alt_n is of course abelian; Alt_4 of order 12, is non-abelian, but is soluble, a normal subgroup being the 4-group $V = \{1, (1\ 2)(3\ 4), (1\ 3)(2\ 4), (1\ 4)(2\ 3)\}$. This is clearly a subgroup and is normal in Sym_4 because it consists of complete conjugacy classes. Thus Sym_4 has a composition series with factors of orders 2, 3, 2, 2.

Exercises

(1) Show that $(G/N)^{(i)} = G^{(i)}N/N$, for any $N \lhd G$.

(2) Let $H, K \lhd G$. Show that if G/H, G/K are soluble, then so is $G/(H \cap K)$.

(3) Show that Alt_5 contains no subgroup of order 15.

(4) A group G is said to be *perfect* if $G' = G$. Verify that Alt_n for $n > 4$ is perfect by expressing each 3-cycle as a commutator. Deduce that Alt_n cannot be soluble for $n > 4$. (Hint. Find the commutator of two 3-cycles with one common symbol.)

(5) Find a composition series for Sym_n.

(6) In any group G, express $(x, y)(y, z)$ as a commutator. (Hint. Equate the given expression to 1 and express the resulting relation as a commutator relation.)

(7) For any subgroups A, B of G define (A, B) to the subgroup generated by all (a, b), where $a \in A$, $b \in B$. If $A, B \lhd G$, show that $(A, B) \subseteq A \cap B$. Deduce that two normal subgroups with trivial intersection commute elementwise (i.e. $ab = ba$ for all $a \in A$, $b \in B$).

(8) If three composition series of a group G differ only in a single term, show that each series has two factors that are not common to the others and that these six factors are all isomorphic. Show moreover that these factors are abelian.

(9) Let $\Sigma(\mathbf{N})$ be the group of permutations on the set $\mathbf{N}$ of positive integers and Sym ($\mathbf{N}$) the subset of permutations moving only finitely many symbols. Show that Sym ($\mathbf{N}$) is a subgroup of $\Sigma(\mathbf{N})$; find a subgroup Alt ($\mathbf{N}$) of index 2 in Sym ($\mathbf{N}$) and show that Alt ($\mathbf{N}$) is simple.

(10) Show that every finite group can be embedded in Sym ($\mathbf{N}$). Show also that each finite group can be embedded in Alt ($\mathbf{N}$).

(11) Show that the correspondence $G \mapsto G'$, where G' is the derived group forms part of a functor D from groups to groups and find its action on morphisms. Show that the inclusion $G' \to G$ is a natural transformation from D to the identity functor.

(12) Show that there can be no natural transformation from the identity functor to D other than the trivial mapping which sends G to the neutral element of G'. (Hint. Apply the square expressing naturality with the infinite cyclic group for G.)

9.6 Direct products

Let A and B be any groups. Then we can define a group on the product set $P = A \times B$ by taking as multiplication†

$$(a, b)(a', b') = (aa', bb').$$

It is easy to see that P becomes a group in this way; the neutral element is $(1_A, 1_B)$ and the inverse of (a, b) is (a^{-1}, b^{-1}). This group P is called the *direct product* of A and B and is again denoted by $A \times B$. We note that the projection mappings $p\colon (a, b) \mapsto a$, $p'\colon (a, b) \mapsto b$ are homomorphisms from P to A, B respectively.

More generally, we can define the direct product of any finite family of groups $A_1, A_2, \ldots, A_n$ as the product set $A_1 \times \cdots \times A_n$ with the group multiplication carried out componentwise:

$$(a_1, \ldots, a_n)(a_1', \ldots, a_n') = (a_1 a_1', \ldots, a_n a_n').$$

The same definition can be used even for an infinite family of groups (A_i) $(i \in I)$. We take as our product set $P = \prod A_i$ and again perform all operations componentwise. As in the case of two factors, the projection mapping on each component is a homomorphism. If all factors are the same, the product becomes a power A^I and we speak of the *direct power* of A. Its elements are A-valued functions on I, with multiplication $xy(i) = x(i)y(i)$.

For a given group G it is useful to know when G can be expressed as a direct product of two groups A, B, for then the structure of G is determined entirely by that of A and B.

THEOREM 1 *Let G be a group and A, B two subgroups of G. Then the following three conditions are equivalent:*

(i) *G is isomorphic to the direct product of A and B: $G \cong A \times B$,*

† Here the brackets indicate pairs, and not commutators as in **9.5**.

(ii) *A and B are normal in G and* $AB = G$, $A \cap B = \mathbf{1}$,

(iii) *A and B commute elementwise and every element x of G is uniquely expressible in the form* $x = ab$, $a \in A$, $b \in B$.

Proof. (i) ⇒ (ii). Let $G \cong A \times B$, then we may identify G with $A \times B$. Denote by p, p' the projections of G on A, B respectively, then $p\colon (a, b) \mapsto a$ is a homomorphism with kernel B, hence $B \lhd G$ and similarly $A \lhd G$. Now any element of G has the form $(a, b) = (a, 1)(1, b)$, which shows that $G = AB$, and $(a, b) \in A \cap B$ iff $a = b = 1$, i.e. $A \cap B = \mathbf{1}$.

(ii) ⇒ (iii). Let $a \in A$, $b \in B$ and form $c = a^{-1}b^{-1}ab$. Since $A \lhd G$, $c = a^{-1}.b^{-1}ab \in A$ and since $B \lhd G$, $c = a^{-1}b^{-1}a.b \in B$, so $c \in A \cap B = \mathbf{1}$, i.e. $c = 1$ and hence $ab = ba$. Now we know that any $x \in G$ can be written $x = ab$; if also $x = a_1b_1$, where $a_1 \in A$, $b_1 \in B$, then $a_1^{-1}a = b_1b^{-1} = d$, say. The first expression for d shows that $d \in A$ and the second shows that $d \in B$. Hence $d \in A \cap B$, but the latter is trivial, and so $d = 1$, i.e. $a^{-1}a = 1$; this shows that $a_1 = a$ and $b_1 = b$. It follows that the expression $x = ab$ is unique.

(iii) ⇒ (i). The mapping $(a, b) \mapsto ab$ $(a \in A, b \in B)$ is a homomorphism from $A \times B$ to G, because A, B commute elementwise, and it is bijective, by hypothesis, hence it is an isomorphism. ∎

In the same way we can prove the

COROLLARY *Let G be a group with subgroups* $G_1, \ldots, G_r$, *then the following three conditions are equivalent*:

(i) *G is isomorphic to the direct product of the* G_i:

$$G \cong G_1 \times \cdots \times G_r,$$

(ii) *each* G_i *is normal in* G, $G = G_1G_2 \ldots G_r$ *and* $G_1 \ldots G_{i-1} \cap G_i = \mathbf{1}$, *for* $i = 2, \ldots, r$,

(iii) G_i *and* G_j *commute elementwise for* $i, j = 1, \ldots, r$, $i \neq j$, *and every element* $x \in G$ *is uniquely expressible in the form* $x = x_1x_2 \ldots x_r$, *where* $x_i \in G_i$. ∎

For an example of a direct product, take two coprime integers r, s, then $\mathbf{C}_{rs} \cong \mathbf{C}_r \times \mathbf{C}_s$. Another example is the group of all non-singular $n \times n$ diagonal matrices over a field F; this is the direct power of n copies of the multiplicative group of F.

For abelian groups, in additive notation, the direct product is also called the *direct sum* and is written $G \oplus H$. An instance of this is the direct sum of vector spaces introduced in **4.4**. For direct sums of abelian groups the criterion of Th. 1 can be stated in a somewhat different form which is sometimes useful. Let $C = A \oplus B$ be a direct sum of abelian groups and denote by $p\colon C \to A$, $p'\colon C \to B$ the projections and by $q\colon A \to C$, $q'\colon B \to C$ the inclusion mappings, given by $aq = (a, 0)$, $bq' = (0, b)$. These mappings may be combined in one diagram:

$$A \underset{p}{\overset{q}{\rightleftarrows}} C \underset{p'}{\overset{q'}{\leftrightarrows}} B$$

and they satisfy the relations:

$$qp = 1_A, \quad q'p' = 1_B, \quad qp' = q'p = 0, \quad pq + p'q' = 1_C. \tag{1}$$

Here 1_A, 1_B, 1_C denote the identity mappings on A, B, C respectively. If we write the projections as a row $P = (p\, p')$ and the injections as a column $Q = \begin{pmatrix} q \\ q' \end{pmatrix}$, these equations may be written more compactly in matrix form as follows:

$$PQ = 1_C, \qquad QP = \begin{pmatrix} 1_A & 0 \\ 0 & 1_B \end{pmatrix} \tag{2}$$

These relations (1) or (2) actually characterize direct sums:

PROPOSITION 2 *Let A, B, C be abelian groups. Then C is isomorphic to the direct sum of A and B if and only if there is a diagram of mappings*

$$A \underset{p}{\overset{q}{\rightleftarrows}} C \underset{p'}{\overset{q'}{\leftrightarrows}} B,$$

satisfying $qp = 1_A$, $q'p' = 1_B$, $qp' = q'p = 0$, $pq + p'q' = 1_C$.

Proof. We have seen that such mappings exist when C is the direct sum of A and B. Conversely, when p, p', q, q' exist satisfying (1), then q, q' are injective because they have right inverses (Lemma 1, **1.3**); hence A and B are isomorphic to their images Aq, Bq' respectively, in C. Any $x \in C$ can be written in the form $x = xpq + xp'q'$ and $xpq \in Aq$, $xp'q' \in Bq'$, hence $C = Aq + Bq'$ and the representation is unique, for if $x = aq + bq'$, where $a \in A$, $b \in B$, then a, b can be expressed in terms of x: $a = xp$, $b = xp'$. ■

The direct product of groups may also be characterized by a universal mapping property. We shall prove this for a general family of groups, (G_i) $(i \in I)$, since that case is no harder than that of two factors.

THEOREM 3 *Let $\{G_i\}$ $(i \in I)$ be a family of groups and $P = \prod G_i$ their direct product, with projections $p_i: P \to G_i$. Then for any family of homomorphisms from a given group G, $f_i: G \to G_i$ there exists a unique homomorphism $f: G \to P$ such that $f_i = fp_i$ for all $i \in I$. Moreover, the group P with the homomorphisms $p_i: P \to G_i$ is determined up to isomorphism by this property.*

We can illustrate this assertion by the accompanying diagram. The claim is that the dotted line can be filled in by a mapping to make the diagram commutative (the diagram should be thought of as a family of triangles,

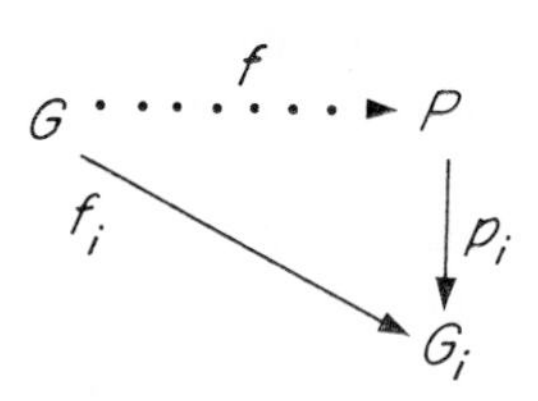

one for each $i \in I$, which have all been sewn together along the horizontal side, marked f). Thus Th. 3 asserts that P with the projections to the groups G_i is universal for families of homomorphisms from a group to the G_i.

Proof. If f exists, it must satisfy $xfp_i = xf_i$ $(i \in I)$ for any $x \in G$. This shows that xf can only be the family (xf_i) and it is clear that the mapping $x \mapsto (xf_i)$ is a homomorphism from G to P with the required property. Thus there exists exactly one such f, as claimed. Now the uniqueness of P follows because P is the solution of a universal problem. ■

As an application we shall determine the structure of the chief factors in a finite group. Of course the composition factors of a group are always simple, but this need not be the case for chief factors.

LEMMA 4 *Any minimal normal subgroup of a finite group is a direct power of a simple group.*

Here a 'minimal normal subgroup' is understood to be 'minimal normal and non-trivial'.

Proof. Let G be a finite group and H a minimal normal subgroup; we must show that H has the form

$$H = K_1 \times \cdots \times K_r, \tag{3}$$

where all the K_i are isomorphic simple groups. Using an induction on $|G|$ we may assume that the result holds for groups of order less than $|G|$. If H is simple the result clearly holds and this is so in particular when $H = G$. Thus we may assume that H is a proper subgroup of G and is not simple. Let K be a minimal normal subgroup of H, then, by the induction hypothesis, K is a direct power of a simple group. Let the distinct conjugates of K in G be $K = K_1, K_2, \ldots, K_s$. Since $H \lhd G$, it follows that $K_i \subseteq H$ for $i = 1, \ldots, s$ and, since $K \lhd H$, each K_i is normal in H. Since $K_1 \neq K_2$, $K_1 \cap K_2$ is a proper subgroup of K_1, again normal in H, and hence $K_1 \cap K_2 = \mathbf{1}$, by the minimality of K. Thus K_1K_2 is a direct product, by Th. 1. Let us renumber the Ks so that for $i \leqslant r$ say we have $K_i \nsubseteq K_1K_2 \ldots K_{i-1}$ while each K_j lies in $K_1K_2 \ldots K_r$. Then $K_1K_2 \ldots K_r = K_1K_2 \ldots K_s = H$, because the product of all the Ks is again a subgroup and, as the product of all conjugates of K, this product is normal in G. Since each K_i is minimal normal in H, we have $K_i \cap K_1 \ldots K_{i-1} = \mathbf{1}$ for $i \leqslant r$. Hence H is the direct product of $K_1, \ldots, K_r$ (by Th. 1, Cor.). Further, each K_i is isomorphic to K which was a direct power of a simple group, P, say. Hence H is a direct power of P. ■

Suppose that G is a soluble group, then any subgroup of G is soluble, and hence any simple subgroup of G is cyclic of prime order. Thus a minimal normal subgroup of a soluble group is a direct power of a cyclic group of prime order; such a group is said to be *elementary abelian.*

Now let G be any finite group and take a chief series for G:

$$G = H_1 \supset H_2 \supset \cdots \supset H_m = \mathbf{1}. \tag{4}$$

The fact that (4) is a chief series can be expressed by saying that each H_i is normal in G and H_{i-1}/H_i is minimal normal in G/H_i. By applying the above lemma to the groups G/H_i we therefore obtain

THEOREM 5 *In a finite group G, each chief factor is a direct power of a simple group; in particular, when G is soluble, each chief factor is elementary abelian.* ■

As an example, consider the symmetric group of degree 4; it has the chief series

$$\mathrm{Sym}_4 \supset \mathrm{Alt}_4 \supset V \supset \mathbf{1},$$

where V is the 4-group. Here $\mathrm{Sym}_4/\mathrm{Alt}_4$ and Alt_4/V are simple, of orders 2, 3 respectively, while V is a direct square of $\mathbf{C}_2$.

Exercises

(1) If $G = H \times K$ and $H_1 \lhd H$, $K_1 \lhd K$, show that $H_1 \times K_1 \lhd G$ and $G/H_1K_1 \cong H/H_1 \times K/K_1$.

(2) If $H, K \lhd G$ and $H \cap K = \mathbf{1}$, show that G is isomorphic to a subgroup of $G/H \times G/K$.

(3) A group is said to be *characteristically simple* if it has no characteristic subgroup apart from itself and the trivial group. Show that any direct power of a simple group is characteristically simple.

(4) Show that any finite characteristically simple group is a direct power of a simple group.

(5) Give an explicit proof that the direct product P is determined up to isomorphism by the property enunciated in Th. 3.

(6) Let H, K be groups with centres C, D respectively. Show that the centre of $H \times K$ is $C \times D$.

(7) Let G be a group. If there is an endomorphism α of G such that $\alpha^2 = \alpha$, and im $\alpha \lhd G$ show that $G = \mathrm{im}\,\alpha \times \ker \alpha$.

(8) If G is complete (cf. Ex. (10), **9.4**) and $G \lhd K$, show that $K = G \times H$ for some $H \lhd K$. (Hint. Show that $G \cong \mathrm{Aut}\, G$ and find an endomorphism α of K such that $\alpha^2 = \alpha$ and im $\alpha = G$; now use Ex. (7).)

(9) Show that for a complete group G, Hol $(G) = L(G) \times R(G)$.

(10) Let A, B, C be abelian groups with homomorphisms $p: C \to A$, $p': C \to B$, $q: A \to C$, $q': B \to C$ such that $qp = 1_A$, $q'p' = 1_B$, $pq + p'q' = 1_C$. Prove that $qp' = q'p = 0$.

(11) If a minimal normal subgroup has more than one decomposition (3), show that it is necessarily elementary abelian. (Hint. If K' is a minimal normal subgroup occurring in another decomposition of H, show that K' commutes elementwise with each of $K_1, K_2, \ldots, K_r$.)

(12) Let S be a simple group and define a group structure G on the union of two copies of the direct square S^2. The elements of G have the form (x_1, x_2), $(x_1, x_2)'$ and the multiplication is defined as

$$(x_1, x_2)(y_1, y_2) = (x_1y_1, x_2y_2), \qquad (x_1, x_2)(y_1, y_2)' = (x_1y_1, x_2y_2)',$$
$$(x_1, x_2)' \,.\, (y_1, y_2) = (x_1y_2, x_2y_1)', \quad (x_1, x_2)' \,.\, (y_1, y_2)' = (x_1y_2, x_2y_1).$$

Find a chief series and a composition series for G.

9.7 Abelian groups

The expression of a group element in terms of a generating set simplifies very much in case the group is abelian. Thus if A is an abelian group, generated by $a_1, \ldots, a_n$, every element x of A can be expressed in the form

$$x = a_1^{\alpha_1} \ldots a_n^{\alpha_n} \qquad \text{where } \alpha_1, \ldots, \alpha_n \in \mathbf{Z}, \tag{1}$$

or when A is an additive group,

$$x = \alpha_1 a_1 + \cdots + \alpha_n a_n. \tag{2}$$

Let us use additive notation, so that we have (2); in general we cannot assert that the coefficients $\alpha_1, \ldots, \alpha_n$ are uniquely determined by x. E.g., if a_1 has finite order m_1, we can change α_1 by a multiple of m_1 and still obtain x. A natural objective would be to find a generating set $\{a_1, \ldots, a_n\}$ for A such that every element x can be expressed in just one way in the form (2), where α_i is restricted to the range $0 \leqslant \alpha_i < m_i$ in case a_i has finite order m_i. Then $a_1, \ldots, a_n$ is called a *basis* for A. Writing A_i for the subgroup generated by a_i, we can express this fact by saying that A is the direct sum of the cyclic groups A_i:

$$A = A_1 \oplus \cdots \oplus A_n. \tag{3}$$

Conversely, if A can be expressed as a direct sum of cyclic groups, this leads to a representation (2) for its elements, which is unique if the α_i are suitably restricted whenever a_i has finite order. Our aim in this section is to prove the basis theorem which asserts the existence of such a decomposition for every finitely generated abelian group.

We recall that in a multiplicative group, $(xy)^n = x^n y^n$ whenever $xy = yx$. Hence in any abelian group, written additively, $n(x+y) = nx + ny$, i.e. the mapping $x \mapsto nx$ is an endomorphism, for any integer n. As a first consequence we have

PROPOSITION 1 *If A is any abelian group, then the subset tA of elements of finite order is a subgroup.*

For if $ra = sb = 0$ (where $a, b \in A, r, s \in \mathbf{Z}$), then $rs(a-b) = rsa - rsb = 0$, hence $a, b \in tA \Rightarrow a-b \in tA$ and $0 \in tA$, so tA is indeed a subgroup of A. ∎

The subgroup tA is called the *torsion subgroup* of A, its elements are the *torsion elements* of A, while the remaining elements of A are *torsion-free*. The group A itself is called a *torsion group* if $tA = A$, and *torsion-free* if $tA = \mathbf{0}$. It is easily verified that tA is a torsion group and A/tA is torsion-free.

A finitely generated torsion group must be finite. For if A is generated by $a_1, \ldots, a_n$ and a_i has finite order m_i, then every element can be written in the form (2) with $0 \leqslant \alpha_i < m_i$, so there are at most $m_1 m_2 \ldots m_n$ different elements in A.

To describe the finitely generated torsion-free groups we make the following

DEFINITION An abelian group A is said to be *free abelian on* $e_1, \ldots, e_n$ if every element of A can be uniquely expressed in the form

$$\lambda_1 e_1 + \cdots + \lambda_n e_n \qquad \text{where } \lambda_1, \ldots, \lambda_n \in \mathbf{Z}.$$

In other words, a free abelian group is a group with a basis consisting of torsion-free elements. Of course A will in general have many different bases, but all have the same number of elements because, on writing $2A = \{2x \mid x \in A\}$, this number is n, where

$$(A : 2A) = 2^n.$$

This number n is called the *rank* of the free abelian group A.

The importance of free abelian groups stems from the following universal mapping property:

THEOREM 2 *Let F_n be the free abelian group on $E = \{e_1, \ldots, e_n\}$. Then any mapping of E into an abelian group A extends to a unique homomorphism $F_n \to A$. In particular, any finitely generated abelian group can be expressed as a homomorphic image of a free abelian group.*

Thus the first assertion is that F_n with the inclusion mapping $E \to F_n$ is universal for mappings of E in abelian groups.

Proof. We must show that for any mapping f from E to an abelian group A, there exists a unique homomorphism $f' : F_n \to A$, whose restriction to E is f. Let f map e_i to $a_i \in A$; if the required homomorphism f' exists, it must map

$$\sum \lambda_i e_i \mapsto \sum \lambda_i a_i.$$

But this is a well-defined mapping, because each element of F_n can be written in the form $\sum \lambda_i e_i$ in just one way. Clearly it is a homomorphism and this proves the first part. Now if A is finitely generated, by $a_1, \ldots, a_n$ say, take F_n and map e_i to a_i. By the first part this extends to a homomorphism $F_n \to A$, the image contains the generating set $\{a_1, \ldots, a_n\}$ of A and hence is the whole of A, thus we have a homomorphism of F_n onto A. ■

Of course the definition of a free abelian group can be extended to the case of infinite rank and every abelian group can be expressed as a homomorphic image of a suitable free abelian group. However, we shall mainly be concerned with finitely generated abelian groups; in that case the free group has a simple characterization which does not extend to the general case:

PROPOSITION 3 *A finitely generated abelian group is free if and only if it is torsion-free.*

Proof. Clearly every free abelian group is torsion-free. Conversely let A be finitely generated, by $a_1, \ldots, a_n$, say, and torsion-free. We use induction on n; for $n = 1$, A is infinite cyclic and the result is clear, so let $n > 1$. If A is free on $a_1, \ldots, a_n$ there is nothing to prove. Otherwise we have a non-trivial relation

$$\lambda_1 a_1 + \cdots + \lambda_n a_n = 0. \tag{4}$$

Here the λs may be taken to be coprime, for if they had a common factor d, say $\lambda_i = d\lambda_i'$ for $i = 1, \ldots, n$, then (4) reads

$$d(\lambda_1' a_1 + \cdots + \lambda_n' a_n) = 0,$$

and here we may cancel d because A is torsion-free. If $\lambda_1 = 1$, we can write

$$a_1 = -(\lambda_2 a_2 + \cdots + \lambda_n a_n).$$

Thus A is already generated by $a_2, \ldots, a_n$ and the result follows by induction; the same applies if $\lambda_1 = -1$. In general we change the generating set of A so as to obtain a relation with a coefficient equal to ± 1. Thus assume $|\lambda_1| \geqslant |\lambda_2| > 0$, say, and replace a_2 by $a_2' = a_2 + \alpha a_1$, where $\alpha \in \mathbf{Z}$. Then the relation (4) becomes

$$(\lambda_1 - \alpha\lambda_2)a_1 + \lambda_2 a_2' + \lambda_3 a_3 + \cdots + \lambda_n a_n = 0.$$

By the division algorithm for $\mathbf{Z}$ we can choose α so that $|\lambda_1 - \alpha\lambda_2| < |\lambda_2|$ and with this choice we obtain a relation with a smaller coefficient of a_1. We continue in this way until one of the coefficients is reduced to ± 1, and then use the relation to eliminate one of the generators, as before. The result now follows by induction on n. ■

As a consequence of this proposition we can express every finitely generated abelian group A as a direct sum of a finite group and a free group. For A/tA is finitely generated torsion-free and hence free. Choose elements $e_1, \ldots, e_n$ in A such that their images in A/tA form a basis of the latter. Then the es satisfy no non-trivial relation mod tA and *a fortiori* they satisfy no non-trivial relation in A, i.e. the subgroup F generated by them is free. We claim that A is the direct sum of tA and F. For since $A/tA \cong F$, every element

$x \in A$ has the form $x = a + \sum \lambda_i e_i$ $(a \in tA)$ and $tA \cap F = \mathbf{0}$, for if $\sum \lambda_i e_i \in tA$, then this element has finite order k say, $\sum k\lambda_i e_i = 0$, hence $k\lambda_i = 0$ because the e_i are free generators and thus $\lambda_i = 0$ for $i = 1, \ldots, n$. This shows that $A = tA \oplus F$; we observe that tA, as homomorphic image of A, is again finitely generated and hence is finite. The result may be stated as a

COROLLARY *Every finitely generated abelian group is the direct sum of a finite group and a free group.* ■

This result shows that in analysing finitely generated abelian groups we can limit ourselves to finite groups. Our task then is to decompose a finite abelian group into a direct sum of cyclic groups. In this connexion we note that $\mathbf{C}_2 \oplus \mathbf{C}_3 \cong \mathbf{C}_6$, but $\mathbf{C}_3 \oplus \mathbf{C}_3 \ncong \mathbf{C}_9$; this example suggests that the problems to be overcome are already present when we deal with a single prime, an observation borne out by the next result.

A group (not necessarily abelian) is said to be *primary* if all its elements are of finite order p^r, where r may vary but the prime p is the same for all elements. Such a group is also called a *p-group*.

Let p be a prime. In each abelian group A, the set of elements of p-power order:

$$A_p = \{x \in A \mid p^r x = 0 \text{ for some } r \geqslant 0\} \tag{5}$$

is a subgroup; for if $p^r x = p^s y = 0$ and $r \geqslant s$, say, then $p^r(x-y) = 0$. Hence $x, y \in A_p \Rightarrow x-y \in A_p$ and clearly $0 \in A_p$. Moreover, A_p is a p-group, the maximal p-group contained in A. When A is finite, $A_p = \mathbf{0}$ unless p divides $|A|$, by Lagrange's theorem. Thus in a finite (or more generally, a finitely generated) abelian group, the maximal p-subgroup is trivial for all except a finite number of primes.

PROPOSITION 4 *Every finite abelian group can be expressed in just one way as a direct sum of p-groups, for different primes p. More precisely, if A is finite and the primes dividing* $|A|$ *are* $p_1, \ldots, p_r$, *then*

$$A = A_{p_1} \oplus \cdots \oplus A_{p_r}, \tag{6}$$

where A_p *is the maximal p-subgroup, defined as in* (5).

Proof. If there is a direct sum representation of the form (6), it is clear that the p-group component must consist of all the elements of p-power order in A and hence the sum is unique if it exists at all. Let $|A| = n = q_1 q_2 \ldots q_r$, where $q_i = p_i^{\alpha_i}$; if we put $P_i = n/q_i$, then $P_1, \ldots, P_r$ are coprime; for if they had a common prime factor p, say, then since each P_i divides n, $p \mid n$ and so p must be one of the p_i, say $p = p_1$, but $p_1 \nmid P_1$, a contradiction. Thus the P_i are coprime and there exist $Q_1, \ldots, Q_r \in \mathbf{Z}$ such that

$$P_1 Q_1 + \cdots + P_r Q_r = 1. \tag{7}$$

For any $a \in A$ write $a_i = P_i Q_i a$, then by (7),

$$a = a_1 + \cdots + a_r, \tag{8}$$

and $q_i a_i = q_i P_i Q_i a = nQ_i a = 0$, hence $a_i \in A_{p_i}$. It remains to prove the uniqueness of (8). If in addition to (8) we also have $a = \sum a_i'$ with $a_i' \in A_{p_i}$, then

$$a_1 - a_1' = a_2' - a_2 + a_3' - a_3 + \cdots + a_r' - a_r. \tag{9}$$

Here the left-hand side has order p_1^β for some β, and the right-hand side has order dividing $(p_2 \ldots p_r)^\gamma$, for some γ; but these two numbers are coprime, hence $up_1^\beta + v(p_2 \ldots p_r)^\gamma = 1$ for some $u, v \in \mathbf{Z}$. This shows that the element (9) has order dividing 1, i.e. it is 0. Thus $a_1' = a_1$ and similarly we find $a_2' = a_2$, $\ldots, a_r' = a_r$, i.e. (8) is unique and so (6) is established. ■

The final step towards our objective is to show that every finitely generated abelian p-group is a direct sum of cyclic groups. By an earlier remark we can limit ourselves to finite groups.

PROPOSITION 5 *Every finite abelian p-group is a direct sum of cyclic groups.*

Proof. Since A is finite, the orders of its elements are bounded. We pick an element $a \in A$ of maximal order p^α, say; write $A_0 = gp\{a\}$ and take a subgroup B of A which is maximal subject to the property $B \cap A_0 = \mathbf{0}$. We assert that

$$A = B \oplus A_0; \tag{10}$$

the result then follows by induction on $|A|$.

To prove (10) we need only show that the subgroup $B + A_0$ coincides with A; so let $A_1 = B + A_0$ and assume that $A_1 \neq A$. Let $x \in A$, $x \notin A_1$, then some power of p maps x into A_1 (because x has finite p-power order) so on replacing x by a multiple if necessary we may assume that $x \notin A_1$ but $px \in A_1$. Hence

$$px = ra + b, \qquad \text{where } r \in \mathbf{Z},\ b \in B.$$

By the definition of α, $0 = p^\alpha x = p^{\alpha-1} ra + p^{\alpha-1} b$, i.e.

$$p^{\alpha-1} b = -p^{\alpha-1} ra. \tag{11}$$

The left-hand side lies in B and the right-hand side in A_0, but these two groups meet in $\mathbf{0}$ (by the definition of B) so the element (11) is 0, i.e. $p^{\alpha-1} ra = 0$. Since a had order p^α, it follows that $p \mid r$, say $r = pr_0$. Then $p(x - r_0 a) = b \in B$, but $x - r_0 a \notin B$ because $x \notin A_1$. By the maximality of B it follows that $gp\{B, x - r_0 a\} \cap A_0 \neq \mathbf{0}$, say

$$sa = t(x - r_0 a) + b' \qquad \text{where } b' \in B.$$

Here $tx = (s + tr_0)a - b' \in A_1$; if $p \mid t$, then $sa \in gp\{B, p(x - r_0 a)\} = B$ and so $sa \in B \cap A_0$, a contradiction. Hence $p \nmid t$, and $tx \in A_1$; since we are in a p-group, it follows that $x \in A_1$ which is again a contradiction. Hence $A_1 = A$ and (10) follows. ■

We observe that Prop. 5 also follows from a more general fact which will be proved in Ch. **10** (Th. 2, **10.6**). The above result shows that every abelian p-group A has the form

$$A = B_1 \oplus \cdots \oplus B_r, \tag{12}$$

where B_i is cyclic of order p^{α_i}, and $0 < \alpha_1 \leqslant \cdots \leqslant \alpha_r$. Unlike the direct sum representation of Prop. 4, the representation (12) is not unique, i.e. there may be different ways of choosing the summands B_i, but the α_i are independent of this choice. To see this we show that the αs can be determined in terms of A alone: we observe that A/pA has the form $\mathbf{C}_p^r$, the direct sum of r copies of the cyclic group of order p, and, generally $p^{i-1}A/p^iA \cong \mathbf{C}_p^{\beta_i}$, where β_i is the number of αs that are at least i. When the βs are given, α_j may be determined as the number of βs that are greater than $r-j$; hence the αs can be found in terms of A alone and so are independent of the particular representation (12). We also say that A is of *type* $(p^{\alpha_1}, p^{\alpha_2}, \ldots, p^{\alpha_r})$ and call $p^{\alpha_1}, \ldots, p^{\alpha_r}$ the *elementary divisors* of A. By combining Prop. 4 and Prop. 5 and remembering Prop. 3, Cor. we obtain

THEOREM 6 (Basis theorem for abelian groups) *Every finitely generated abelian group A can be written as a direct sum of cyclic groups:*

$$A = B_1 \oplus \cdots \oplus B_r$$

where each B_i is either infinite or of prime power order, and the orders which occur are uniquely determined. ■

The orders of the cycles occurring here are again called the *elementary divisors* of the abelian group A. They may also be described as the orders of the indecomposable summands of A, i.e. the summands which cannot be further decomposed.

There is another decomposition into cyclic groups that is of interest: the decomposition into 'largest' cyclic summands. If A has elementary divisors

$$p^{\alpha_1}, p^{\alpha_2}, \ldots, p^{\alpha_u}, q^{\beta_1}, q^{\beta_2}, \ldots, q^{\beta_v}, \ldots, r^{\gamma_1}, \ldots, r^{\gamma_w},$$

where $p, q, \ldots, r$ are distinct primes and for each prime the exponents are in ascending order, then on grouping the largest cyclic subgroup for each prime together (and remembering that the direct product of cyclic groups of coprime orders is again cyclic) we find that A has a cyclic direct summand of order $p^{\alpha_u}q^{\beta_v} \ldots r^{\gamma_w}$, another of order $p^{\alpha_{u-1}}q^{\beta_{v-1}} \ldots r^{\gamma_{w-1}}$ (where terms with negative subscripts are to be omitted) and so on. These numbers are called the *invariant factors* of A. E.g., if A has elementary divisors $2, 2, 2^2, 2^4$; $3^2, 3^5$; $7, 7^2, 7^2$, then the invariant factors in descending order are $2^4 . 3^5 . 7^2$, $2^2 . 3^2 . 7^2, 2 . 7, 2$. In **10.6** we shall again look at the problem of decomposing an abelian group, from a more general point of view.

We have already met groups of type $(p, p, \ldots, p)$; they are the elementary abelian p-groups. In such a group every non-zero element has order p. i.e. $pA = \mathbf{0}$. Conversely, any finite abelian group A satisfying $pA = \mathbf{0}$ is an elementary abelian p-group. This follows of course from Prop. 5, but it can also be proved directly: we regard A as a vector space over $\mathbf{F}_p$. Since A is finitely generated over $\mathbf{F}_p$ (it is even finite), it has a basis $a_1, \ldots, a_n$, say, and any such basis corresponds to a decomposition of A as a direct sum of cyclic groups, necessarily of order p.

Elementary abelian p-groups are of importance in finite group theory because they occur as chief factors of soluble groups (Th. 5, **9.6**). By the vector space theory of Ch. **4**, an endomorphism of an elementary abelian p-group of order p^n is given by an $n \times n$ matrix over $\mathbf{F}_p$ (relative to a given basis in A) and the automorphisms constitute the general linear group of degree n over $\mathbf{F}_p$, denoted by $\mathbf{GL}_n(\mathbf{F}_p)$ or also $\mathbf{GL}(n, p)$, so that we have

$$\mathrm{Aut}\,(\mathbf{F}_p^n) \cong \mathbf{GL}(n, p).$$

Each automorphism of $A \cong \mathbf{F}_p^n$ can be specified by the basis to which it maps a given basis, hence the number of automorphisms of A equals the number of ordered bases. The first basis element e_1 can be any non-zero element of A; this can be chosen in $p^n - 1$ ways. For the next basis element, e_2, we can take any element not a multiple of e_1 and we thus have $p^n - p$ choices. In choosing e_3 we must avoid the 2-dimensional subspace spanned by e_1, e_2 and so have $p^n - p^2$ choices. Continuing in this way we find that the number of bases in A, and hence the order of its automorphism group, is

$$(p^n - 1)(p^n - p) \ldots (p^n - p^{n-1}).$$

Exercises

(1) Let A be an abelian group and tA its torsion subgroup. Show that (i) tA is a torsion group, (ii) A/tA is torsion-free and (iii) tA is the only subgroup of A satisfying (i) and (ii).

(2) Show that there is a functor t which associates with each abelian group its torsion subgroup and describe its effect on homomorphisms. Show that the inclusion $tA \to A$ is a natural transformation from t to the identify functor. Is there a non-trivial natural transformation from the identity to t?

(3) Show that an abelian group of order n contains a subgroup of order m iff $m \mid n$. If A is a p-group of type $(p^{\alpha_1}, \ldots, p^{\alpha_r})$, show that A contains a subgroup of type $(p^{\beta_1}, \ldots, p^{\beta_s})$ iff $s \leqslant r$ and $\beta_s \leqslant \alpha_r$, $\beta_{s-1} \leqslant \alpha_{r-1}, \ldots, \beta_1 \leqslant \alpha_{r-s+1}$.

(4) Find all abelian groups of orders 30, 36, 90, 360. In each case, find how many elements there are of each order.

(5) Show that the number of abelian groups of order p^n is the number of partitions of n into positive integers.

(6) If two finite abelian groups have the same number of elements of any given order, show that they are isomorphic.

(7) Show that in $\mathbf{GL}_2(\mathbf{Z})$, the group of invertible 2×2 matrices over $\mathbf{Z}$, the torsion elements do not form a subgroup.

(8) Show that the additive group of rational numbers is torsion-free but not free.

(9) Let $\mathbf{Q}^*$ be the multiplicative group of non-zero rational numbers. Show that the torsion subgroup is of order 2, and that $\mathbf{Q}^*$ is the direct product of its torsion subgroup and the free abelian group on $p_1, p_2, \ldots$, the different primes.

(10)* Let A be a finite abelian group and let $A = A_1 \oplus \cdots \oplus A_r$ be the decomposition into largest cyclic summands. Show that any automorphism of A which commutes with all automorphisms, maps each A_i into itself. Deduce that the centre of Aut A has order $\varphi(m)$, where m is the largest invariant factor of A.

9.8 The Sylow theorems

In looking for subgroups of finite groups we found Lagrange's theorem a useful aid, since it provides a necessary condition on the order of a subgroup. However, this condition is not sufficient: The alternating group Alt_4 of degree 4 has order 12, but it has no subgroup of order 6. A sufficient condition for a subgroup to exist was found by Sylow in 1872: If n divides $|G|$, then G has a subgroup of order n provided that n is a prime power. This result is one of the basic tools in modern finite group theory. It is proved in Th. 1 and Th. 3 below; the proof makes use of the following remark. Let G be a p-group acting on a set X and denote by X_0 the set of points fixed by G, i.e. the 1-point orbits. The number of points in any orbit is a power of p, by the orbit formula, and hence is either 1 or divisible by p. Thus modulo p we can ignore orbits with more than one point and so find that

$$|X_0| \equiv |X| \pmod{p}. \tag{1}$$

If G is a group of order $n = p^\alpha n'$, where $p \nmid n'$, any subgroup of order p^α is called a *Sylow p-subgroup*. Clearly if such a subgroup exists at all (and Sylow's theorem asserts that it always does), then it is maximal among p-subgroups of G.

THEOREM 1 (Sylow) *Let G be a finite group and p a prime. Then*

(i) *G has Sylow p-subgroups and every p-subgroup of G is contained in a Sylow p-subgroup,*

(ii) *all Sylow p-subgroups of G are conjugate,*

(iii) *the number of Sylow p-subgroups is congruent to* 1 (mod p).

Proof (*Wielandt*). We begin by showing that Sylow p-subgroups exist. Let us write $|G| = n = p^\alpha n'$, where $p \nmid n'$. Of course if $\alpha = 0$, the assertion holds trivially, so we may assume that $\alpha > 0$.

Let $M = \{X_1, X_2, \ldots, X_k\}$ be the collection of all subsets of G with exactly p^α elements. The number of these subsets is the number of ways of choosing p^α elements from a total of n:

$$k = |M| = \binom{n}{p^\alpha} = \frac{n(n-1)(n-2)\ldots(n-p^\alpha+1)}{p^\alpha.\ 1.\quad 2.\quad \ldots\quad (p^\alpha-1)},$$

and this is not divisible by p, since corresponding factors in the numerator and denominator (as written) are divisible by the same power of p.

We let G act on M by right multiplication. Each $a \in G$ defines a permutation $X \mapsto Xa$ of M. Since $p \nmid k$, at least one orbit has a number of elements prime to p. Take X in this orbit and let $S = \{a \in G \mid Xa = X\}$ be its stabilizer, then S has index prime to p and hence has order divisible by p^α. Let us pick an element $u \in X$; if $a \in S$, then $Xa = X$, hence $ua \in X$, and so $a \in u^{-1}X$. This shows that $S \subseteq u^{-1}X$ and therefore $|S| \leqslant |X| = p^\alpha$. It follows that $|S| = p^\alpha$ and so S is a Sylow p-subgroup.

Let S be the Sylow p-subgroup just found and H any p-subgroup of G. We next show that H lies in a conjugate of S. To do this, we let H act on the set of left cosets of S in G:

$$Sx \mapsto Sxa \qquad (a \in H).$$

The total number of cosets, $(G : S) = n'$ is prime to p and H is a p-group, hence by (1), some coset Sx is fixed by H, i.e. $SxH = Sx$. Hence $xH \subseteq Sx$, i.e. $H \subseteq x^{-1}Sx$. Now $x^{-1}Sx$ has the same order as S and so is again a Sylow subgroup. This proves that H is contained in a Sylow p-subgroup of G, i.e. (i). If H is itself a Sylow p-subgroup, then $H = x^{-1}Sx$ and so we see that all Sylow p-subgroups are conjugate to a single one, i.e. (ii).

Finally let $\Sigma = \{S_1, \ldots, S_r\}$ be the set of all Sylow p-subgroups, and let S_1 act on Σ by conjugation: $S_i \mapsto a^{-1}S_ia$ $(a \in S_1)$. The only fixed point under this action is S_1; for if $a^{-1}S_ia = S_i$ for some $i \neq 1$ and all $a \in S_1$, then $S_1S_i = S_iS_1$ is a subgroup and S_i is normal in S_1S_i. But S_1 and S_i are both Sylow p-subgroups of S_1S_i, hence S_1 is a conjugate of S_i in S_1S_i; this is impossible because S_i is normal in S_1S_i but different from S_1. Thus Σ has just one fixed point under the action of S_1 and the formula (1) shows that $|\Sigma| \equiv 1 \pmod{p}$, which establishes (iii). ∎

COROLLARY *Let G be a finite group and S a Sylow p-subgroup. Then S is the only Sylow p-subgroup if and only if $S \lhd G$. In that case S is characteristic in G.*

For the Sylow p-subgroups of G are just the conjugates of S and they reduce to a single one iff $a^{-1}Sa = S$ for all $a \in G$, i.e. when $S \lhd G$. Moreover, any automorphism α of G transforms S into another subgroup of the same order as S, hence $S\alpha$ is again a Sylow subgroup and so $S\alpha = S$, i.e. S is characteristic in G. ∎

As an illustration of the theorem consider a group G of order pq, where p,q are distinct primes. The only abelian group of this order is $\mathbf{C}_{pq} = \mathbf{C}_p \times \mathbf{C}_q$, the cyclic group of order pq. If G is non-abelian, let S be a Sylow p-subgroup and let H be the normalizer of S in G:

$$H = \{x \in G \mid x^{-1}Sx = S\}.$$

Clearly $S \subseteq H \subseteq G$, so either $H = G$ or $H = S$, because $(G:S) = q$ is a prime. If $H = G$, $S \lhd G$; otherwise S has $q = (G:H)$ conjugates and so by Th. 1 (iii), $q \equiv 1 \pmod{p}$. This shows that $p \mid (q-1)$ and, in particular, $p < q$. Thus when G has order pq, the Sylow subgroup corresponding to the larger prime must be normal in G. Taking q to be the larger prime, we see that unless G is abelian, the Sylow p-subgroup is not normal and so has q conjugates. We also note that G has $q-1$ elements of order q, one element of order 1 and $q(p-1)$ elements of order p. An example is Sym_3: here $p = 2$, $q = 3$.

The Cor. of Th. 1 shows that the normalizer of a Sylow subgroup S of a group G cannot be contained in a proper normal subgroup N of G; for S would be characteristic in N and hence normal in G. There is a strengthening of this remark which is sometimes useful.

PROPOSITION 2 *Let G be a finite group, S a Sylow subgroup of G and T its normalizer. Then any subgroup containing T is its own normalizer.*

Proof. Let $H \supseteq T$ and assume that $a^{-1}Ha = H$ for some $a \in G$, then $a^{-1}Sa \subseteq a^{-1}Ha = H$. Hence S and $a^{-1}Sa$ are both Sylow subgroups of H and so are conjugate in H: $a^{-1}Sa = b^{-1}Sb$ for some $b \in H$. Hence $ab^{-1} \in T \subseteq H$, and so $a \in Hb = H$. ∎

We still have to justify the claim made in the first paragraph of this section that a finite group G has a subgroup of any prime power dividing $|G|$. By Sylow's theorem it is enough to prove the result for p-groups; this is fairly easy (and was known earlier). We collect this and some other results on p-groups together:

THEOREM 3 *Let G be a non-trivial p-group, then*

(i) *G has a non-trivial centre,*

(ii) *for each divisor r of $|G|$, G has a subgroup of order r,*

(iii) *G is soluble.*

Proof. (i) Let G act on itself by conjugation, then the fixed points constitute the centre C of G, hence by (1),

$$|C| \equiv |G| \equiv 0 \pmod{p},$$

Thus C has order divisible by p and so is non-trivial.

We now prove (ii) and (iii) for any p-group, by induction on $|G|$, using (i). When $|G| = 1$, both assertions are obviously true, so let $|G| = p^n > 1$. Then by (i) the centre is a non-trivial abelian p-group and so it contains an element c of order p. Let C_1 be the subgroup generated by c, then C_1 has order p and $C_1 \lhd G$, because $x^{-1}cx = c$ for all $x \in G$. Hence we can form G/C_1; this is a group of order p^{n-1}, by induction it has subgroups of all orders, which by the third isomorphism theorem correspond to subgroups of all orders p^r $(0 < r \leqslant n)$ of G. This proves (ii); to prove (iii), we may assume by induction that G/C_1 is soluble, then since C_1 is soluble, it follows that G is soluble. ■

Groups in which all Sylow subgroups are normal have a simple structure that is noteworthy:

PROPOSITION 4 *A finite group G is the direct product of its Sylow subgroups if and only if all its Sylow subgroups are normal. Such a group is always soluble.*

Proof. Every direct factor is normal, so the condition is necessary. When it holds, there is just one Sylow subgroup for each prime; different Sylow subgroups meet in **1**, and hence commute elementwise. Moreover, each $x \in G$ can be written uniquely as a product of elements of distinct prime power orders: simply apply Prop. 4, **9.7**, (the easy part of the basis theorem for abelian groups) to the cyclic group generated by x. Now the result follows from the characterization of direct products given in Th. 1, **9.6**, Cor.

Each Sylow subgroup is soluble, by Th. 3 (iii), and hence so is their direct product. ■

Exercises

(1) Find the Sylow subgroups in Sym_3 and Sym_4.

(2) Show that every group of order 45 is abelian.

(3) Show that a group of order p^2q, where p, q are distinct primes, contains a normal Sylow subgroup.

(4) Show that the subgroups of order 2 in Sym_4 are not all conjugate.

(5) Determine all groups of order less than 16. (There are 28.)

(6) Show that no group of order 200 is simple.

(7) For any finite group G show that Inn G, the group of inner automorphisms of G, has as many Sylow p-subgroups as G itself.

(8) Show that $\mathbf{GL}_n(\mathbf{F}_p)$, the group of invertible $n \times n$ matrices over $\mathbf{F}_p$, has a Sylow p-subgroup consisting of triangular matrices. Find its order.

(9)* Let G be a group, H a subgroup and S a Sylow p-subgroup of G. Using a doublet coset decomposition of G with respect to H and S (cf. Ex. (15), Further Exercises Ch. **3**), show that S has a conjugate S' such that $(H: H \cap S')$ is prime to p. Deduce that $H \cap S'$ is a Sylow p-subgroup of H.

(10) Let G be a finite group of order n and p a prime. Show that a faithful G-action can be defined on $\mathbf{F}_p^n$ and hence embed G in $\mathbf{GL}_n(\mathbf{F}_p)$. Use Ex. (8) and Ex. (9) to deduce the existence of a Sylow p-subgroup of G.

(11) Let G be a finite group, N a normal subgroup, S a Sylow subgroup of N, and T the normalizer of S in G. Show that $G = NT$. (Hint. Any conjugate of S under G lies in N. Now use Th. 1, (ii).)

(12) Let G be a finite group, S a Sylow subgroup and let $N \lhd G$. Show that $S \cap N$ is a Sylow subgroup of N and SN/N is a Sylow subgroup of G/N.

(13) Let G be a finite group and p a prime dividing $|G|$. (i) If every conjugacy class of G has order either 1 or divisible by p, use the class equation:

$$G = \bigcup C_\lambda,$$

where C_λ runs over the conjugacy classes, to show that the centre of G has order divisible by p. (ii) If G has a conjugacy class of order greater than 1 and prime to p, show that G has a proper subgroup of order divisible by p. Deduce (without using Sylow's theorem) that a finite group of order divisible by a prime p contains an element of order p. (Cauchy)

(14) Let $f: G \to H$ be a homomorphism of finite groups, then for any $x \in G$, show that the order of xf divides the order of x. Moreover, if $z \in Gf$ has p-power order, show that there exists $x \in G$ such that $xf = z$ and x also has p-power order.

(15) Show that a finite group with non-trivial cyclic Sylow 2-subgroup has a subgroup of index 2. (Hint. The regular representation contains odd permutations.)

9.9 Generators and defining relations; free groups

In order to describe a group we usually take one of its naturally occurring representations, e.g. the symmetric group on n symbols, or the rotation group in 3 dimensions. But the group may be given abstractly, in which case we may describe it by writing down a list of its elements and its multiplication table. This description can still be abbreviated: in place of a complete list of all its elements it is enough to have a generating set of the group and from the multiplication table we need only write down enough products to enable us to deduce the rest. Of course these products must now be expressed in terms of the given generating set. A set of equations between products, which suffice to construct the multiplication table is called a *set of defining relations* for the group and the description we have given is called a *presentation of the group by generators and defining relations.*

To give an example, the cyclic group $\mathbf{C}_n$ of order n can be generated by a single element c with the defining relation

$$c^n = 1. \tag{1}$$

Once we are given (1), we know that $\mathbf{C}_n$ has at most n elements, which may be written as $1, c, c^2, \ldots, c^{n-1}$ and using (1) we can reconstruct the multiplication table:

$$c^i c^j = \begin{cases} c^{i+j} & \text{if } i+j < n, \\ c^{i+j-n} & \text{if } i+j \geqslant n. \end{cases}$$

We shall write this presentation for $\mathbf{C}_n$ in the form

$$\mathbf{C}_n = gp\{c \mid c^n = 1\}. \tag{2}$$

Generally, if a group G is presented by a generating set $a_1, \ldots, a_r$ and defining relations $f_1 = \cdots = f_s = 1$, where the f_j are *words* (i.e. formal products) in the *a*s and their inverses, we write

$$G = gp\{a_1, \ldots, a_r \mid f_1 = \cdots = f_s = 1\}. \tag{3}$$

The left-hand side of a relation $f = 1$ is also called a *relator* of G. A relation in a group will have the general form $g = h$, where g and h are words in the generators and their inverses but we can reduce the right-hand side to 1 by writing the relation as $gh^{-1} = 1$, where of course $(c_1 \ldots c_t)^{-1} = c_t^{-1} \ldots c_1^{-1}$.

A presentation is clearly a very economical way of describing an abstract group, which is often more perspicuous than the description in terms of the multiplication table. This is very clear in the above example of the cyclic group $\mathbf{C}_n$. There is just one thing that the presentation (2) of $\mathbf{C}_n$ does not tell us immediately: it is easy to see from the presentation that $\mathbf{C}_n$ cannot have more than n elements, but not so easy to see that the order is exactly n. To convince ourselves of this fact we must show that the expressions $1, c, \ldots, c^{n-1}$ all represent distinct elements; we give two proofs.

Firstly we have the representation method. Let us define a mapping of $\mathbf{C}_n$ into the complex numbers by

$$c^\nu \mapsto \exp(2\pi i\nu/n).$$

This is a well-defined mapping because the relation $c^n = 1$ goes over into the correct relation $\exp(2\pi i) = 1$, and it is clearly a homomorphism of $\mathbf{C}_n$ into the multiplicative group of non-zero complex numbers. Now $1, c, \ldots, c^{n-1}$ are distinct because they have distinct images.

The second proof also constructs a representation of $\mathbf{C}_n$, namely its right regular representation, or at least what would be its right regular representation if we knew that $1, c, \ldots, c^{n-1}$ were distinct. Of course we must avoid the apparent circularity of this informal description. We take a set of n symbols $E = \{e_1, \ldots, e_n\}$ and define a permutation γ on E by the rule

$$e_i\gamma = \begin{cases} e_{i+1} & \text{if } i \leqslant n-1, \\ e_1 & \text{if } i = n. \end{cases}$$

Thus γ is just a cycle of length n. Again $c^\nu \mapsto \gamma^\nu$ defines a homomorphism, because $\gamma^n = 1$, and $1, c, \ldots, c^{n-1}$ are distinct, because this is true of their images.

We observe that both proofs use the fact that (2) defines a group. Of course this is not in question; we take the given generating set and all consequences of the defining relations. What we then have to prove is that certain relations such as $c^\nu = 1$ for $0 < \nu < n$ do *not* follow from the defining relations. This problem, of deciding from a given presentation when two words represent the same element is called the *word problem* for groups. Below we shall give more examples of presentations for which the word problem can be solved. But this is often a difficult problem and in general it is undecidable; this means that there are group presentations for which no algorithm exists for deciding when two words represent the same element.

In the case of finitely generated abelian groups the word problem can be completely solved. We saw in **9.7** that every finitely generated abelian group A has a presentation

$$A = \mathrm{gp}\{a_1, \ldots, a_r \mid a_i a_j = a_j a_i, \quad a_1^{\alpha_1} = \cdots = a_r^{\alpha_r} = 1\},$$

where $\alpha_i \mid \alpha_{i+1}$ (and some of the αs may be 0, corresponding to infinite cyclic factors). It follows that every element of A can be uniquely expressed in the form $a_1^{\beta_1} \ldots a_r^{\beta_r}$, where $0 \leqslant \beta_i < \alpha_i$; thus we have a normal form for the elements of A. To find out if two words represent the same element we need only bring them both into normal form and compare these normal forms.

To give a non-abelian example, consider Sym_3; it has the presentation

$$\mathrm{Sym}_3 = gp\{a, b \mid a^3 = b^2 = (ab)^2 = 1\}, \tag{4}$$

and the reader should verify that this presentation defines a group of order 6.

Let us see how the method works when the group being presented is not known in advance. Consider the presentation

$$G = gp\{a, b \mid a^r = b^s = 1, \quad b^{-1}ab = a^{-1}\},$$

where r, s are integers greater than 1. We begin by taking all words in a, b, a^{-1}, b^{-1} and seeing how they can be simplified. In the first place, the relation $a^r = 1$ can be used to reduce all exponents of a to the range $0, 1, \ldots, r-1$, and similarly the exponents of b can be reduced to the range $0, 1, \ldots, s-1$. Next the commutation rule, in the form $ba = a^{-1}b$ $(= a^{r-1}b)$ can be used to move all occurrences of a to the left, so that each word can be reduced to the form

$$a^i b^j \qquad (0 \leqslant i < r, \quad 0 \leqslant j < s). \tag{5}$$

This already shows that G has order at most rs. We now show that the order is exactly rs by showing that the rs expressions (5) represent distinct elements of G. Let us first note that $b^2a = ba^{-1}b = ab^2$, thus a commutes with b^2. Since a also commutes with $b^s(=1)$, it follows that if s is odd, say $s = 2k+1$, then a commutes with $b^s(b^2)^{-k} = b$ and hence the group is then abelian. But this is immaterial in what follows (it just means that for odd s the given presentation can still be simplified). To represent G we take a set $E = \{e_{ij} \mid 0 \leqslant i < r, 0 \leqslant j < s\}$ in bijective correspondence with (5) and define two mappings of E into itself:

$$e_{ij}\alpha = \begin{cases} e_{i+1\,j} & \text{if } j \text{ is even and } i < r-1, \\ e_{i+1-r\,j} & \text{if } j \text{ is even and } i = r-1, \\ e_{i-1\,j} & \text{if } j \text{ is odd and } i > 0, \\ e_{r-1\,j} & \text{if } j \text{ is odd and } i = 0. \end{cases}$$

$$e_{ij}\beta = \begin{cases} e_{i\,j+1} & \text{if } j < s-1, \\ e_{i0} & \text{if } j = s-1. \end{cases}$$

It is only necessary to verify that $\alpha^r = \beta^s = 1$ and $\beta^{-1}\alpha\beta = \alpha^{-1}$, a task which may be left to the reader. Once this is done, we see that the correspondence $a \mapsto \alpha$, $b \mapsto \beta$ defines a G-action on E and now the elements (5) are distinct because they map e_{00} to distinct elements: $e_{00}a^ib^j = e_{ij}$. We observe that the presentation (4) of Sym_3 is the special case $r = 3$, $s = 2$.

The definition of a group by a presentation is an instance of a universal mapping property. This is an obvious but useful observation, which has already been used informally in the above proofs.

THEOREM 1 (Dyck) *Given a group G with the presentation*

$$G = gp\{a_1, \ldots, a_r \mid f_1 = \cdots = f_s = 1\}, \tag{6}$$

if H is any group with a family of elements $\{b_1, \ldots, b_r\}$ *such that the relations* $f_j(b) = 1$ $(j = 1, \ldots, s)$ *hold in H, then the mapping* $a_i \mapsto b_i$ $(i = 1, \ldots, r)$ *defines a unique homomorphism from G to H.*

Proof. Let $g(a)$, $h(a)$ be any words in the as and their inverses. If $g(a) = h(a)$ in G, then this is a consequence of the defining relations $f_j = 1$, i.e. by inserting occurrences of f_j, or removing such occurrences, together with the usual group laws (which tell us that we can insert or remove occurrences of $a_ia_i^{-1}$ or of $a_i^{-1}a_i$) we can transform $g(a)$ into $h(a)$. If we carry out the corresponding process on $g(b)$ we reach $h(b)$ and, since the relations $f_j(b) = 1$ hold in H, it follows that in H we have $g(b) = h(b)$. Therefore the correspondence

$$g(a) \mapsto g(b)$$

between elements of G and elements of H is actually a mapping. It is trivial to check that this is a homomorphism. To prove that it is unique, assume that φ, φ' are two homomorphisms from G to H which agree on $a_1, \ldots, a_r$. The set $\{x \in G \mid x\varphi = x\varphi'\}$ is easily seen to be a subgroup (cf. Lemma 3, **6.4**); it contains a generating set of G and therefore is the whole of G, i.e. $\varphi = \varphi'$. ■

Of particular interest are the groups with no defining relations, i.e. with a presentation of the form

$$F_r = gp\{x_1, \ldots, x_r \mid \}.$$

F_r is called the *free group* on $x_1, \ldots, x_r$. We observe that $F_r^{ab} = F_r/F_r'$ is the free abelian group of rank r. This remark shows that the integer r is uniquely determined by F_r, for if $F_r \cong F_s$ then $F_r^{ab} \cong F_s^{ab}$ and by the uniqueness of the rank of a free abelian group it follows that $r = s$; this integer r is again called the *rank* of F_r.

If in Th. 1 we take G to be free, we obtain the universal mapping property for free groups; it shows that free groups play the same role in group theory as free abelian groups do in the theory of abelian groups:

COROLLARY *If F is the free group on a set X, then any mapping from X to a group G can be extended in just one way to a homomorphism from F to G.*

In particular, by taking X large enough to map to a generating set of G, we see that every group G is the homomorphic image of a free group.

The last part follows because the image of the homomorphism is a subgroup of G containing the image of X, so if the latter generates G, the homomorphism is surjective. ■

Since we have only defined free groups of finite rank, we should, in the last sentence of the Corollary, restrict ourselves to finitely generated groups. However, with appropriate definitions of a free group of infinite rank, the Corollary holds generally; in fact the same proof goes through without difficulty.

To end this section we describe a normal form for the elements of a free group which enables us to solve the word problem for these groups. We recall that by a *word* in $x_1, \ldots, x_r$ and $x_1^{-1}, \ldots, x_r^{-1}$ we understand a formal product

$$x_{i_1}^{\varepsilon_1} \ldots x_{i_n}^{\varepsilon_n} \qquad (\varepsilon_\nu = \pm 1). \tag{7}$$

The number of letters, n, is its *length*. E.g., $x_1x_1x_2^{-1}x_1$ and $x_1x_3^{-1}x_3x_1x_2x_2^{-1}x_2^{-1}x_1$ are words (which happen to represent the same element). If we think of (7) as representing an element in a free group F, there may be certain simplifications we can make, even though the free group has no defining relations. Thus if $x_ix_i^{-1}$ or $x_i^{-1}x_i$ occurs in (7) we can omit such an occurrence without affecting the element of F represented. By carrying out all such simplifications we may assume that (7) contains no occurrence of $x_ix_i^{-1}$ or $x_i^{-1}x_i$ (of course x_i must be immediately followed or preceded by x_i^{-1}, thus $x_1^{-1}x_2x_1$ admits no simplification). Then (7) is said to be in reduced form; put differently, the word (7) is *reduced* if for no $\nu = 1, \ldots, n-1$, $i_\nu = i_{\nu+1}$ and $\varepsilon_\nu + \varepsilon_{\nu+1} = 0$.

THEOREM 2 *Let F be a free group on $x_1, \ldots, x_r$. Then each element of F has a unique expression as a reduced word in the xs and their inverses.*

Proof. The remarks preceding the theorem show that every element of F can be expressed as a reduced word in at least one way; we must show that two reduced words cannot represent the same element unless they are equal. It comes to the same to show that no reduced word except the empty word represents 1; for if u, v are any words, then $u = v$ is equivalent to $uv^{-1} = 1$ and, if u and v are distinct reduced words, it is easily seen that uv^{-1} has a non-empty reduced form.

Thus let (7) be a non-empty reduced word which represents 1, then under any homomorphism of F into any group, the element (7) maps to 1. Let G be the symmetric group of degree $n+1$ and map the x_i to permutations of 1, 2, $\ldots, n+1$ in any way such that x_{i_ν} maps ν to $\nu+1$ if $\varepsilon_\nu = 1$ and it maps $\nu+1$ to ν if $\varepsilon_\nu = -1$. This is possible, for, given x_i, we have assigned an image to ν under x_i if either $i = i_\nu$ and $\varepsilon_\nu = 1$ (then $\nu x_i = \nu+1$), or if $i = i_{\nu-1}$ and $\varepsilon_\nu = -1$ (then $\nu x_i = \nu-1$). These possibilities cannot both be realized (fortunately!) because (7) was assumed in reduced form, so $i_\nu = i_{\nu-1}$ and $\varepsilon_{\nu-1}+\varepsilon_\nu$

$= 0$ is impossible. Similarly, if $\mu \neq \nu$, then $\mu x_i \neq \nu x_i$ so that a permutation with the required properties exists. In this realization of F consider the effect of (7) on the symbol 1. Each $x_{i_\nu}^{\varepsilon_\nu}$ maps ν to $\nu+1$, hence (7) maps 1 to $n+1$. In particular, it is not the identity permutation and so the word (7) is not equal to 1 in F, as we wished to show. ∎

Exercises

(1) Verify that the group defined by the presentation (4) has order 6.

(2) Show that $\text{Sym}_4 = \text{gp}\{a, b \mid a^2 = b^3 = (ab)^4 = 1\}$.

(3) Show that $\text{Alt}_4 = \text{gp}\{a, b \mid a^2 = b^3 = (ab)^3 = 1\}$.

(4) Show that $\text{gp}\{a, b \mid a^4 = b^4 = 1, b^{-1}ab = a^{-1}\}$ has order 8 and is non-abelian, but all its subgroups are normal. (This group is called the *quaternion group*.)

(5) Find a presentation for the affine group $\mathbf{Af}_1(\mathbf{F}_p)$.

(6) Show that the affine group over $\mathbf{Z}$, $\mathbf{Af}_1(\mathbf{Z})$, has the presentation $\text{gp}\{a, b \mid a^2 = b^2 = 1\}$. Show that the natural homomorphism $\mathbf{Z} \to \mathbf{Z}/n$ induces a homomorphism $\mathbf{Af}_1(\mathbf{Z}) \to \mathbf{Af}_1(\mathbf{Z}/n)$ and find a presentation for the image.

(7) Prove Th. 2 by defining an action of the free group on the set of reduced words by right multiplication.

(8) Show that two elements of a free group commute iff they can be written as powers of the same element.

(9) Show that the group $\text{gp}\{a, b, c \mid a^{-1}ba = b^2,\ b^{-1}cb = c^2, c^{-1}ac = a^2\}$ is trivial. (B. H. Neumann)

(10) If a finite group G has a presentation in which all the defining relations have even length, show that every relation has even length; deduce that G has even order. If moreover, G is a permutation group, show that every permutation of odd order is even.

Further exercises on Chapter 9

(1) Prove that if N_1, N_2 are normal subgroups of a group G, then so is N_1N_2.

(2) Show that a group is abelian and simple iff it has prime order.

(3) Let G be a group, H a subgroup and N its normalizer in G. Show that H has exactly $(G\colon N)$ conjugates in G. Prove that if H is a proper subgroup, then G contains an element not in any conjugate of H.

(4) Let R be a ring, regarded as a category with a single object. A functor F from R to abelian groups is said to be *additive* if $(a+b)F = aF+bF$ for all $a, b \in R$. Show

that a covariant additive functor from R to abelian groups is an abelian group A with a homomorphism $R \to \text{End}(A)$, i.e. a right R-module. Similarly show that a contravariant additive functor is a left R-module.

(5)* Let G be a finite group and α an automorphism of G which leaves only the unit-element of G fixed. Show that the mapping $x \mapsto x^{-1}x^{\alpha}$ is a permutation of G. If for some $c \in G$, c^{α} is conjugate to c, show that there is a conjugate of c fixed by α, and deduce that $c = 1$. If moreover, $\alpha^2 = 1$, show that $xx^{\alpha} = 1$ for all $x \in G$ and deduce that G is abelian.

(6) Show that the holomorph of a cyclic group of odd order is complete.

(7)* Let G be a finite group and α an automorphism of order prime to $|G|$. If α maps each conjugacy class into itself, show that the subgroup H of elements fixed by α meets each conjugacy class and deduce that $\alpha = 1$. (W. Gaschütz)

(8) Let G be a finite group. Show that the automorphisms which map each conjugacy class into itself form a normal subgroup of Aut G whose order contains only prime factors occurring in $|G|$.

(9)* Show that $\mathbf{Af}_1(\mathbf{Z}/8)$ has order 32. If $\tau: x \mapsto ax+b$, define $\tau^{\alpha}: x \mapsto ax+b+\frac{1}{2}(a^2-1)$. Show that α is an automorphism which on τ agrees with conjugation by the element $\sigma_a: x \mapsto x+\frac{1}{2}(a+1)$. Deduce that α maps each conjugacy class into itself, but show that α is not inner. (G. E. Wall)

(10) Let G be any group and Z its centre. If G/Z is cyclic, show that G is abelian; deduce that every group of order p^2, where p is a prime, is abelian.

(11) Show that any proper subgroup of a finite group is contained in a maximal proper subgroup. If G is a p-group which contains at most one subgroup of any given order, show that G is cyclic.

(12) In any group G, show that an element x commutes with all its conjugates iff $((y, x), x) = 1$ for all $y \in G$.

(13) Show that the direct product of cyclic groups with coprime orders is cyclic.

(14) Show that the direct product of soluble groups is soluble.

(15)* Show that the group of order 168 in Ex. (8), **3.6**, is isomorphic to $\mathbf{GL}_3(\mathbf{F}_2)$. (Hint. Interpret the triples as lines in $\mathbf{F}_2^3$.)

(16) Show that a finite abelian group is cyclic iff it does not contain a direct square as subgroup.

(17) Show that the additive group of an entire ring is either torsion-free or every non-zero element has the same prime order.

(18) Show that in any group an element of finite order can be written as a product of commuting elements with prime power orders.

(19) If an abelian group A has the form $A = T+F$, where T is a torsion group and F is free abelian, show that $T = tA$ and that the sum is direct.

(20) Show that the centre of a group is not a functor, i.e. there is no functor from groups to abelian groups which assigns to each group its centre.

(21) Let A be an abelian torsion group and I an infinite set. Show that the direct power A^I is a torsion group iff there is a positive integer n such that $nA = 0$.

(22) Show that a proper subgroup of a p-group cannot be its own normalizer. Deduce that every subgroup of index p in a p-group is normal.

(23)* If a finite soluble group G has composition series in which the factors occur in arbitrary prescribed order, show that G is a direct product of its Sylow subgroups.

(24) A word is said to be *cyclically reduced* if it is not conjugate to a shorter word. Show that (i) every word is conjugate to a cyclically reduced word and (ii) two cyclically reduced words are conjugate iff one is a cyclic permutation of the other. (This solves the *conjugacy problem* for free groups, i.e. the problem of deciding in a finite number of steps whether a given pair of words represent conjugate elements.)

10

Rings and Modules

10.1 Ideals and quotient rings

In this section all rings considered may be arbitrary (not necessarily commutative). We shall analyse homomorphisms of rings in much the same way as was done for groups in **9.1**. Let

$$f: R \to S \tag{1}$$

be any homomorphism of rings. Thus f is a mapping satisfying $(x+y)f = xf+yf$, $(xy)f = xf \,.\, yf$, $1f = 1$. The last requirement cannot be omitted; e.g. the zero mapping $xf = 0$ (for all $x \in R$) satisfies the first two conditions, but not the third, unless $S = 0$. In any homomorphism (1), the image of f is a subring of S, as is easily seen. To study the kernel we shall need the following

DEFINITION An *ideal* $\mathfrak{a}$ in a ring R is a subgroup of the additive group of R such that $R\mathfrak{a} \subseteq \mathfrak{a}$, $\mathfrak{a}R \subseteq \mathfrak{a}$.

We usually denote ideals by small gothic letters and write $\mathfrak{a} \lhd R$ to indicate that $\mathfrak{a}$ is an ideal in R.

The definition shows in particular that an ideal is closed under addition, subtraction and multiplication, but it will not in general contain 1, and so will not be a subring of R. In fact if an ideal $\mathfrak{a}$ contains 1, then it contains any $x \in R$, for $x = x \,.\, 1 \in \mathfrak{a}$; hence $\mathfrak{a} = R$. Any ideal different from R is said to be *proper*.

Let us return to the kernel of the homomorphism (1). By definition this is just the set

$$\ker f = \{x \in R \mid xf = 0\}.$$

We claim that $\ker f$ is an ideal in R. Clearly it is a subgroup of the additive group of R; if $a \in \ker f$, then $af = 0$ and hence for any $x \in R$, $(ax)f = af \,.\, xf = 0$, $(xa)f = xf \,.\, af = 0$, so ax, $xa \in \ker f$, which shows that $\ker f$ is indeed an ideal.

Consider a ring R and an ideal $\mathfrak{a}$ in R; denote by $R/\mathfrak{a}$ the set of cosets of $\mathfrak{a}$ in R. As we have seen in **9.1**, this set is again a group, abelian like the additive group of R, and the natural homomorphism $\lambda: R \to R/\mathfrak{a}$ is a homomorphism of groups. We now show that a multiplication can be defined on these cosets so that $R/\mathfrak{a}$ becomes a ring and the mapping λ a ring-homomorphism. In fact there is precisely one way of doing this.

Let $a, b \in R$, then $a\lambda, b\lambda$ are cosets of $\mathfrak{a}$ whose product must satisfy $a\lambda \,.\, b\lambda = (ab)\lambda$, i.e.

$$(a+\mathfrak{a})(b+\mathfrak{a}) = ab+\mathfrak{a}. \tag{2}$$

This equation tells us how to proceed: given cosets α, β, we choose $a \in \alpha$, $b \in \beta$ and take as the product of α and β the coset $ab+\mathfrak{a}$. To ensure that this multiplication is well-defined, we must show that it is independent of the choice of a and b in their respective cosets. Let a', b' be other choices, then $a' = a+u$, $b' = b+v$, where $u, v \in \mathfrak{a}$, and hence

$$a'b' = ab+av+ub'.$$

Since $\mathfrak{a}$ is an ideal, $av+ub' \in \mathfrak{a}$ and so $a'b'$ lies in the same coset as ab; hence the product (2) is indeed well-defined. Now it is an easy matter to verify the associative law and the distributive laws. Thus $R/\mathfrak{a}$ is a ring; if R is commutative, then so is $R/\mathfrak{a}$. The kernel is of course $\mathfrak{a}$, so we have (in analogy to Th. 1, **9.1**),

THEOREM 1 *Given a ring homomorphism $f: R \to S$, its image is a subring of S and its kernel is an ideal of R. Conversely, given any ideal $\mathfrak{a}$ of R, a ring structure can be defined on the set of cosets $R/\mathfrak{a}$ in such a way that the natural mapping from R to $R/\mathfrak{a}$ is a homomorphism with kernel $\mathfrak{a}$. Moreover, if R is commutative, then so is $R/\mathfrak{a}$.* ■

The ring $R/\mathfrak{a}$ is called the *residue class ring* or *quotient ring* of R by the ideal $\mathfrak{a}$. For example, if $R = \mathbf{Z}$, each positive integer m defines an ideal (m) consisting of all the multiples of m, and $\mathbf{Z}/(m) \cong \mathbf{Z}/m$ is the ring of integers mod m, already encountered in **2.3**.

We have analogues of the factor theorem and the isomorphism theorems for groups. The proofs in the case of rings are essentially the same, and are therefore left to the reader.

FACTOR THEOREM FOR RINGS *Given a homomorphism of rings f: $R \to S$ and an ideal $\mathfrak{a}$ of R satisfying $\mathfrak{a} \subseteq \ker f$, there exists a unique mapping f': $R/\mathfrak{a} \to S$ such that the accompanying triangle commutes. Moreover, f' is a ring-homomorphism, injective if and only if $\mathfrak{a} = \ker f$.* ■

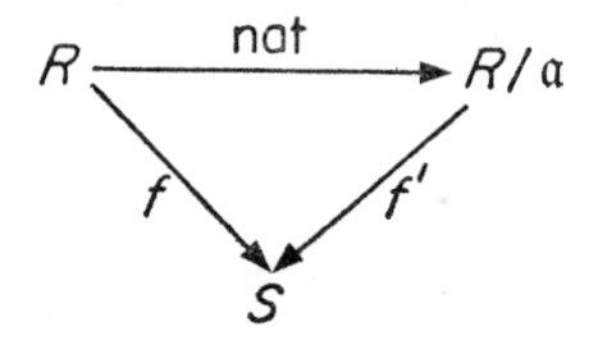

FIRST ISOMORPHISM THEOREM *Given any ring homomorphism f: $R \to S$, there is a factorization $f = \alpha f_1 \beta$, where α: $R \to R/\ker f$ is the natural homomorphism, β: $\operatorname{im} f \to S$ the inclusion mapping, and f_1: $R/\ker f \to \operatorname{im} f$ is an isomorphism.* ■

SECOND ISOMORPHISM THEOREM *Let R be a ring, S a subring and $\mathfrak{a} \lhd R$, then $\mathfrak{a} \cap S \lhd S$ and there is an isomorphism*

$$S/(S \cap \mathfrak{a}) \cong (S+\mathfrak{a})/\mathfrak{a}. \blacksquare$$

THIRD ISOMORPHISM THEOREM *Let R be a ring and $\mathfrak{a} \lhd R$. Then subrings (and ideals) of $R/\mathfrak{a}$ correspond in a natural way to subrings (and ideals) of R that contain $\mathfrak{a}$ and, if $\mathfrak{b} \supseteq \mathfrak{a}$ is an ideal of R, corresponding to the ideal $\mathfrak{b}/\mathfrak{a}$ of $R/\mathfrak{a}$, then*

$$R/\mathfrak{a}/\mathfrak{b}/\mathfrak{a} \cong R/\mathfrak{b}. \blacksquare$$

As an example consider the relation between maximal ideals and simple rings. An ideal of R is said to be *maximal* if it is maximal among all the proper ideals. A ring R which is non-trivial and has no ideals apart from R and $\mathbf{0}$ is said to be *simple*. Now the third isomorphism theorem shows the truth of

THEOREM 2 *An ideal $\mathfrak{a}$ in a ring R is maximal if and only if $R/\mathfrak{a}$ is simple.* $\blacksquare$

In the commutative case we can say rather more: Let K be a commutative simple ring. Given $a \in K$, the set aK of all multiples of a is an ideal in K and so is either $\mathbf{0}$ or K. If $a \neq 0$, the first alternative is ruled out and $aK = K$, hence $ab = 1$ for some $b \in K$. Thus every non-zero element of K has an inverse and K is then a field. This proves the

COROLLARY *In a commutative ring R, an ideal $\mathfrak{a}$ is maximal if and only if $R/\mathfrak{a}$ is a field.* $\blacksquare$

In the non-commutative case there are many simple rings besides fields. We shall meet some of them in the exercises.

Let us return to a general ring R. Given any family $\{\mathfrak{a}_i\}$ of ideals in R, the intersection $\bigcap \mathfrak{a}_i$ is again an ideal, as is easily verified. In particular, if X is a subset of R, the intersection of all ideals containing X is an ideal, the least ideal containing X. This ideal is obtained in a more explicit fashion as the set of all finite sums

$$a_1x_1b_1 + \cdots + a_rx_rb_r \qquad \text{where } x_i \in X,\ a_i, b_i \in R.$$

It may be denoted by RXR and is called the ideal *generated by* X. If R is commutative and X is finite, say $X = \{x_1, \ldots, x_r\}$, the ideal generated by X consists of all expressions $x_1a_1 + \cdots + x_ra_r$ $(a_i \in R)$. In that case it may be denoted by XR or $\sum x_iR$, or also $(x_1, \ldots, x_r)$. In particular, the ideal generated by $x \in R$ is written (x).

Exercises

(1) Prove the isomorphism theorems and the factor theorem for rings.

(2) Find all ideals in $\mathbf{Z}$; likewise in $\mathbf{Z}/m$.

(3) If $\mathfrak{a}, \mathfrak{b} \lhd R$, show that $\mathfrak{a}+\mathfrak{b} = \{a+b \mid a \in \mathfrak{a},\ b \in \mathfrak{b}\}$, $\mathfrak{ab} = \{\sum a_i b_i \mid a_i \in \mathfrak{a}, b_i \in \mathfrak{b}\}$ and $\mathfrak{a} \cap \mathfrak{b}$ are again ideals in R.

(4) Let R be a ring and S a set with two operations $x+y$, xy. If $f\colon R \to S$ is a surjective mapping such that $(x+y)f = xf+yf$, $(xy)f = xf \cdot yf$, show that S is a ring with zero element $0f$ and unit-element $1f$. Moreover if R is commutative, then so is S. Deduce that the structure defined on $R/\mathfrak{a}$ in the text satisfies the laws for rings.

(5) Show that the additive group $\mathbf{Q}/\mathbf{Z}$ has no ring structure.

(6) Show that there is a homomorphism $\mathbf{Z}/n \to \mathbf{Z}/m$ iff $m \mid n$.

(7) Show that the category of rings and homomorphisms has an initial object and a final object, but these two are not isomorphic.

(8) Let R be a ring. If the set of all non-units is an ideal $\mathfrak{m}$, show that $R/\mathfrak{m}$ is a skew field. Show that this is so if for each $x \in R$, either x or $1-x$ is a unit.

(9) Show that the two ways of defining the least ideal containing a set X described in the text give the same answer.

(10) Let R be a ring and R_n the ring of all $n \times n$ matrices over R. If E_{ij} is the matrix with (i, j)-entry 1 and the other entries 0, show that for any matrix $A = (a_{ij})$, $E_{rs}AE_{uv} = a_{su}E_{rv}$. Deduce that the ring of all $n \times n$ matrices over a field (even skew) is simple.

10.2 Modules over a ring

In Ch. **9** we encountered abelian groups with a ring R as operator domain, or briefly, R-modules, as a common generalization of abelian groups and vector spaces. Modules are a useful tool for studying rings and in this and the following sections we describe some general properties of modules, as well as some more precise results that hold for modules over special rings. The reader is advised to keep in mind the case of vector spaces, remembering however that a submodule of a module will not in general be complemented.

We begin by listing some examples of modules, partly to illustrate the concept and also for later use.

(i) Every abelian group may be regarded as a $\mathbf{Z}$-module; to operate with $n(\geqslant 0)$ on a, we add a to itself n times:

$$na = a+a+\cdots+a \qquad (n \text{ terms } a), \tag{1}$$

nd for $n < 0$, we define na to be $-(-n)a$. The reader should verify that this gives indeed a $\mathbf{Z}$-module structure. We observe that 'submodule' and 'subgroup' mean the same thing in this case: any submodule is clearly also a subgroup and, conversely, a subgroup contains with any element a, also na and $-na$, and so is a $\mathbf{Z}$-submodule. Thus any result proved for modules will automatically apply to abelian groups.

(ii) If R is any ring, we may regard R as a right R-module with respect to the multiplication in R. To avoid confusion, this right R-module is often written R_R. The submodules of R_R are called the *right ideals* of R; thus a right ideal $\mathfrak{a}$ is a subgroup of the additive group of R satisfying $\mathfrak{a}R \subseteq \mathfrak{a}$. An ideal of R, as defined in **10.1**, is necessarily a right ideal, but the converse need not hold. To take an example, let $R = k_n$ be the ring of all $n \times n$ matrices over a field k. The only ideals of R are 0 and R (thus R is a simple ring, cf. Ex. (10), **10.1**), but R has many right ideals, for example the set $\mathfrak{r}_1$ of all matrices whose only non-zero entries are in the first row. If $\mathfrak{r}_i$ $(i = 2, \ldots, n)$ is the corresponding set for the ith row, we have a direct sum of abelian groups:

$$R = \mathfrak{r}_1 \oplus \cdots \oplus \mathfrak{r}_n.$$

Of course R has other right ideals (obtained e.g. by changing the matrix basis in R), but we shall see later that each of them is isomorphic, as R-module, to the direct sum of a number of copies of $\mathfrak{r}_1$.

We can also regard R as a left R-module under the multiplication in R. This is written ${}_RR$ and the submodules of ${}_RR$ are called *left ideals*. Again every ideal is a left ideal; moreover a subset of R is an ideal precisely if it is both a left and right ideal. For emphasis ideals are sometimes called *two-sided*, in contrast to left or right ideals, which are one-sided. Of course in a commutative ring the distinction between left, right and two-sided ideals disappears and we may without risk of ambiguity simply speak of ideals.

(iii) Let R be any ring and R° the *opposite ring*: this is a ring with the same additive group structure as R, but with multiplication

$$a \circ b = ba.$$

For a commutative ring there is of course no difference between R and R°, they are identical under the mapping which identifies the underlying sets. Even for some non-commutative rings R° may be isomorphic to R, e.g. when R is a full matrix ring over a field, but now R and R° are no longer identical. In general, however, R° need not be isomorphic to R.

Given any ring R, a left R-module M may always be regarded as a right R°-module in a natural way: we define $x \,.\, a = ax$ $(x \in M,\ a \in R)$; then we have $(x.\, a) \,.\, b = b(ax) = (ba)x = x \,.\, (ba) = x \,.\, (a \circ b)$, which shows that M is indeed a right R°-module. In the same way right R-modules may be considered as left R°-modules.

In practice this remark is often used in the other direction: by using both left and right modules we need never consider modules over the opposite of a ring. Thus given a ring R and a right R°-module M, we may consider M as a left R-module.

(iv) Let R be a ring, A an abelian group with endomorphism ring End (A) and suppose that we are given a ring-homomorphism $\rho\colon R \to \text{End}\,(A)$.

This means that each $a \in R$ defines an endomorphism ρ_a of A, subject to the rules:

$$\rho_{a+b} = \rho_a + \rho_b, \qquad \rho_{ab} = \rho_a \rho_b, \qquad \rho_1 = 1. \tag{2}$$

Let us define a mapping $A \times R \to A$ by the rule $(x, a) \mapsto x\rho_a$. If we write xa instead of $x\rho_a$, the above rules become

$$x(a+b) = xa + xb, \qquad x(ab) = (xa)b, \qquad x1 = x, \tag{3}$$

while the fact that ρ_a is an endomorphism of A is expressed by the equation

$$(x+y)a = xa + ya. \tag{4}$$

Thus we see that A has been made into a right R-module. It is worth noting that all right R-modules arise in this way; for if M is a right R-module, then the mapping

$$\rho_a : x \mapsto xa \quad (x \in M)$$

is an endomorphism of the additive group of M, by (4), so that $a \mapsto \rho_a$ is a mapping of R into End (M), and now (3) shows that this mapping is a homomorphism.

Since a module over a ring is essentially an abelian group with operators, all the results proved for groups with operators in Ch. **9** still hold for modules. In fact the results are simpler for modules, since we do not have to be concerned with normality: every submodule is normal. We shall restate the isomorphism theorems for modules and use the opportunity to present an important piece of notation: exact sequences.

Let R be a ring which may be quite arbitrary, but will be fixed in what follows. If M, N are any right R-modules, a homomorphism $f: M \to N$ is a homomorphism of abelian groups satisfying

$$(xa)f = (xf)a \qquad (x \in M, a \in R).$$

It is easily verified that $\ker f$, $\operatorname{im} f$ are submodules of M and N respectively. A sequence of R-modules and homomorphisms

$$\cdots \to M_{i-1} \xrightarrow{f_{i-1}} M_i \xrightarrow{f_i} M_{i+1} \xrightarrow{f_{i+1}} \cdots$$

is said to be *exact at* M_i if $\ker f_i = \operatorname{im} f_{i-1}$; the sequence is called *exact* if it is exact at every module.

For example, a homomorphism $f: M \to N$ is injective iff

$$0 \to M \xrightarrow{f} N$$

is exact; it is surjective iff

$$M \xrightarrow{f} N \to 0$$

is exact, while the exactness of

$$0 \to M \xrightarrow{f} N \to 0$$

means that f is an isomorphism. We next take a less trivial case, which is of importance in studying module extensions. If we are given a 3-term exact sequence:

$$0 \to M' \xrightarrow{\lambda} M \xrightarrow{\mu} M'' \to 0, \tag{5}$$

also called a *short exact sequence*, we see that M' is isomorphic to a submodule of M with quotient M''; thus M is an extension of M' by M''. In fact this is a convenient way of writing extensions of modules. This was not in evidence when we studied vector spaces because there every submodule has a complement, so that any extension has the form of a direct sum and is therefore trivial. If in (5) $\text{im}\,\lambda = \ker \mu$ is a direct summand of M, the exact sequence is said to *split* or also to be *split exact*. As an example of a non-split exact sequence we have

$$0 \to \mathbf{Z} \xrightarrow{k} \mathbf{Z} \to \mathbf{C}_k \to 0,$$

where k denotes multiplication by the positive integer k, thus $\text{im}\,k = k\mathbf{Z}$, and the quotient is $\mathbf{C}_k$, the cyclic group of order k.

Given a homomorphism $f: M \to N$, in addition to the kernel and image of f, let us define the *coimage* as $\text{coim}\,f = M/\ker f$ and the *cokernel* of f as $\text{coker}\,f = N/\text{im}\,f$. Then we have the following diagram, in which the square commutes and the row is exact:

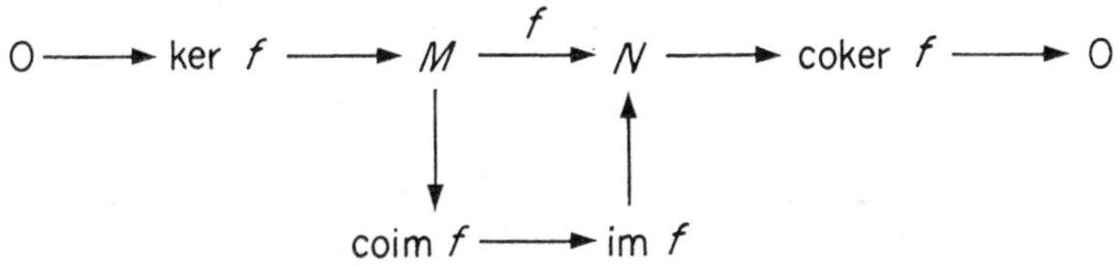

$$\begin{array}{ccccccccc}
0 & \longrightarrow & \ker f & \longrightarrow & M & \xrightarrow{f} & N & \longrightarrow & \text{coker}\,f \longrightarrow 0 \\
 & & & & \downarrow & & \uparrow & & \\
 & & & & \text{coim}\,f & \longrightarrow & \text{im}\,f & &
\end{array}$$

The commutative square already occurred in the statement of the first isomorphism theorem (Th. 3, **9.1**), which also tells us that the mapping $\text{coim}\,f \to \text{im}\,f$ is an isomorphism. For this reason it is not really necessary to have a separate name for the coimage; nevertheless the gain in symmetry makes it desirable to have a special name.

The second isomorphism theorem (parallelogram law) states that for any submodules N_1, N_2 of N,

$$(N_1 + N_2)/N_1 \cong N_2/(N_1 \cap N_2). \tag{6}$$

This may be expressed in the following commutative diagram with exact rows:

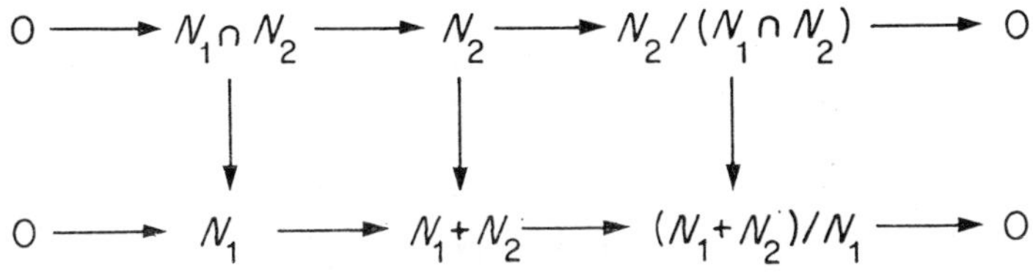

$$\begin{array}{ccccccccc}
0 & \longrightarrow & N_1 \cap N_2 & \longrightarrow & N_2 & \longrightarrow & N_2/(N_1 \cap N_2) & \longrightarrow & 0 \\
 & & \downarrow & & \downarrow & & \downarrow & & \\
0 & \longrightarrow & N_1 & \longrightarrow & N_1 + N_2 & \longrightarrow & (N_1 + N_2)/N_1 & \longrightarrow & 0
\end{array}$$

where the last column is an isomorphism. This in turn is a special case of the following commutative diagram with exact rows and columns:

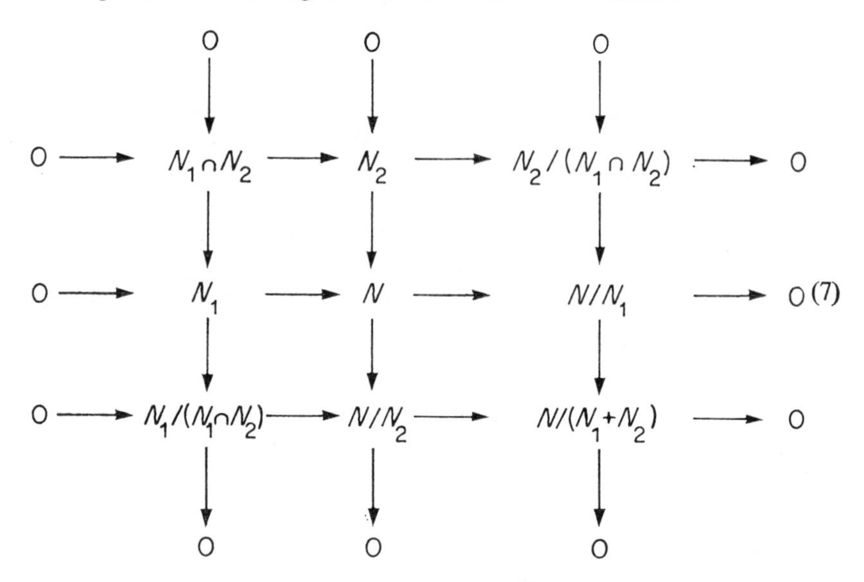

The exactness of the first two rows and columns is clear; for the third row (and column) we need to use the isomorphism (6); thus the image of the mapping $N_2/(N_1 \cap N_2) \to N/N_1$ is $(N_1+N_2)/N_1$, from which the result follows. In the diagram the mappings are not labelled since they are in all cases easily identifiable. But in general it will be necessary to define the mappings explicitly and to verify the exactness and commutativity by working out their effect on individual elements. This method of proof (of which we shall soon meet examples) is called 'diagram chasing'.

The third isomorphism theorem may be stated as follows: Given a module N and submodules $N_1 \subseteq N_2 \subseteq N$, we have the commutative diagram with exact rows:

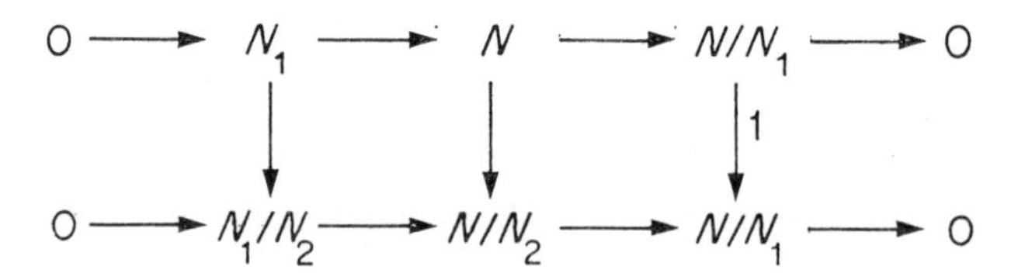

where the last column is the identity mapping. We observe that this diagram may be obtained by considering the last two rows of the previous diagram in the case $N_2 \subseteq N_1$.

We conclude this section with a result which is often useful and whose proof illustrates the technique of diagram chasing.

LEMMA 1 (The 5-lemma) *Given a commutative diagram*

$$\begin{array}{ccccccccc}
A_1 & \xrightarrow{\alpha_1} & A_2 & \xrightarrow{\alpha_2} & A_3 & \xrightarrow{\alpha_3} & A_4 & \xrightarrow{\alpha_4} & A_5 \\
\downarrow{\scriptstyle f_1} & & \downarrow{\scriptstyle f_2} & & \downarrow{\scriptstyle f_3} & & \downarrow{\scriptstyle f_4} & & \downarrow{\scriptstyle f_5} \\
B_1 & \xrightarrow{\beta_1} & B_2 & \xrightarrow{\beta_2} & B_3 & \xrightarrow{\beta_3} & B_4 & \xrightarrow{\beta_4} & B_5
\end{array}$$

with exact rows, if f_1, f_2, f_4, f_5 are isomorphisms, then so is f_3. More precisely,

(i) *if f_1 is surjective and f_2, f_4 are injective, then f_3 is injective,*

(ii) *if f_5 is injective and f_2, f_4 are surjective, then f_3 is surjective.*

Proof. We shall prove (i); the proof of (ii) is quite similar and the first part then follows by combining (i) and (ii).

Let $x \in A_3$ and $xf_3 = 0$, we must show that $x = 0$. We have $x\alpha_3 f_4 = xf_3\beta_3 = 0$, hence $x\alpha_3 = 0$ (because f_4 is injective). By exactness there exists $y \in A_2$ such that $x = y\alpha_2$. Moreover, $0 = xf_3 = y\alpha_2 f_3 = yf_2\beta_2$, hence by exactness there is $z \in B_1$ such that $yf_2 = z\beta_1$. But f_1 is surjective, so we can write $z = tf_1$ for some $t \in A_1$. Now $yf_2 = z\beta_1 = tf_1\beta_1 = t\alpha_1 f_2$ and f_2 is injective, hence $y = t\alpha_1$. It follows that $x = y\alpha_2 = t\alpha_1\alpha_2 = 0$ (by exactness), and this shows f_3 to be injective, as claimed. ∎

Exercises

(1) Prove in detail that an abelian group is a **Z**-module, under the definition given in (1).

(2) Given a commutative diagram of short exact sequences

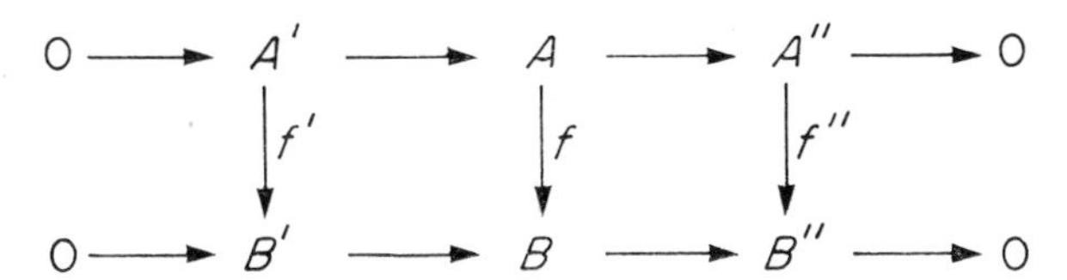

prove without using the 5-lemma that if two of f, f', f'' are isomorphisms, then so is the third.

(3) Given two modules A, B, we can form the category whose objects are extensions E of A by B, while the morphisms are homomorphisms $f\colon E \to E'$ such that the diagram

$$\begin{array}{ccccccccc}
 & & & \nearrow & E & \searrow & & & \\
0 & \longrightarrow & A & & \downarrow & & B & \longrightarrow & 0 \\
 & & & \searrow & E' & \nearrow & & &
\end{array}$$

commutes. Show that every morphism in this category is an isomorphism.

(4) Prove the part of the 5-lemma left to the reader.

(5) Show that a short exact sequence (5) splits iff λ has a right inverse (or, equivalently, μ has a left inverse).

(6) Let $n = rs$; show that the short exact sequence of $\mathbf{Z}/n$-modules

$$0 \to r\mathbf{Z}/n \xrightarrow{\lambda} \mathbf{Z}/n \xrightarrow{\mu} s\mathbf{Z}/n \to 0$$

where λ is the inclusion mapping and μ multiplication by s, splits iff $(r, s) = 1$.

(7) Given a short exact sequence of vector spaces

$$0 \to V' \to V \to V'' \to 0$$

show that $\dim V = \dim V' + \dim V''$.

(8) Show that a long exact sequence $\cdots \to M_i \xrightarrow{fi} M_{i+1} \to \cdots$ is equivalent to the conjunction of short exact sequences

$$0 \to \ker f_i \to M_i \to \ker f_{i+1} \to 0.$$

(9) Given an exact sequence of vector spaces

$$0 \to V_0 \to V_1 \to \cdots \to V_n \to 0,$$

show that $\sum (-1)^\nu \dim V_\nu = 0$.

(10) (3×3 lemma) Given any commutative 3×3 diagram (7) with exact rows and columns, in which the arrows in the first column are missing, there is just one way of filling in this column so as to keep the diagram commutative and the column so obtained is then exact.

10.3 Direct products and direct sums

We have already met the direct product of groups in **9.6** and it is clear how this notion applies to modules: Given a family (M_i) $(i \in I)$ of modules, all over the same ring R, we define a module structure on the set-theoretical Cartesian product $P = \prod M_i$, by performing all the operations componentwise. Thus if $x = (x_i)$ is a typical element of P, then

$$x + y = (x_i + y_i), \qquad xr = (x_i r) \quad (r \in R),$$

and the canonical projection $\varepsilon_i\colon x \mapsto x_i$ is a homomorphism from P to M_i. In the case of a finite family of modules this is essentially the direct sum of spaces or of abelian groups, as defined earlier, but we shall still speak of a product; the precise connexion between sums and products will become clear later in this section.

We recall Th. 3, **9.6**, which states that the direct product $\prod M_i$ with the canonical projections is universal for mappings into the family (M_i). This naturally leads one to ask whether a dual construction exists, i.e. a module S with mappings $M_i \to S$ which is universal for mappings from the family (M_i).

In detail we ask for a module with the following property (obtained by reversing the arrows in Th. 3, **9.6**).

Given a family (M_i) $(i \in I)$ *of R-modules, there exists an R-module S with homomorphisms* $\mu_i: M_i \to S$ *such that for any family of homomorphisms* $f_i: M_i \to N$ *there is a unique homomorphism* $f: S \to N$ *such that*

$$f_i = \mu_i f \qquad \text{for all } i \in I. \tag{1}$$

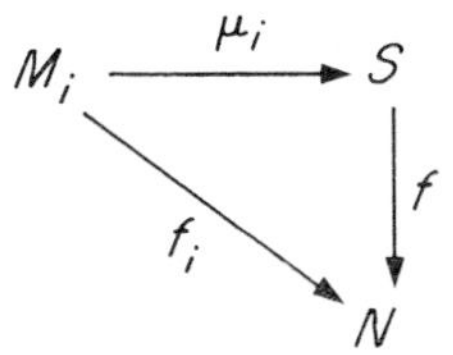

It is again clear that such a module S, if it exists, must be unique up to isomorphism. To find such a module, our first try might be to take the direct product $P = \prod M_i$ and, as homomorphism μ_j, take the mapping which sends $c \in M_j$ to the element $x = (x_i)$, where $x_j = c$ and $x_i = 0$ for $i \neq j$. This is an embedding of M_j in P. However, finding f leads to difficulties unless the index-set I is finite (the reader is invited to try this for himself). In the case of a finite product it becomes possible to define f by the rule

$$(x_i)f = \sum x_i f_i, \tag{2}$$

and it is not hard to show that this mapping f satisfies the above conditions in the case of a finite index-set I.

Thus the direct product satisfies the above condition for the module S in the case of finite index-sets but not in general; however, this case provides us with a clue. In general, let us define S as the submodule of P generated by the im μ_i for all $i \in I$. This is the submodule of P consisting of all (x_i) with only finitely many non-zero coefficients. It is called the *direct sum* (or also sometimes the *coproduct*) of the M_i and is written $\oplus M_i$ or $\coprod M_i$ while the injections $\mu_i: M_i \to S$ are called the *canonical injections*. We observe that S coincides with P whenever I is finite (or more generally, whenever $M_i = \mathbf{0}$ except for a finite number of suffixes i). Now (2) makes sense for every element of S, because on the right we are adding only finitely many non-zero elements, and the mapping f, so defined, clearly satisfies (1). In this sense the direct sum and direct product of modules, as defined here, are dual notions, although the actual constructions are not dual. In fact, as we have seen, they coincide in the finite case.

If (M_i) is a family of submodules of a module M, the submodule generated by all the M_i is written $\sum M_i$ and is called their *sum*. By applying the universal property of direct sums to the family (M_i), with the inclusion mappings $M_i \to \sum M_i$, we find a homomorphism $\oplus M_i \to \sum M_i$, which is clearly surjective. If it is an isomorphism, the M_i are said to be an *independent* family of submodules of M. We observe that this will be the case precisely when any family (x_i), where $x_i \in M_i$, and where all but a finite number of the x_i are zero, has a sum $\sum x_i$ which is different from zero unless all the x_i are zero. Thus we obtain the following criterion for independence, generalizing Th. 1, **9.6**, Cor.

PROPOSITION 1 *Let R be any ring and M an R-module. Then a family (M_i) of submodules of M is independent if and only if for any family of elements $x_i \in M_i$, of which at most finitely many are non-zero, $\sum x_i = 0$ implies $x_i = 0$ for all $i \in I$.* ∎

Let us now consider the case of a finite family of modules $M_1, \ldots, M_n$ where, as we saw, the direct sum and direct product coincide and both may be denoted by $M = \oplus M_i$. We have the canonical injections $\mu_i \colon M_i \to M$ and projections $\varepsilon_i \colon M \to M_i$ and it is easily seen that they satisfy the equations

$$\sum \varepsilon_i \mu_i = 1_M, \qquad \mu_i \varepsilon_j = \delta_{ij} 1_{M_i}. \tag{3}$$

If we write the εs and μs as a row and column respectively,

$$\varepsilon = (\varepsilon_1, \ldots, \varepsilon_n), \qquad \mu = (\mu_1, \ldots, \mu_n)^T, \tag{4}$$

we can write the equations (3) more briefly in matrix form:

$$\varepsilon\mu = \begin{pmatrix} 1_{M_1} & & & 0 \\ & \cdot & & \\ & & \cdot & \\ & & & \cdot \\ 0 & & & 1_{M_n} \end{pmatrix}, \qquad \mu\varepsilon = 1_M. \tag{5}$$

This leads to another characterization of the direct sum of a finite family of modules, corresponding to Prop. 2, **9.6**.

PROPOSITION 2 *Let R be any ring and $M, M_1, \ldots, M_n$ any R-modules, with homomorphisms $\mu_i \colon M_i \to M$, $\varepsilon_i \colon M \to M_i$, and let us define ε and μ as in* (4). *Then M is a direct product* (*or equivalently, a direct sum*) *of the M_i if and only if ε, μ satisfy* (5).

Proof. We have seen that the equations (5) necessarily hold in a direct product. Conversely, assume that they hold; we shall complete the proof by verifying that M with the mappings ε_i satisfies the universal mapping property. Let $f_i \colon N \to M_i$ be any family of mappings and write $f = (f_1, \ldots, f_n)$, then $F = f\mu$ is a mapping from N to M which satisfies $F\varepsilon = f\mu\varepsilon = f$, while if also $F'\varepsilon = f$, then $F' = F'\varepsilon\mu = f\mu = F$, which proves the uniqueness of F. ∎

Of course a similar proof could be given using the defining property of direct sums. We shall write $\oplus M_i$ or $\prod M_i$ or also $M_1 \oplus \cdots \oplus M_n$ for the direct sum of $M_1, \ldots, M_n$. If all summands are the same, say $M_i = M$, we denote the direct sum by M^n. Sometimes it will be convenient to use the notation ^nM; this is used if the elements are to be regarded as rows, while M^n is used when the elements are to be thought of as columns. When there are infinitely many terms M, indexed by I, we shall denote the direct product (also called *direct power*) by M^I and the direct sum by ^IM.

Exercises

(1) Prove Prop. 2 by verifying the characteristic property of direct sums.

(2) Let (M_i) be a family of R-modules and define a *direct family* of homomorphisms as a module N with homomorphisms $\alpha_i\colon M_i \to N$, $\beta_i\colon N \to M_i$ such that $\alpha_i\beta_j = 1_{M_i}$ if $i = j$ and 0 otherwise. If N' with corresponding homomorphisms α'_i, β'_i is a second family, a *morphism* between them is a homomorphism $f\colon N \to N'$ such that $\alpha_i f = \alpha'_i$, $\beta_i = f\beta'_i$. Show that the direct families and morphisms form a category and describe the initial and final objects in this category.

(3) Show that Hom $(M, \prod N_i) \cong \prod$ Hom (M, N_i), Hom $(\oplus M_i, N) \cong \prod$ Hom (M_i, N).

(4) For a fixed set I, show that the correspondence $M \mapsto M^I$ is an additive functor which preserves exact sequences; similarly for $M \mapsto {}^IM$. (A functor which preserves exact sequences is called *exact*.)

(5) A category $\mathscr{A}$ is said to possess direct products if with every family (B_i) of $\mathscr{A}$-objects an $\mathscr{A}$-object $\prod B_i$ is associated such that

$$\mathrm{Hom}_{\mathscr{A}}(A, \prod B_i) \cong \prod \mathrm{Hom}_{\mathscr{A}}(A, B_i)$$

where $\cong$ denotes natural equivalence in the category of sets (or the category of abelian groups, in case $\mathrm{Hom}_{\mathscr{A}}(A, B)$ is an abelian group). Show that $\prod B_i$ may be obtained as the solution of a universal problem and hence is unique up to isomorphism. Similarly discuss the *coproduct* $\coprod B_i$, defined (when it exists) by

$$\mathrm{Hom}_{\mathscr{A}}(\coprod B_i, A) \cong \prod \mathrm{Hom}_{\mathscr{A}}(B_i, A).$$

(6) Let $A_i\ (i = 1, \ldots, r)$, $B_\lambda\ (\lambda = 1, \ldots, s)$ be R-modules and $f_{i\lambda}\colon A_i \to B_\lambda$ be homomorphisms. Show that the mapping $f = (f_{i\lambda})\colon \oplus A_i \to \oplus B_\lambda$ defined by $(x_i)f = (\sum x_i f_{i\lambda})$ is a homomorphism; further, show that every homomorphism from $\oplus A_i$ to $\oplus B_\lambda$ is of this form.

(7)* Given an exact sequence

$$0 \to C \xrightarrow{f} A \oplus A' \xrightarrow{g} B \oplus B' \xrightarrow{h} D \to 0,$$

where $g = \begin{pmatrix} \alpha & \beta \\ \gamma & \delta \end{pmatrix}$ and $\delta\colon A' \to B'$ is an isomorphism, obtain an exact sequence

$$0 \to C \xrightarrow{f'} A \xrightarrow{g'} \mathrm{B} \xrightarrow{h'} D \to 0,$$

where $f' = fp$ (and p is the projection $A \oplus A' \to A$), $h' = ih$ (and i is the injection $B \to B \oplus B'$) and $g' = \alpha - \beta\delta^{-1}\gamma$.

10.4 Free modules

The study of vector spaces over a field is made easy by the fact that every vector space has a basis. For modules over a general ring this is no longer so, but it is still of interest to examine the modules which have a basis; they are the free modules which form the topic of this section.

Let R be any ring and M a right R-module. If we are given a subset X of M, then by the submodule *generated* by X we understand the least submodule containing X, denoted by XR or $\langle X \rangle$. In fact it is not hard to see that the set of all finite linear combinations in X,

$$XR = \{\sum x_i a_i \mid x_i \in X, a_i \in R\}$$

is a submodule containing X and clearly any submodule of M containing X also contains XR, so that XR is indeed the least submodule containing X. If $XR = M$, we call X a *generating* or *spanning* set of M and also say that X *spans* M.

Next we define linear independence, as in the case of vector spaces. A sequence of elements $x_1, \ldots, x_n$ of an R-module M is said to be *linearly independent* if for any $a_1, \ldots, a_n \in R$,

$$\sum x_i a_i = 0 \quad \text{implies} \quad a_1 = \cdots = a_n = 0.$$

The sequence is *linearly dependent* if it is not linearly independent; however unlike the case of vector spaces over a field, in a module it is not generally true that in a linearly dependent set some element can be expressed in terms of the rest.

An infinite family (x_i) in a module is called *linearly independent* if every finite subfamily is linearly independent. This amounts to the requirement that for any family of elements (a_i) in R, such that all but a finite number of the a_i vanish,

$$\sum x_i a_i = 0 \quad \text{implies } a_i = 0 \text{ for all } i.$$

Here the sum makes sense because only finitely many non-zero terms occur, by the restriction imposed on the a_i.

By a *basis* of an R-module M one understands a family of elements which is linearly independent and spans M. Of course not every module has a basis, e.g. a cyclic group, regarded as a $\mathbf{Z}$-module, has a basis iff it is infinite. A module M which has a basis X is called a *free* module, more precisely, it is free on the basis X. For $R = \mathbf{Z}$ the free $\mathbf{Z}$-module is just the free abelian group introduced in **9.7**.

As in the case of vector spaces, a given free module will in general have many different bases, and the change of basis is most easily described by matrices. Thus if a free right R-module M has two finite bases $e_1, \ldots, e_m$; $f_1, \ldots, f_n$, each can be expressed uniquely in terms of the other:

$$f_\lambda = \sum e_i a_{i\lambda}, \; e_i = \sum f_\lambda b_{\lambda i}, \tag{1}$$

where the latin alphabet is used for the range $1, \ldots, m$ and greek characters for the range $1, \ldots, n$. Of course we cannot at this stage assume that $m = n$; soon we shall prove that in a free module over a commutative ring, all bases have the same number of elements. Although this result can be extended to many of the non-commutative rings usually encountered, it does not hold without exception.

From (1) we obtain by elimination of the *f*s,

$$e_i = \sum e_j a_{j\lambda} b_{\lambda i},$$

hence by the linear independence of the *e*s, we find $\sum a_{j\lambda} b_{\lambda i} = \delta_{ji}$, i.e. on writing the coefficients as matrices: $A = (a_{i\lambda})$, $B = (b_{\lambda i})$, we have

$$AB = I_m,$$

where I_m is the unit matrix of order m. By symmetry we find that $BA = I_n$. Thus, assuming that $m = n$, we see that the change from one basis of a free module to another is described by a pair of mutually inverse matrices. We now prove that m and n are in fact equal.

PROPOSITION 1 *Let M be a finitely generated free module over a non-trivial commutative ring R. Then all bases of M are finite, with the same number of elements.*

Proof. By hypothesis M is free, on a basis (e_i), say; it is also finitely generated, by $u_1, \ldots, u_n$ say. If we express each u_ν in terms of the e_i, only finitely many e_i occur in the expression for any one u_ν and so only finitely many of the e_i are needed for all the u_ν. But the u_ν span M, hence M is also spanned by a finite subfamily of the e_i, and since the *e*s form a basis, the whole family (e_i) must be finite.

Now let $e_1, \ldots, e_m$ and $f_1, \ldots, f_n$ be two bases of M and as in (1) write A, B for the matrices of transformation between them. Further, suppose that $m > n$ and let us partition A, B as follows:

$$A = \begin{pmatrix} A_1 \\ A_2 \end{pmatrix} \qquad B = (B_1 \quad B_2),$$

where A_1, B_1 are both $n \times n$ matrices, and so A_2 is $(m-n)\times n$ and B_2 is $n\times(m-n)$. By what we have seen, $AB = I$, hence

$$\begin{pmatrix} I & 0 \\ 0 & I \end{pmatrix} = \begin{pmatrix} A_1 \\ A_2 \end{pmatrix}(B_1 \quad B_2) = \begin{pmatrix} A_1B_1 & A_1B_2 \\ A_2B_1 & A_2B_2 \end{pmatrix}. \tag{2}$$

In particular, $A_1B_1 = I$. Hence $\det(A_1)\det(B_1) = \det(A_1B_1) = 1$, therefore A_1 is invertible, with (2-sided) inverse B_1. Now multiply both sides of (2) by $\begin{pmatrix} B_1 & 0 \\ 0 & I \end{pmatrix}$ on the left and by its inverse $\begin{pmatrix} A_1 & 0 \\ 0 & I \end{pmatrix}$ on the right, then we obtain

$$\begin{pmatrix} I & 0 \\ 0 & I \end{pmatrix} = \begin{pmatrix} I \\ A_2 \end{pmatrix}(I \quad B_2) = \begin{pmatrix} I & B_2 \\ A_2 & A_2B_2 \end{pmatrix}.$$

This shows that $A_2 = 0$, $B_2 = 0$, $A_2B_2 = I$, which is a contradiction unless $1 = 0$, i.e. $R = 0$. But that case has been excluded, hence $m \leqslant n$, and by symmetry, $n \leqslant m$, i.e. $m = n$. ■

This result can be extended to free modules that are not finitely generated and, for such modules, it holds over all rings without exception (see Vol. 2).

Given any set X, we can construct a free module on X, simply by taking a direct sum of copies of R, as right R-module, indexed by X. Thus if X is finite, say $X = \{x_1, \ldots, x_n\}$, then the free module on X is isomorphic to R^n, with the n-tuple $(a_1, \ldots, a_n)$ of R^n corresponding to $\sum x_i a_i$. In particular, R itself is free on one generator, the unit-element of R, while for $n = 0$, the zero module may be regarded as the free module on $\varnothing$ as free generating set.

Given any set X, we recall that R^X is the direct product of copies of R indexed by X, i.e. the module whose elements are families (a_x) of elements of R indexed by X, with componentwise operations. Further, ${}^XR$ is the direct sum of copies of R indexed by X; we can embed X in R^X by identifying $y \in X$ with the element whose x-component is 1 if $x = y$ and 0 otherwise, i.e. the element (δ_{yx}). Then ${}^XR$ is just the submodule of R^X spanned by X; moreover, X is linearly independent and hence is a basis of ${}^XR$. Thus ${}^XR$ is the free module on X. From the universal property of the direct sum we derive the following property of free modules, which can also be used to characterize them:

PROPOSITION 2 *Given a set X, let ${}^XR$ be the free R-module on X and $\lambda: X \to {}^XR$ the embedding defined above, then for any mapping of X into an R-module M there exists a unique homomorphism from ${}^XR$ into M such that the accompanying diagram commutes.*

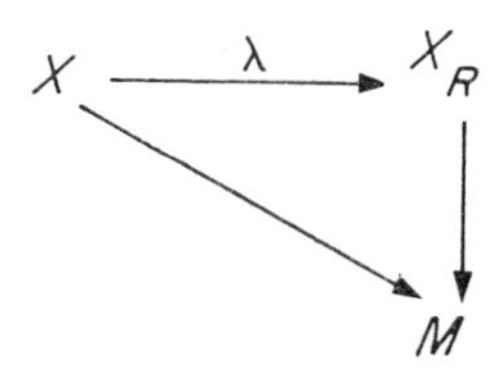

This result may be derived from the characterization of direct sums in **10.3** or proved directly, by an argument similar to that used in proving Prop. 2, **10.3**. The details are left to the reader. ■

We shall often regard X as a subset of ${}^XR$, via the above embedding λ; then this Proposition states that any mapping $X \to M$ can be extended to a unique homomorphism ${}^XR \to M$.

A trivial but important property of free modules is expressed in

THEOREM 3 *Every R-module can be expressed as a quotient of a free module.*

Proof. Let M be the given module and X a spanning set of M (e.g. M itself). By Prop. 2, the inclusion $X \to M$ can be extended to a homomorphism

$f\colon {}^X R \to M$; its image is a submodule containing X and hence is the whole of M, thus f maps ${}^X R$ onto M and we have expressed M as a quotient of a free module. ■

Let us take the homomorphism $f\colon {}^X R \to M$ obtained in the proof of Th. 3 and write $K = \ker f$, then we have the exact sequence

$$0 \to K \to {}^X R \to M \to 0.$$

The module K will not in general be free, but it can itself be written as a homomorphic image of a free module ${}^Y R$, say. Thus we get an exact sequence

$${}^Y R \to {}^X R \to M \to 0.$$

This is called a *presentation* of M; it is analogous to the presentation of a group by generators and defining relations.

Exercises

(1) In an entire ring R, show that $aR = bR$ iff a and b are right associated, i.e. $a = bu$ for some unit u in R.

(2) Show that $\mathbf{Q}$ as additive abelian group cannot be finitely generated and any generating set X of $\mathbf{Q}$ has a proper subset which also generates $\mathbf{Q}$ (thus $\mathbf{Q}$ has no minimal generating set).

(3) Let M be a finitely generated R-module. Show that each generating set X of M contains a finite subset X_0 which already generates M; deduce that M has a minimal generating set and every minimal generating set of M is finite.

(4) Show that if F is a free R-module of rank n, then so is $\mathrm{Hom}_R(F, R)$.

(5) Given the diagram with exact rows:

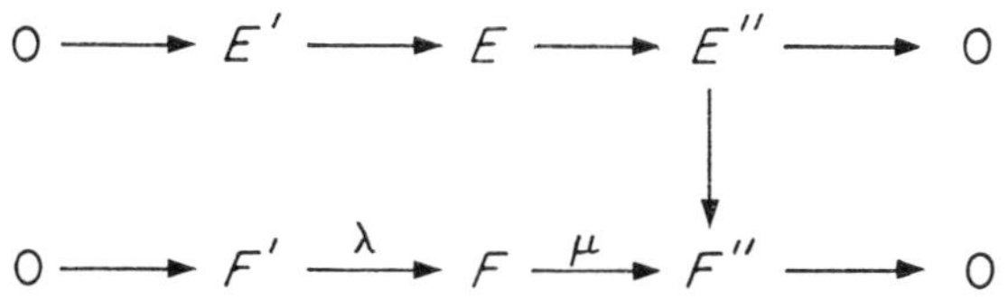

$$\begin{array}{ccccccccc} 0 & \longrightarrow & E' & \longrightarrow & E & \longrightarrow & E'' & \longrightarrow & 0 \\ & & & & & & \downarrow & & \\ 0 & \longrightarrow & F' & \xrightarrow{\lambda} & F & \xrightarrow{\mu} & F'' & \longrightarrow & 0 \end{array}$$

if E is free, find homomorphisms $E \to F$ and $E' \to F'$ such that the resulting diagram is commutative. If f_1, f_2 are two such homomorphisms from E to F with this property, show that there exists $h\colon E \to F'$ such that $f_1 - f_2 = h\lambda$.

10.5 Principal ideal domains

The first examples of rings we encountered were the ring of integers and the ring of polynomials (in one variable over a field). We found that both could be treated by the same method, using the Euclidean algorithm. It might therefore seem natural to embark on a study of Euclidean domains. In fact, we shall cast our net a little wider and discuss the class of principal ideal domains, to be defined soon. We shall see that every Euclidean domain is principal, but not conversely. On the other hand, many important properties of Euclidean domains are shared by principal ideal domains and the latter have the advantage that their definition is intrinsic, unlike the definition of Euclidean domains, which makes reference to a norm function. It is possible to define principal ideal domains even in the non-commutative case and much of the discussion can be carried out in this more general setting; but the commutative case is significantly simpler and, since this is enough for many important applications, we shall in this section limit ourselves to commutative rings (apart from some initial definitions).

In any ring R, a right (or left) ideal is said to be *principal* if it can be generated by a single element. Thus a principal right ideal may be written in the form aR, where $a \in R$; similarly a principal left ideal has the form Ra. By a *principal ideal ring* we understand a ring in which every right (and every left) ideal is principal; such a ring is also called *principal.* As already indicated, we shall confine our attention here to the commutative case, and we shall mainly be concerned with the situation where there are no zero-divisors, thus we shall be dealing with *principal ideal domains.* We first prove a result which ensures an adequate supply of principal ideal domains.

THEOREM 1 *Every Euclidean domain is principal.*

Proof. Let R be a Euclidean domain and $\mathfrak{a}$ an ideal in R. If $\mathfrak{a} \neq \mathbf{0}$, there are non-zero elements in $\mathfrak{a}$ and we pick one, a, say, of least norm $\varphi(a)$. Then $(a) \subseteq \mathfrak{a}$; we claim that equality holds. For if $(a) \neq \mathfrak{a}$, we can find $b \in \mathfrak{a}$, $b \notin (a)$. By hypothesis, $b = aq+r$, where $\varphi(r) < \varphi(a)$. However, $r = b-aq \in \mathfrak{a}$ and so, by the minimality of $\varphi(a)$, it follows that $r = 0$, i.e. $b = aq \in (a)$, a contradiction. This proves that $\mathfrak{a} = (a)$, hence every ideal of R is principal, as claimed. ∎

Thus, for example the integers $\mathbf{Z}$ form a principal ideal domain; every positive integer generates an ideal of $\mathbf{Z}$ and together with $\mathbf{0}$ these are all the ideals in $\mathbf{Z}$. Likewise $k[X]$, the ring of polynomials in an indeterminate over a field k is principal; here the generators of the non-zero ideals are the monic polynomials. On the other hand, the polynomials over $\mathbf{Z}$ do not form a principal ideal domain, neither do the polynomials in two or more variables over a field. Examples of principal ideal domains that are not Euclidean are a little harder to construct; we shall meet some in Vol. 2.

An important property of principal ideal domains is that any two elements have an HCF (highest common factor) and an LCM (least common multiple), and that, moreover, the HCF of a and b can be written as a linear combination of a and b, as in a Euclidean domain (see Th. 4, **6.5**). For, given a principal ideal domain R and $a, b \in R$, the ideal generated by a and b is principal, say

$$(a, b) = (d). \tag{1}$$

This means that

$$d = au+bv, \tag{2}$$

where

$$a = da_1, \qquad b = db_1. \tag{3}$$

If we exclude the trivial case where $a = b = 0$, we see that d is a common factor of a and b, by (3), and by (2), any common factor of a and b is also a factor of d; thus d is indeed an HCF of a and b. Now the existence of an LCM can be deduced as in **6.5**: in fact $m = ab/d$ is the required LCM.

We observe that in this discussion we have not used the full force of the definition. Let us define a *Bezout domain* as an integral domain in which any ideal generated by two elements is principal, then it is clear that all that has been said applies to Bezout domains. In fact they may be characterized as integral domains in which the HCF of any pair of elements exists and can be written as a linear combination of them. Thus we have

THEOREM 2 *In a Bezout domain, any two elements a, b have an HCF d and there exist u and v such that*

$$d = au+bv.$$

Conversely, any integral domain with this property is a Bezout domain.

Moreover, in a Bezout domain R any finitely generated ideal is principal, more precisely, $(a_1, \ldots, a_n) = (d)$, where d is an HCF of $a_1, \ldots, a_n$. Thus any finite set of elements of R has an HCF, and likewise it has an LCM.

Only the last paragraph still needs proof. We first show, by induction on n, that every ideal generated by n elements is principal. For $n = 1$ there is nothing to prove and for $n = 2$ this is true by definition. Now let $n > 2$ and assume that every ideal on $n-1$ generators is principal. Given $a_1, \ldots, a_n \in R$, we have $(a_1, a_2) = (b)$, hence $(a_1, a_2, \ldots, a_n) = (b, a_3, \ldots, a_n)$ and, by induction on n, the latter ideal is principal.

Given $a_1, \ldots, a_n$, not all zero, let

$$d = \sum a_i b_i, \tag{4}$$

and

$$a_i = da_i' \qquad (i = 1, \ldots, n). \tag{5}$$

By (5), d is a common factor of $a_1, \ldots, a_n$ and by (4), any common factor of the a_i is a factor of d, so d is an HCF of $a_1, \ldots, a_n$.

To find the LCM of $a_1, \ldots, a_n$ we may assume that no a_i vanishes (otherwise the LCM would be zero). Write $p = a_1 a_2 \ldots a_n$, then we can also write $p = a_i A_i$ for $i = 1, \ldots, n$. Let c be an HCF of $A_1, \ldots, A_n$, say,

$$c = \sum A_i u_i. \tag{6}$$

Clearly $c \mid p$; we assert that $m = p/c$ is the required LCM. For by hypothesis, $c \mid A_1$, hence $ca_1 \mid a_1 A_1 = p = cm$, therefore $a_1 \mid m$ and similarly $a_i \mid m$ for $i = 2, \ldots, n$. This shows that m is a common multiple of the a_i; if m' is another common multiple, say $m' = a_i v_i$ $(i = 1, \ldots, n)$, then by (6), $m'c = m' . \sum A_i u_i = \sum a_i v_i A_i u_i$ (where we have put $m' = a_i v_i$ in the ith term). Hence $m'c = \sum p v_i u_i = cm . \sum v_i u_i$; this shows that $m \mid m'$, so m is indeed a least common multiple, as claimed. ∎

Suppose that the a_i are pairwise coprime, i.e. $(a_i, a_j) = 1$ † for $i, j = 1, \ldots, n$, $i \neq j$. Then the A_i have no common factor and so $c = 1$ in (6); conversely when $c = 1$, the a_i must be coprime in pairs, and so we obtain the

COROLLARY *The LCM of any finite set of non-zero elements in a Bezout domain is their product if and only if they are pairwise coprime.* ∎

It is clear that any principal ideal domain is Bezout and it is easy to give examples of Bezout domains that are not principal: Consider the subring of **Q**$[x]$ consisting of all polynomials with integral constant term. It is not hard to verify that this is a Bezout domain, but the ideal generated by $x/2, x/2^2, \ldots$ is not principal.

This example suggests that Bezout domains differ from principal ideal domains in lacking a certain finiteness condition. This is made explicit in the following description of principal ideal domains, which is often useful as a test, and at the same time tells us that they are unique factorization domains. Let us call an integral domain *atomic*, if every element not zero or a unit is expressible as a product of atoms.

THEOREM 3 *For any ring R the following are equivalent:*

(i) *R is a principal ideal domain,*

(ii) *R is a Bezout domain which is atomic,*

(iii) *R is Bezout and a unique factorization domain.*

Proof. (i) ⇒ (ii). We need only show that a principal ideal domain is atomic (since it is clearly Bezout). Let us first observe that every ascending chain of ideals in a principal ideal domain must terminate:

$$(a_1) \subset (a_2) \subset \cdots$$

For the union is an ideal, generated by b, say, and if $b \in (a_k)$, then $\bigcup(a_n) = (b) = (a_k)$, i.e. $(a_k) = (a_{k+1}) = \ldots$. It follows that every non-unit is divisible by an atom. For if a_1 is a non-unit, either a_1 is an atom or it has a proper non-unit factor a_2, likewise a_2 is either an atom or it has a proper non-unit factor a_3; continuing in this way we obtain a strictly ascending chain, which must terminate, by what we have seen. But this can only happen when we reach an atom which is a factor of a_1.

† Observe that this expression has been defined in different ways in **6.5**, and **10.1** which, however, mean the same thing in a Bezout domain.

Now we repeat the process, but taking a_1/a_2, $a_2/a_3, \ldots$ to be atoms, as we may by what has been proved. Thus we obtain

$$a_1 = p_2 a_2 = p_2 p_3 a_3 = \ldots,$$

where $p_2, p_3, \ldots$ are atoms. Again this process terminates and this can only happen when a_k is a unit; but then we have an atomic factorization $a_1 = p_2 p_3 \ldots p_k a_k$ of a_1.

(ii) ⇒ (iii). This follows closely the proof of Th. 5, **6.5**; in fact we need only show that every atom is prime, and this follows from the Bezout equation (2), as in Th. 5, **6.5.**

(iii) ⇒ (i). Let R be a Bezout domain and a UFD. We recall that in a UFD, each non-zero element has a *length*, which is the number of factors in a complete factorization into atoms. Given an ideal $\mathfrak{a}$ in R, we must show that $\mathfrak{a}$ is principal. This is clear if $\mathfrak{a} = \mathbf{0}$, otherwise we take an element a of least length in $\mathfrak{a}$, then $(a) \subseteq \mathfrak{a}$. If the inequality is strict we can find an element b in $\mathfrak{a}$ but not in (a), then $(a, b) = (d) \supset (a)$, hence d is a proper factor of a and so has shorter length. But $d \in \mathfrak{a}$ and so we have reached a contradiction. This shows that $\mathfrak{a} = (a)$ and hence every ideal of R is principal, as claimed. ■

We conclude this section with a result on the reduction of matrices over a principal ideal domain, the *PAQ-reduction*, also called *Smith's normal form*. This will be useful in **10.6** when we come to generalize the basis theorem for abelian groups to modules over a principal ideal domain.

THEOREM 4 *Let A be an $m \times n$ matrix over a principal ideal domain R; then there exist invertible matrices P, Q of orders m, n respectively over R, such that*

$$PAQ = \begin{pmatrix} d_1 & 0 & 0 & \ldots & 0 \\ 0 & d_2 & 0 & \ldots & 0 \\ \cdot & \cdot & \cdot & \cdot & \cdot \\ 0 & \ldots & d_r & \ldots & 0 \\ \cdot & \cdot & \cdot & \cdot & \cdot \\ 0 & 0 & & \ldots & 0 \end{pmatrix} \tag{7}$$

and $d_i \mid d_{i+1}$ $(i = 1, \ldots, r-1)$.

Proof. We first take the case $m = 1$, $n = 2$. Thus we have a 1×2 matrix $(u \ v)$ say, and we are looking for an invertible matrix Q of order 2 such that

$$(u \ v)Q = (t \ 0). \tag{8}$$

If $u = v = 0$, there is nothing to prove. Otherwise we can find $a, b, c, d \in R$ such that $ud - vc = t$, $u = ta$, $v = tb$, then $t \neq 0$ and on dividing the relation $tad - tbc = t$ by t, we obtain

$$ad - bc = 1. \tag{9}$$

It follows that the matrix $\begin{pmatrix} a & b \\ c & d \end{pmatrix}$ is invertible, with inverse $Q = \begin{pmatrix} d & -b \\ -c & a \end{pmatrix}$ and it is easily checked that (8) holds with this value for Q.

Now consider the general case. We shall apply elementary operations to the columns of A, as in Ch. **5**. These correspond to multiplying A on the right by certain invertible matrices. The operations are

(i) *interchange of two columns,*

(ii) *multiplying a column by an invertible element of R,*

(iii) *adding a multiple of one column to another.*

They are as in Ch. **5**, except that there we allowed any non-zero factor in (ii). But this is just a special case of the above, because in Ch. **5** we were dealing with a field, where every non-zero element is invertible.

In addition we now have a further (non-elementary) operation:

(iv) *alter two columns in such a way that the first elements in each are replaced by their HCF and 0 respectively.*

For this can be accomplished by right multiplication by a matrix constructed as Q was in (8).

Of course corresponding operations can also be carried out on the rows of A.

We can now proceed with the reduction. If $A = 0$, there is nothing to prove. Otherwise we permute the rows and the columns, by operation (i), so as to bring a non-zero element into the (1, 1)-position. Next we apply operation (iv) to replace a_{11} and a_{12} by their HCF and 0 respectively; we repeat this process to replace the new a_{11} and a_{13} by their HCF and 0 respectively, and so on, until the first row has only zeros after the first place. We note that if $a_{11} \mid a_{1i}$, say, then operation (iv) is not needed; we can then reduce a_{1i} to 0 by subtracting a multiple of the first column from the ith, by operation (iii). If on the other hand $a_{11} \nmid a_{1i}$, so that we have to use the operation (iv), then the length of a_{11} is reduced in the process.

We now carry out a corresponding process on the first column; thus if $a_{11} \mid a_{21}$, we reduce a_{21} to 0 by subtracting an appropriate multiple of the first row from the second and, similarly, for a_{i1} ($i = 3, \ldots, m$). All this will leave the first row unaffected. However, if $a_{11} \nmid a_{21}$, we must use operation (iv) to replace a_{11}, a_{21} by their HCF and 0 respectively. This may introduce new non-zero elements into the first row, but at the same time it reduces the length of a_{11}. Thus by appropriate row operations, we can reduce the elements in the first column after a_{11} to 0, and *either* the elements in the first row remain unchanged (i.e. all apart from a_{11} are 0), *or* the length of a_{11} is reduced. In the second case we repeat the process by which the elements in the first row are reduced to 0. Since a_{11} has finite length, this process (of alternately reducing the first row and the first column) must come to an end.

When it does, we have reduced A to the form

$$P_0AQ_0 = \begin{pmatrix} a_1 & 0 \quad 0 \quad \dots \quad 0 \\ 0 & \\ \cdot & \\ \cdot & A_1 \\ \cdot & \end{pmatrix} \tag{10}$$

where A_1 is an $(m-1)\times(n-1)$ matrix. By induction on $m+n$ we can reduce A_1 to diagonal form, say

$$P_1A_1Q_1 = \text{diag}\,(a_2, a_3, \dots, a_r, 0, \dots, 0),$$

hence on writing

$$P_1' = \begin{pmatrix} 1 & 0 \quad \dots \quad 0 \\ 0 & \\ \cdot & \\ \cdot & P_1 \\ \cdot & \\ 0 & \end{pmatrix} \qquad Q_1' = \begin{pmatrix} 1 & 0 \quad \dots \quad 0 \\ 0 & \\ \cdot & \\ \cdot & Q_1 \\ \cdot & \\ 0 & \end{pmatrix}$$

we obtain the equation

$$PAQ = \text{diag}\,(a_1, a_2, \dots, a_r, 0, \dots, 0),$$

where $P = P_1'P_0$, $Q = Q_0Q_1'$.

Now consider the 2×2 matrix formed by the first two rows and columns of PAQ. We have

$$\begin{pmatrix} 1 & 1 \\ 0 & 1 \end{pmatrix}\begin{pmatrix} a_1 & 0 \\ 0 & a_2 \end{pmatrix} = \begin{pmatrix} a_1 & a_2 \\ 0 & a_2 \end{pmatrix}$$

and unless $a_1 \mid a_2$, we can again reduce the length of a_1. Thus by further reductions we may assume that $a_1 \mid a_2$ and, similarly, $a_1 \mid a_i$ $(i = 3, \dots, r)$ By repeating this procedure for a_2 in place of a_1, etc we finally reach a situation where $a_i \mid a_{i+1}$ for $i = 1, \dots, r-1$. ■

When R is Euclidean, the proof becomes a little simpler; in particular, we then do not need the operation (iv). We can then use the norm instead of the length. Thus we first bring a non-zero element of least norm into the (1, 1)-position. If the new a_{1j} is not a factor of every a_{1j} (and every a_{i1}), we obtain an element of smaller norm by subtracting a suitable multiple of the first column (or row). When a_{11} divides all a_{i1} and a_{1j} we can reduce the latter to zero by operation (iii), and so reach (10).

Now the same argument as before can be used to ensure that each diagonal element divides the next. Thus we obtain the

COROLLARY *In* Th. 4, *if R is Euclidean, then P and Q can be taken to be products of elementary and diagonal matrices.* ■

It remains to be seen how far the form (7) obtained is unique. Of course P and Q are far from unique, but as we shall now show, the d_i are unique up to associates. We recall (from Ex. (8), Further Exercises, Ch. **7**) that each kth order minor of AQ is a linear combination of the kth order minors of A. Similarly each kth order minor of PAQ is a linear combination of the kth order minors of AQ and hence of the kth order minors of A itself. Thus if $D = PAQ$, then each kth order minor of D is a linear combination of the kth order minors of A. When P and Q are invertible, we can also write $A = P^{-1}DQ^{-1}$ and it follows that the kth order minors of A and those of D generate the same ideal in R, (Δ_k), say, where the generator Δ_k is determined up to a unit factor by the matrix A, and $\Delta_0 = 1$. But by (7), we have $\Delta_k = d_1 d_2 \ldots d_k$, hence the ds can be expressed in terms of the Δs:

$$d_i = \Delta_i/\Delta_{i-1} \qquad (i = 1, \ldots, r). \tag{11}$$

This shows that the d_i are unique up to unit factors. They are called the *invariant factors* of the matrix A.

The PAQ-reduction was first carried out by H. J. S. Smith in 1861 for integer matrices. Previously the reduction $A \to AQ$ had been used over a field to obtain the echelon form by Gauss and others. The extension to principal ideal domains is more recent, but the most important case remains the well known case of Euclidean domains, mainly on account of its applications to matrices, cf. Ch. **11.**

Exercises

(1) Show that every homomorphic image of a principal ideal ring is again principal.

(2) Show that over a Euclidean domain every matrix of determinant 1 is a product of elementary matrices.

(3) Show that in Th. 4, P and Q can be chosen to have determinant 1. Show that the theorem remains true if we only use operations (i), (iii), (iv), or only (ii), (iii), (iv). Moreover, using only (iii), (iv) and (*i'*): interchange of two columns (or rows) and change of sign in one of them, show that Th. 4 can still be proved, with P, Q chosen to have determinant 1.

(4) Show that over a principal ideal domain, every matrix is right associated to a triangular matrix (*Hermite reduction*). Does this remain true for Bezout domains?

(5) Let $a_1, \ldots, a_n$ be elements of a Bezout domain R with HCF d. Show that there is an $n \times n$ matrix over R with first row $a_1, \ldots, a_n$ and determinant d.

(6) If R is a principal ideal domain, it can be shown that every right ideal of the full matrix ring R_n is principal. Prove this statement for finitely generated right ideals. (Hint. Given $A_1, \ldots, A_r$ apply the PAQ-reduction to the matrix of order rn whose first n rows are $(A_1, \ldots, A_r)$).

(7) Let R be a Bezout domain and K its field of fractions. Show that every element of K can be written in the form a/b, where a and b are coprime. To what extent is this representation unique?

(8) Let R, K be as in Ex. (7) and (c_λ) any finite family of elements of K. Show that the c_λ have a HCF d and $d = \sum a_\lambda c_\lambda$, where $a_\lambda \in R$.

(9) Let R be a commutative ring. If every submodule of a free module is free, show that R is a principal ideal domain. If every finitely generated submodule of a free module is free show that R is a Bezout domain.

10.6 Modules over a principal ideal domain

Our endeavour in this section is to extend the results on abelian groups obtained in **9.7** to modules over a principal ideal domain. In particular, we shall give an independent proof of the main result (Th. 2) and this will provide another proof of the basis theorem for abelian groups. The more general form of this result will find application in Ch. **11**.

Let R be an integral domain. Given any R-module M, an element $u \in M$ is called a *torsion element* if $ua = 0$ for some non-zero element a of R. We observe that uR is free on u precisely when u is *not* a torsion element. As for abelian groups, the set tM of all torsion elements forms a submodule, the *torsion submodule* of M, and we shall call M a *torsion module* or *torsion-free* according as $tM = M$ or $tM = \mathbf{0}$. As for abelian groups, we see that tM is a torsion module and M/tM is torsion-free and tM can again be characterized by these two properties.

The basic result of **9.7** was the fact that every finitely generated abelian group is a direct sum of cyclic groups. Here we shall prove the analogue for modules over principal ideal domains; by a *cyclic* module in this case we mean a module which can be generated by a single element. Our first task will be to show that every finitely generated module over a principal ideal domain is finitely presented. This will follow from

PROPOSITION 1 *Let R be a principal ideal domain; then every submodule of R^n is free, of rank at most n.*

Proof. We shall use induction on n; for $n = 0$, R^0 is interpreted as $\mathbf{0}$ and this is free on the empty set, so we may assume that $n > 0$ and let M be a submodule of R^n. We shall denote the canonical projections $R^n \to R$ by $\varepsilon_1, \ldots, \varepsilon_n$ and their restrictions to M by $\varepsilon_1', \ldots, \varepsilon_n'$. Thus ε_1' is a homomorphism from M to R. If $\varepsilon_1' = 0$, then $M = \ker \varepsilon_1' \subseteq R^{n-1}$ and the result follows by induction on n. If $\varepsilon_1' \neq 0$, its image is a non-zero ideal in R, say $\varepsilon_1'(M) = (a)$, where $a \neq 0$. We choose $u \in M$ such that $\varepsilon_1'(u) = a$ and claim that

$$M = uR \oplus \ker \varepsilon_1'. \tag{1}$$

Let us first show how the result follows from (1). By induction hypothesis, $\ker \varepsilon_1'$ is free of rank at most $n-1$ and uR is free on u. For if $uc = 0$ for some $c \in R$, then $ac = \varepsilon_1'(u)c = \varepsilon_1'(uc) = 0$ and hence $c = 0$ (because R is entire), so u is not a torsion element. Thus uR is free of rank 1, and now (1) shows that M is free of rank at most n.

To establish (1), let $x \in M$ and write $\varepsilon_1'(x) = ab$, then $\varepsilon_1'(x-ub) = ab-ab = 0$, hence $x = ub+(x-ub) \in uR+\ker \varepsilon_1'$. Moreover, the sum is direct, for if $uc \in \ker \varepsilon_1'$ for some $c \in R$, then $0 = \varepsilon_1'(uc) = ac$, hence $c = 0$ and so $uc = 0$. This establishes (1) and completes the proof. ∎

Now let M be a finitely generated module over a principal ideal domain R. Then M is a homomorphic image of R^n, for some n, and the kernel of the mapping $R^n \to M$ is a free module of rank at most n, by Prop. 1. Hence we obtain

COROLLARY 1 *Let R be a principal ideal domain. Then every finitely generated R-module M has a presentation*

$$0 \to R^m \to R^n \to M \to 0, \tag{2}$$

where $m \leqslant n$. ∎

We record another useful consequence of Prop. 1.

COROLLARY 2 *Let R be a principal ideal domain and M any finitely generated R-module, then any submodule of M is again finitely generated. More precisely, if M can be generated by n elements, then any submodule can also be generated by n elements.*

For we can write $M = F/G$, where F is free of rank n, say. By the third isomorphism theorem, any submodule M' of M can be expressed as F'/G, where F' is a submodule of F containing G. By Prop. 1, F' is free of rank at most n, hence M' (as a homomorphic image of F') can be generated by n (or fewer) elements. ∎

We now come to the analogue of the basis theorem for abelian groups.

THEOREM 2 *Let R be a principal ideal domain. Then any finitely generated R-module is a direct sum of cyclic modules. More precisely, if M is a finitely generated R-module, then*

$$M \cong R^s \oplus R/a_1R \oplus \cdots \oplus R/a_rR, \tag{3}$$

where the a_i are non-zero non-units and

$$a_i \mid a_{i+1}, \quad i = 1, \ldots, r-1; \tag{4}$$

the decomposition (3), *subject to* (4), *is unique, up to isomorphism.*

This follows by taking a presentation of the form (2) for M. Here the mapping $f\colon R^m \to R^n$ is represented by an $m \times n$ matrix $A = (a_{ij})$: if $e_1, \ldots, e_m$ and $f_1, \ldots, f_n$ are standard bases in R^m and R^n respectively, then $e_i f =$

$\sum a_{ij}f_j$ as in the case of vector spaces (cf. **4.5**), and if we change the bases in R^m, R^n, the matrix A is replaced by PAQ^{-1}, where P, Q are invertible matrices of orders m and n respectively (cf. **4.6**). Thus by a suitable choice of bases we can take A in the diagonal form of Th. 4 and this leads to the form (3) for M, where $s = n-m$ if we omit terms corresponding to units. ∎

The number s is called the *rank* of M and $a_1, \ldots, a_r$ are called the *invariant factors* of M. For abelian groups (the case $R = \mathbf{Z}$), this definition agrees with that given in **9.7** and we observe that Th. 2 provides an independent proof of the basis theorem for Abelian groups (Th. 6, **9.7**). As in the case of groups we have the

COROLLARY *A finitely generated module over a principal ideal domain is free if and only if it is torsion-free.*

This is clear from the expression (3) for M. If $r > 0$, M has torsion elements, while for $r = 0$, M is free. ∎

More generally, Th. 2 shows that any finitely generated module M is a direct sum of its torsion submodule tM and a free module. However, here the torsion module need not be finite.

In addition to Th. 2 there is a second kind of decomposition into primary cyclic torsion modules, analogous to Prop. 4, **9.7**. We recall that in a principal ideal domain atoms and primes are essentially the same. Let p be a prime in a principal ideal domain R, then a *p-primary module* is a module each of whose elements is annihilated by some power of p (but not necessarily the same power of p will do for each element). By a *primary* module we understand a p-primary module, for some prime p.

THEOREM 3 *Let R be a principal ideal domain and M a torsion module over R. Then M can be expressed as a direct sum of p-primary modules, for the different primes p of R, in a unique way:*

$$M = \bigoplus M_p. \tag{5}$$

If M is finitely generated, only finitely many terms on the right are different from zero.

Proof. For any prime p define

$$M_p = \{x \in M \mid xp^n = 0 \text{ for some } n \geqslant 0\}.$$

Then it is easily checked that M_p is a p-primary module, and the definition makes it clear that it is the largest p-primary submodule of M. To establish (5) we argue much as in the proof of Prop. 4, **9.7**, but using a 'local' rather than a 'global' argument, since M need not be finite, nor even finitely generated.

Let $x \in M$; since M is a torsion module, there is a non-zero element $a \in R$ annihilating x: $xa = 0$. We take a complete factorization of a:

$$a = q_1 q_2 \ldots q_r, \qquad \text{where } q_i = p_i^{\alpha_i} \text{ and the } p_i \text{ are distinct primes.}$$

Put $P_i = a/q_i$, then since the q_i are pairwise coprime, the P_i have no common factor and so there exist $Q_1, \ldots, Q_r \in R$ such that

$$P_1Q_1 + \cdots + P_rQ_r = 1.$$

Now put $x_i = xP_iQ_i$ $(i = 1, \ldots, r)$, then $x = \sum x_i$ and $x_iq_i = xP_iQ_iq_i = xaQ_i = 0$, hence $x_i \in M_{p_i}$ and it follows that $M = \sum M_p$. To show that the sum is direct, assume that $x \in M_{p_0}$ and $x \in \sum' M_p$, where the sum $\sum'$ ranges over all primes $p \neq p_0$. Then $x = x_1 + \cdots + x_r$ say, where x_i is annihilated by $q_i = p_i^{\alpha_i}$ say $(p_i \neq p_0)$. Hence x is annihilated by $a = q_1 \ldots q_r$, but x also lies in M_{p_0} and so is annihilated by p_0^β. It follows that x is annihilated by the HCF of p_0^β and a, but these elements are coprime, so x is annihilated by 1, i.e. $x = 0$. This shows that (5) is direct.

Finally, if M is finitely generated, by $u_1, \ldots, u_r$ say, assume that $u_ia_i = 0$, then $b = a_1 \ldots a_r$ annihilates every element of M. But clearly b can annihilate a non-zero p-primary module only if $p \mid b$. Therefore the only non-zero terms on the right of (5) are those in which $p \mid b$. ∎

If we apply Th. 3 to a cyclic module M we obtain a direct sum of primary modules which are themselves cyclic (as homomorphic images of M). Thus we have the

COROLLARY *Any finitely generated module over a principal ideal domain R can be written as a direct sum of a free module and modules $R/p_i^{\alpha_{ij}}R$, where $\alpha_{i1} \leqslant \alpha_{i2} \leqslant \cdots \leqslant \alpha_{ir_i}$.* ∎

From the uniqueness of the decompositions of Th. 2 and Th. 3 it follows that the elements $p_i^{\alpha_{ij}}$ are again unique up to unit factors. They are called the *elementary divisors* of the module. As in the case of abelian groups, they may be characterized as the generators of the ideals used to describe a direct decomposition into cyclic modules with the largest number of terms, while the invariant factors arise from direct decompositions in which the cyclic terms are each as 'large' as possible.

Th. 2 and Th. 3 show that every finitely generated module (over a principal ideal domain) is a direct sum of cyclic modules. It remains to see when a direct sum of cyclic modules is cyclic. If we turn to abelian groups, we find that a direct sum of cyclic groups is cyclic iff their orders are finite and coprime in pairs. We shall find that a corresponding result holds for modules over principal ideal domains, once we have defined the analogue of order.

Let M be a module over a principal ideal domain R. We recall that an element $x \in M$ is a torsion element iff the ideal (c) of elements annihilating x is non-zero. In that case we call c the *order* of x. Of course the order of an element is only determined up to a unit factor. The module M is said to be *bounded* if there is a non-zero element of R annihilating all of M; any such element is called an *annihilator* of M. The set of all these annihilators is an ideal (a), and a is called the *bound* of M; clearly it is the LCM of all the orders of elements of M. In a cyclic module the bound (if it exists) is the order of the generator, e.g. $R/(c)$, where $c \neq 0$, has the bound c.

A finitely generated module is bounded iff it is a torsion module. For every finitely generated module M can be written

$$M \cong R^s \oplus R/a_1R \oplus \cdots \oplus R/a_rR, \qquad a_i \mid a_{i+1},$$

and this will be unbounded unless $s = 0$. When $s = 0$, M is bounded; in fact its bound is then a_r. We observe that a_r is also the order of an element of M. Thus we have proved

PROPOSITION 4 *A finitely generated module over a principal ideal domain is bounded if and only if it is a torsion module, and there is then an element whose order is the bound of the module.* ∎

Next we determine the relation between the bound of a direct sum and that of its summands.

PROPOSITION 5 *Let M be a module over a principal ideal domain, and suppose that*

$$M = M_1 \oplus \cdots \oplus M_r. \tag{6}$$

Then M is bounded if and only if each M_i is bounded, and the bound of M is then the LCM of the bounds of the M_i.

Proof. If $Mc = 0$, then $M_ic = 0$ and, conversely, if $M_ic_i = 0$, then $Mc = 0$, where $c = c_1c_2 \ldots c_r$. This shows M to be bounded iff $M_1, \ldots, M_r$ are. Now let $c, c_1, \ldots, c_r$ be the exact bounds of $M, M_1, \ldots, M_r$ respectively and write c' for the LCM of $c_1, \ldots, c_r$. Since $M_ic = 0$, $c_i \mid c$ for $i = 1, \ldots, r$, hence $c' \mid c$. On the other hand, $M_ic' = 0$ because $c_i \mid c'$, and so $Mc' = 0$. This shows that $c \mid c'$, therefore c' is an associate of c and this proves the assertion (since the LCM is only determined up to associates). ∎

We can now decide when a direct sum of cyclic modules is cyclic:

THEOREM 6 *Over a principal ideal domain, a direct sum of finitely many cyclic modules is cyclic if and only if the summands are all bounded and their bounds are pairwise coprime.*

Proof. Let M be given by (6), where M_i is cyclic with generator u_i and bound q_i. Suppose that the q_i are pairwise coprime and write $c = q_1 \ldots q_r$, $P_i = c/q_i$, then the P_i have no common factor, hence there exist $Q_1, \ldots, Q_r \in R$ such that

$$P_1Q_1 + \cdots + P_rQ_r = 1.$$

We assert that $u = u_1 + \cdots + u_r$ is a generator of M. For we have $u_iP_j = 0$ when $i \neq j$, hence $u_i = \sum_j u_iP_jQ_j = u_iP_iQ_i$. Therefore

$$uP_iQ_i = \textstyle\sum_j u_jP_iQ_i = u_iP_iQ_i = u_i.$$

This shows that the module uR contains each u_i and hence each M_i. It follows that $uR = M$, i.e. M is cyclic as claimed.

Conversely, let us take two cyclic modules $A = R/aR$ and $B = R/bR$, where $a = da_1$, $b = db_1$. If a, b are not both zero, we take d to be their HCF and assume that this is a non-unit; in case $a = b = 0$ we take d to be any element not 0 or a unit (this is possible unless R is a field, in which case the assertion can easily be proved directly). Now A and B each have R/dR as a homomorphic image, therefore if $A \oplus B$ is cyclic, so is its quotient, $R/dR \oplus R/dR$. But this is clearly false: if

$$R/dR \oplus R/dR \cong R/cR,$$

then by Th. 2 $R/dR = 0$ which is a contradiction. Thus if $A \oplus B$ is cyclic, the bounds of A and B must be finite and coprime. Now if the module M, given by (6), is cyclic, so is any module $M_i \oplus M_j$ $(i \neq j)$, as quotient of M, and hence the bounds of M_i and M_j are finite and coprime. ■

Exercises

(1) An abelian group has generators a, b, c, d and defining relations $3b+2c+8d = 0$, $5a+b-4c+8d = 0$, $-2a+b+4c-8d = 0$, $-a+3b+2c+8d = 0$. Express the group as a direct sum of cyclic groups and find the invariant factors.

(2) Let M be a module over a principal ideal domain R, with a presentation

$$0 \to R^m \to R^n \to M \to 0.$$

Show that the integer $r(M) = n-m$ is independent of the presentation and coincides with the rank of M, as defined after Th. 2.

(3) For any short exact sequence of finitely generated modules over a principal ideal domain show that $r(M) = r(M')+r(M'')$. (Hint. Use the 3×3 lemma.)

(4) Let R be a principal ideal domain and u an element of the field of fractions. By applying Th. 2 and Th. 3 to the module $(R+uR)/R$, obtain a representation $u = u_1 + \cdots + u_t$, where

$$u_i \equiv a_{i0} + a_{i1}/p_i + a_{i2}/p_i^2 + \cdots + a_{ir_i}/p_i^{r_i},$$

and where $p_1, \ldots, p_t$ are the distinct (i.e. non-associated) primes dividing the denominator of u. (Partial fraction decomposition for u.)

(5) In Ex. (4), if $R = \mathbf{Z}$, show that the expression for u is unique if a_{ij} satisfies $0 \leqslant a_{ij} < p_i$. Obtain partial fraction decompositions for 23/36, 1000/510, 1/2400.

(6) In Ex. (4), if $R = k[x]$, show that the expression for u is unique if the p_i are monic and $\deg a_{ij} < \deg p_i$. Obtain the partial fraction decompositions over $\mathbf{R}$ and $\mathbf{C}$ for $(x^4+1)/(x^4-1)$, $1/(x^6-2x^5+6x^4-8x^3+10x^2-6x+3)$.

(7) Decompose $\dfrac{(x+1)\ldots(x+n)-n!}{x(x+1)\ldots(x+n)}$ into partial fractions. Deduce the formula $\sum \dfrac{(-1)^{\nu+1}}{\nu}\dbinom{n}{\nu} = 1+1/2+\cdots+1/n.$

(8) If $f(x)$ has distinct zeros $\alpha_1, \ldots, \alpha_n$, show that

$$\frac{f'(x)}{f(x)} = \sum \frac{1}{x-\alpha_i}.$$

Further exercises on Chapter 10

(1) If R is any commutative ring, find the centre of the full matrix ring R_n.

(2) Let R be a ring and $\mathfrak{c}$ the ideal generated by all elements $xy-yx$. Show that the quotient ring $R^{ab} = R/\mathfrak{c}$ has the following universal mapping property: R^{ab} is a commutative ring and the natural homomorphism $R \to R^{ab}$ is universal for homomorphisms from R to a commutative ring.

(3) Show that the additive group of a simple ring is either torsion-free or, for some prime p, every non-zero element has order p.

(4) Show that an endomorphism θ of an R-module M is an automorphism iff ker $\theta = \mathbf{0}$ and im $\theta = M$. If M is a simple module (i.e. $M \neq \mathbf{0}$ and M has no submodules apart from $\mathbf{0}$, M), show that $\mathrm{End}_R(M)$ is a skew field. (This result is known as *Schur's lemma.*)

(5) Given a split exact sequence $\mathscr{E}$ and any module M, show that the sequences Hom $(\mathscr{E}, M)$, Hom $(M, \mathscr{E})$ are again split exact. What can be said about the new sequences if we merely know that $\mathscr{E}$ is exact? Find these sequences when $R = \mathbf{Z}$, $M = \mathbf{Z}/2$ and $\mathscr{E}$ is the non-split exact sequence $0 \to \mathbf{Z}/2 \to \mathbf{Z}/4 \to \mathbf{Z}/2 \to 0$.

(6) For any square matrix A over a commutative ring R define tr $(A) = \sum a_{ii}$ as the sum of the diagonal elements. If A, B are matrices such that AB, BA are both defined, though not necessarily of the same order, show that tr $(AB) =$ tr (BA). If the unit-element in R has infinite additive order (i.e. $n1 \neq 0$ for $n \neq 0$), deduce another proof of Prop. 1, **10.4**, for R.

(7) Let R be a principal ideal domain and A a skew symmetric matrix over R. Show that A is congruent to a matrix diag $(d_1J, \ldots, d_rJ, 0)$ where $J = \begin{pmatrix} 0 & 1 \\ -1 & 0 \end{pmatrix}$ and $d_1|d_2| \ldots |d_r$ (Frobenius).

(8) Show that a system of equations over a principal ideal domain: $Ax = b$, has a solution iff for each $i = 1, 2, \ldots, r = \rho(A, b)$, the HCF of the ith order minors of A and (A, b) agree.

(9) A polynomial over a UFD is said to be *primitive* if its coefficients have no common prime factor. Let K be a UFD and f, g polynomials in t over K; if deg $g = n$ and g is primitive, show that $t^{n+1}f+g$ is again primitive. Using Gauss's lemma, show that the ring of all fractions f/g, where g is primitive, is a Bezout domain.

(10) Let F be a free module over a principal ideal domain R, and (e_λ) a basis. Given $a \in F$, if $a = \Sigma a_\lambda e_\lambda$ say, show that the ideal $dR = \Sigma a_\lambda R$ depends only on a, not on

the e_λ. The element a is said to be *primitive* if $\Sigma a_\lambda R = R$. Show that an element a of F occurs in a basis of F iff it is primitive. (Hint. Apply Th. 2 to F/aR.)

(11) Show that every finitely generated torsion module over a principal ideal domain has a composition series. What does the Jordan–Hölder theorem tell us in this case?

(12) Let R be a Bezout domain. Show that any finitely generated torsion-free R-module is free.

11

Normal forms for matrices

In this chapter we shall discuss a seemingly special topic, which, however, is of great importance both historically and in present day applications of the theory. We have met the matrix as an auxiliary concept, describing an entity such as a linear mapping or a quadratic form; in each case the actual entries of the matrix depended on the particular coordinate system. Our endeavour therefore will be to choose coordinate systems in which the matrix representing a given entity takes on a particularly simple form. What we need to know in order to find this form is not the nature of the entities represented, but the transformation law to which the matrix is subject. Our goal will be to find a *normal form* or *canonical form* for the matrix, i.e. a class of expressions such that each matrix can be transformed to precisely one matrix within the given class.

For example, linear mappings from one vector space to another, $\theta: U \to V$ (over a field k) are represented by $m \times n$ matrices, where dim $U = m$, dim $V = n$, subject to the transformation rule

$$A' = PAQ^{-1}. \tag{1}$$

Two $m \times n$ matrices A, A' are said to be *associated*† if they are related as in (1), where P, Q are invertible matrices of orders m, n respectively. It is clear that being associated is an equivalence relation and over a field each $m \times n$ matrix A of rank r is associated to $\begin{pmatrix} I_r & 0 \\ 0 & 0 \end{pmatrix}$ (Th. 1, **4.7**, Cor.). This is a complete solution of the normal form problem for matrices subject to the relation of being associated, at least over a field (over a principal ideal domain a solution is given by the Smith normal form, see **10.5**). This form shows that the rank of A is an invariant and that for the transformation law (1) over a field, the rank is the only invariant.

We shall be particularly concerned with the normal form problem in the following case. Consider a linear mapping of a vector space into itself, i.e. an endomorphism $\theta: V \to V$. Such a mapping is represented by a square

† Sometimes such matrices are called 'equivalent'. It seems better not to use that overworked term but to follow the above terminology which agrees with current usage in ring theory.

matrix A subject to the transformation law

$$A' = PAP^{-1}. \tag{2}$$

Two square matrices A, A' related as in (2) are said to be *similar* and one of our main tasks in this chapter will be to find a normal form under similarity. We remark here that we have already met a special case of this problem in Ch. **8**. For, given a quadratic form, or more generally, a hermitian form, this is described by a symmetric or hermitian matrix A with the transformation law

$$A' = PAP^{\mathrm{H}}. \tag{3}$$

This gives rise to the relation of *congruence*, discussed in **8.2**, and we found a normal form there over the real or complex numbers (Th. 3 and Th. 4, **8.2**). But if the underlying space is Euclidean or unitary, we can only allow transformations preserving the metric, i.e. P will then have to be orthogonal or unitary. Thus we now have the transformation law

$$A' = PAP^{\mathrm{H}}, \qquad \text{where } PP^{\mathrm{H}} = I. \tag{4}$$

This is a special case of similarity (since $P^{\mathrm{H}} = P^{-1}$), called *unitary similarity*, and the normal form in this case was obtained in **8.4**. It is rather simpler than the general case (2), although the notion of eigenvalue plays an important role in both. We therefore begin by discussing eigenvalues in a rather more general context.

11.1 Eigenvalues and eigenvectors

Consider an endomorphism of a vector space V (over a field k):

$$\theta: V \to V. \tag{1}$$

In order to find a normal form for its matrix we look for any invariants, i.e. characteristics of θ which can be read off from its matrix but do not depend on the coordinate system. Such invariants are e.g. the fixed points of θ, i.e. vectors $x \in V$ satisfying $x\theta = x$ and, more generally, the vectors $x \in V$ such that

$$x\theta = \lambda x \qquad \text{for some } \lambda \in k. \tag{2}$$

A vector x satisfying (2) is called an *eigenvector of* θ, and if $x \neq 0$, the element λ occurring in (2) is called an *eigenvalue* of θ. Many other names are in use, such as 'proper value', 'latent root' or 'characteristic value' (the latter is 'Eigenwert' in German, from which 'eigenvalue' is formed).

If A is a square matrix of order n, we may regard it as an endomorphism of k^n (referred to the standard basis) and this allows us to speak of eigenvalues and eigenvectors for A. Thus an *eigenvector* for A is a row ξ such that

$$\xi A = \lambda \xi \qquad \text{for some } \lambda \in k, \tag{3}$$

and if $\xi \neq 0$, λ is an *eigenvalue* of A. The equation (3) may be rewritten in the form

$$\xi(\lambda I - A) = 0.$$

From the criterion for the solubility of homogeneous linear equations (cf. **5.1**) we obtain

THEOREM 1 *Let A be a square matrix over a field k. An element λ of k is an eigenvalue of A if and only if λ satisfies the equation in x:*

$$\det (xI - A) = 0. \quad \blacksquare \tag{4}$$

In particular, this shows that the definition of eigenvalue (and eigenvector) given here agrees with that given in **8.4** in the case of symmetric or hermitian matrices. As in that case we shall call (4) the *characteristic equation* and its left-hand side the *characteristic polynomial* of A. Clearly (4) is a monic equation of degree n; thus a matrix of order n has at most n eigenvalues (Th. 3, **6.6**, Cor.). The set of these eigenvalues is also called the *spectrum* of A.

Explicitly the characteristic equation reads

$$\begin{vmatrix} x-a_{11} & -a_{12} & \cdots & \cdots & -a_{1n} \\ -a_{21} & -a_{22} & \cdots & \cdots & -a_{2n} \\ \cdot & \cdot & \cdot & \cdot & \cdot \\ -a_{n1} & -a_{n2} & \cdots & \cdots & x-a_{nn} \end{vmatrix} = 0.$$

On expansion this takes the form

$$\varphi(x) = x^n + c_1 x^{n-1} + \cdots + c_n = 0. \tag{5}$$

We note in particular the values of the last and first coefficients: $c_n = (-1)^n \det A$ and $c_1 = -\operatorname{tr} A$, where $\operatorname{tr} A$, the *trace* of A, is defined by

$$\operatorname{tr} A = a_{11} + \cdots + a_{nn}. \tag{6}$$

THEOREM 2 *Similar matrices have the same characteristic polynomial and hence the same trace and determinant.*

For if $B = PAP^{-1}$, then $xI - B = P(xI - A)P^{-1}$ and hence $\det (xI - B) = (\det P)(\det (xI - A))(\det P^{-1}) = (\det P)(\det (xI - A))(\det P)^{-1} = \det (xI - A)$ $\blacksquare$

The terms *characteristic equation* and *characteristic polynomial* are also applied to endomorphisms of vector spaces. Thus let $\theta: V \to V$ be an endomorphism and denote its matrix relative to a given basis of V by A, then $\det (xI - A)$ is called the *characteristic polynomial* of θ and the equation (4) its *characteristic equation.* Relative to another basis of V, θ has a matrix of the form PAP^{-1}, and by Th. 2, this matrix leads to the same characteristic polynomial. Hence the latter is independent of the choice of basis in V.

There is a remarkable connexion between a matrix and its characteristic equation, which results when we replace the unknown by the matrix:

THEOREM 3 (Cayley–Hamilton theorem) *Every matrix satisfies its own characteristic equation, i.e. if* $\det(xI-A) = \varphi(x)$, *then* $\varphi(A) = 0$.

Some students on meeting this result for the first time are apt to regard it as trivial: 'Simply put $x = A$ in $\det(xI-A)$ and obtain $\det(A-A) = \det 0$'. Of course this argument is inadmissible because, in evaluating the determinant, x is to be treated as a scalar, not a matrix. What must be proved is that

$$A^n+c_1A^{n-1}+\cdots+c_nI = 0,$$

if φ is given by (5).

Nevertheless, there is a rigorous proof which retains something of the naive approach just quoted; it uses the remainder theorem proved in **6.6**.

Write $K = k[A]$ for the ring of polynomials in the matrix A and form the polynomial ring $K[x]$. In the latter ring, let us divide $\varphi(x)I$ by $xI-A$:

$$\varphi(x)I = (xI-A)Q(x)+R, \tag{7}$$

where $Q(x)$ and R are matrices; here R does not involve x because it is of lower degree in x than $xI-A$, and with this proviso $Q(x)$ and R are uniquely determined (Th. 1, **6.6**). To obtain the value of R we apply the remainder theorem (Th. 2, **6.6**):

$$R = \varphi(A). \tag{8}$$

Now for any matrix C, $\det C = C \,.\, \operatorname{adj} C$, hence on writing $C = xI-A$, we find

$$\varphi(x)I = (xI-A)\,.\,\operatorname{adj}(xI-A).$$

On comparing this expression with (7) we see that $R = 0$, and hence by (8), $\varphi(A) = 0$, as we wished to show. ■

We observe that this proof holds for a matrix over any commutative ring.

Exercises

(1) Find the eigenvalues of αI_n, $\begin{pmatrix} 0 & 1 \\ -1 & 0 \end{pmatrix}$, $\begin{pmatrix} \cos\theta & -\sin\theta \\ \sin\theta & \cos\theta \end{pmatrix}$, $\begin{pmatrix} 1 & 1 & 1 \\ 9 & -7 & -9 \\ -13 & 4 & 6 \end{pmatrix}$,

$\begin{pmatrix} 3 & 2 & 2 \\ 1 & 4 & 1 \\ -2 & -4 & -1 \end{pmatrix}$.

(2) Let A be a square matrix with all elements below the main diagonal zero. Show that the eigenvalues are just the diagonal elements.

(3) If A is a matrix with eigenvalues $\lambda_1, \ldots, \lambda_n$, show that the eigenvalues of A^r are $\lambda_1^r, \ldots, \lambda_n^r$. Deduce that $\operatorname{tr} A^i = s_i$ are the power sums of the eigenvalues.

(4) Show that $\operatorname{tr}(A_1A_2\ldots A_r) = \operatorname{tr}(A_iA_{i+1}\ldots A_{i-1})$ for any r matrices of order n.

(5) If $A \in {}^m\mathbf{R}^n$, $\mathbf{B} \in {}^nR^m$, where $m < n$, how are the eigenvalues of AB and BA related? (Hint. Recall Ex. (6), Further Exercises, Ch. **10**.)

(6) Show that any real 2×2 matrix $A \neq I$ satisfying $A^3 = I$ has trace -1.

(7) Show that $\begin{pmatrix} 0 & 0 \\ 1 & 0 \end{pmatrix}$ is similar to $\begin{pmatrix} 0 & 1 \\ 0 & 0 \end{pmatrix}$.

(8) If A, B are square matrices over $\mathbf{C}$, show that $AX = XB$ has a non-zero solution X iff A and B have a common eigenvalue. (Hint. Observe that $f(A)X = Xf(B)$ for any polynomial f.)

(9) In the matrix $\begin{pmatrix} 4 & 2y \\ -y & x \end{pmatrix}$ find integers x and y so that the eigenvalues are 2 and 3.

(10) If a matrix A has characteristic polynomial $\varphi = \prod(x-\alpha_i)$ and f is any polynomial, show that $f(A)$ has characteristic polynomial $\prod(x-f(\alpha_i))$ and $\operatorname{tr} f(A) = \sum f(\alpha_i)$, $\det f(A) = \prod f(\alpha_i)$.

(11)* Let $F(x) = \sum C_i x^i$ be a polynomial in x with coefficients in k_n (the $n\times n$ matrices over the field k) and write $\det F(x) = f(x)$. If $A \in k_n$ satisfies $F(A) = \sum C_i A^i = 0$, show that $f(A) = 0$. (Hint. Show that $xI-A$ divides $F(x)$.)

(12) Show that a 2×2 matrix with positive entries has distinct eigenvalues and at least one is positive.

11.2 The $k[x]$-module defined by an endomorphism

In a vector space V over a field k, let $\theta\colon V \to V$ be an endomorphism and write $\mathfrak{o}$ for the polynomial ring $k[x]$. Then V may be defined as a right $\mathfrak{o}$-module, by taking the action of x on V to be given by θ. Thus the general element of $\mathfrak{o}$ is a polynomial $f = \sum a_i x^i$ and its effect on a vector $u \in V$ is defined to be

$$u\,.\,f = \sum a_i\,.\,u\theta^i. \tag{1}$$

It is easily verified that with this definition V becomes an $\mathfrak{o}$-module. The submodules of V are the subspaces admitting θ, i.e. the subspaces U of V such that $U\theta \subseteq U$; they are also called θ-subspaces. The endomorphisms of V as $\mathfrak{o}$-module are the linear mappings of V into itself which commute with θ. For a module endomorphism is a linear mapping ω such that

$$(uf)\omega = (u\omega)f \qquad \text{for all } f \in \mathfrak{o}, \text{ and all } u \in V. \tag{2}$$

Now if ω satisfies (2), then $\theta\omega = \omega\theta$ (by taking $f = \theta$) and, when this holds, (2) follows.

If we write $\operatorname{End}_k(V)$ for the ring of all endomorphisms of V as vector space over k, then another way of describing the action of $\mathfrak{o}$ on V is by saying that we have a ring homomorphism

$$\mathfrak{o} \to \operatorname{End}_k(V), \tag{3}$$

given by $x \mapsto \theta$. This homomorphism is not injective, because the powers of x are linearly independent over k, whereas the powers of θ are linearly dependent, by the Cayley–Hamilton theorem. Thus the kernel of the mapping (3) is a non-zero ideal of $\mathfrak{o}$. Its generator is a polynomial which will be determined uniquely if we take it to be monic. This monic polynomial, μ say, is called the *minimal polynomial* of θ (or of any matrix representing θ). It is characterized by the fact that for any polynomial $f \in \mathfrak{o}$, $f(\theta) = 0$ iff $\mu \mid f$. In particular, μ divides the characteristic polynomial of θ, whence $\deg \mu \leqslant n$. Later (in **11.4**) we shall find the precise relation between the minimal polynomial and the characteristic polynomial of a given endomorphism or a matrix. For the moment we record the properties of μ found so far.

PROPOSITION 1 *Let A be an $n \times n$ matrix over a field, then there is a monic polynomial μ of degree at most n such that A satisfies any polynomial f if and only if $\mu | f$. In particular, μ divides the characteristic polynomial of A.* ■

We shall find this method of regarding V as an $\mathfrak{o}$-module useful, because $\mathfrak{o}$ is a principal ideal domain and we are able to use the theory of modules over such rings developed in **10.6**. As a first application we examine the conditions under which a matrix can be transformed to diagonal form.

Let us say that a matrix A, or an endomorphism represented by A, is *diagonalizable* if A is similar to a diagonal matrix. E.g., any diagonal matrix is diagonalizable and, as we shall soon see, so are many other matrices. On the other hand, consider the matrix $A = \begin{pmatrix} 0 & 1 \\ -1 & 0 \end{pmatrix}$; it is easily seen that A cannot be transformed to diagonal form over $\mathbf{R}$, thus A is not diagonalizable over $\mathbf{R}$, although it is so over $\mathbf{C}$. For an example of a matrix which is not diagonalizable over any field, take $\begin{pmatrix} 0 & 1 \\ 0 & 0 \end{pmatrix}$.

We note the form taken by the matrix of an endomorphism θ when we choose a basis adapted to a θ-subspace. Thus let θ be an endomorphism of V and let U be a θ-subspace of V. If $e_1, \ldots, e_n$ is a basis of V adapted to U, say $e_1, \ldots, e_m$ is a basis of U, then relative to this basis we have

$$e_i\theta = \sum a_{ij}e_j, \tag{4}$$

where $e_i\theta \in U$ for $i \leqslant m$, hence $a_{ij} = 0$ for $i \leqslant m < j$. This shows that the matrix $A = (a_{ij})$ has the form

$$A = \begin{pmatrix} A_1 & 0 \\ A_3 & A_4 \end{pmatrix}. \tag{5}$$

If V can be written as a direct sum of θ-subspaces: $V = U \oplus U'$, where U, U' are θ-subspaces, and we take a basis adapted to this decomposition, we find that in (4) a_{ij} also vanishes for $j \leqslant m < i$, so A now takes the form

$$A = \begin{pmatrix} A_1 & 0 \\ 0 & A_4 \end{pmatrix}. \tag{6}$$

The extreme case is that where V has a basis of eigenvectors for θ; on referring the matrix of θ to this basis we obtain a diagonal matrix. Conversely, if θ is represented by a diagonal matrix, the basis must consist of eigenvectors. Thus θ is diagonalizable iff V is spanned by eigenvectors of θ, for when θ is spanned in this way, we can pick out a linearly independent spanning set, i.e. a basis. This proves

PROPOSITION 2 *An endomorphism of a vector space V is diagonalizable if and only if its eigenvectors span V.* ∎

Prop. 2 provides a usable criterion for diagonalizability. Thus, given a matrix A, we find its eigenvalues (by solving the characteristic equation) and the eigenvectors (by solving the associated systems of linear equations), to find out if there is a spanning set among them. Here it is useful to know that eigenvectors corresponding to different eigenvalues are linearly independent:

PROPOSITION 3 *Let θ be an endomorphism of a vector space V. If $\lambda_1, \ldots, \lambda_r$ are distinct eigenvalues of θ and u_i is a non-zero eigenvector corresponding to λ_i $(i = 1, \ldots, r)$, then $u_1, \ldots, u_r$ are linearly independent.*

Proof. Suppose that there is a non-trivial linear relation between $u_1, \ldots, u_r$, then one of them can be expressed as a linear combination of earlier ones. Let u_s be the first eigenvector linearly dependent on earlier ones, say

$$u_s = \alpha_1 u_1 + \cdots + \alpha_{s-1} u_{s-1}. \tag{7}$$

Applying θ we find

$$\lambda_s u_s = \alpha_1 \lambda_1 u_1 + \cdots + \alpha_{s-1} \lambda_{s-1} u_{s-1}; \tag{8}$$

now multiply (7) by λ_s and subtract it from (8):

$$\alpha_1(\lambda_1 - \lambda_s)u_1 + \cdots + \alpha_{s-1}(\lambda_{s-1} - \lambda_s)u_{s-1} = 0.$$

Here the differences $\lambda_i - \lambda_s$ are all non-zero and some α_i will be non-zero, by (7), because $u_s \neq 0$; so we have a non-trivial relation between $u_1, \ldots, u_{s-1}$. Therefore one of these must be linearly dependent on earlier ones, which is a contradiction. ∎

If V is n-dimensional and θ has n distinct eigenvalues, we can find a non-zero eigenvector for each and hence, by Prop. 3, a basis. This proves the

COROLLARY *An endomorphism of an n-dimensional space with n distinct eigenvalues is diagonalizable.* ∎

There is another criterion for diagonalizability which is sometimes useful:

THEOREM 4 *An endomorphism θ of a vector space is diagonalizable if and only if its minimal polynomial can be written as a product of distinct linear factors.*

Proof. We define V as an $\mathfrak{o}$-module via the action of θ and take its minimal polynomial in the form $\mu = p_1, \ldots p_m$, where the $p_i = x - \alpha_i$ are distinct linear factors. Since $\mu(\theta) = 0$, $\mu(x)$ annihilates every vector in V; by Th. 3, **10.6**, we can write V as a direct sum of p_i-primary modules

$$V = V_1 \oplus \cdots \oplus V_m \tag{9}$$

where V_i is annihilated by $p_i = x - \alpha_i$ by Prop. 5, **10.6**. Hence $u\theta = \alpha_i u$ for all $u \in V_i$, and so if we take a basis of V adapted to the decomposition (9), θ will be represented by a diagonal matrix; the distinct diagonal elements are $\alpha_1, \ldots, \alpha_m$ and α_i occurs dim V_i times. Conversely, let D be a diagonal matrix whose distinct diagonal elements are $\alpha_1, \ldots, \alpha_m$, then it is easily checked that the minimal polynomial of D is $\prod (x - \alpha_i)$. Any matrix similar to D has the same minimal polynomial and this shows that the condition is necessary. ∎

An endomorphism θ (or also the corresponding matrix) is called *idempotent* if $\theta^2 = \theta$, and *nilpotent* if $\theta^r = 0$ for some $r \geqslant 1$.

COROLLARY *Any idempotent matrix A is similar to a matrix* $\begin{pmatrix} I & 0 \\ 0 & 0 \end{pmatrix}$; *moreover, if the underlying field has characteristic* 0, *then the rank of A is* tr A.

Here the last part follows because both rank and trace are similarity invariants. ∎

By contrast a nilpotent matrix is not diagonalizable unless it is 0.

Exercises

(1) Show that over an algebraically closed field every square matrix is similar to a triangular matrix. (Hint. Use induction on the order.) Over the complex numbers show that the transformation can be accomplished by a unitary matrix.

(2) Prove the Cayley–Hamilton theorem for an $n \times n$ matrix with n distinct eigenvalues and deduce the general result by the principle of irrelevance of algebraic inequalities. Why does this method not show that (over an algebraically closed field) every matrix is similar to a diagonal matrix?

(3) Let θ be an endomorphism of V and define V as a $k[x]$-module by (1). If $v_1, \ldots, v_n$ is a basis of V and A the matrix of θ relative to this basis, then, on writing $\mathbf{v} = (v_1, \ldots, v_n)$, show that

$$\mathbf{v}(\theta I - A) = 0.$$

Deduce that $\mathbf{v}\varphi(\theta) = 0$, where φ is the characteristic polynomial of A, and hence obtain another proof of the Cayley–Hamilton theorem.

(4) Show that every eigenvalue of a matrix A is a root of the minimal equation of A.

(5) Let A be an $n \times n$ matrix over $\mathbf{C}$. If $A^r = I$, show that tr A is a sum of rth roots of 1.

(6) Show that over a field of characteristic 0, a matrix A of order n is nilpotent iff tr $(A^\nu) = 0$ for $\nu = 1, \ldots, n$.

(7) Show that the minimal equation of a hermitian matrix has distinct roots.

(8) Let $A_1, \ldots, A_r$ be matrices whose minimal polynomials are pairwise coprime. Show that for any polynomials $f_1, \ldots, f_r$ there exists a polynomial f such that $f(A_i) = f_i(A_i)(i = 1, \ldots, r)$. (Hint. Use the Chinese remainder theorem.)

11.3 Cyclic endomorphisms

An endomorphism θ of a vector space V is called *cyclic*, if V as a θ-space can be generated by a single vector. Let v be a generator of a cyclic space V, and write $v_i = v\theta^{i-1}$, then V is spanned by $v_1, v_2, \ldots$ over k. Let v_{r+1} be the first of these vectors linearly dependent (over k) on the preceding ones. Then $v_1, \ldots, v_r$ are linearly independent, while

$$v_{r+1} = v_1 a_1 + \cdots + v_r a_r \qquad (a_i \in k). \tag{1}$$

We claim that every v_i is linearly dependent on $v_1, \ldots, v_r$. For $i \leqslant r$ this is clear and for $i = r+1$ it follows by (1), so we take $i > r+1$ and use induction on i. Applying θ^{i-r-1} to (1) we obtain

$$v_i = v_{i-r} a_1 + \cdots + v_{i-1} a_r,$$

and by the induction hypothesis the right-hand side is linearly dependent on $v_1, \ldots, v_r$; hence so is v_i. Thus $v_1, \ldots, v_r$ form a linearly independent spanning set, i.e. a basis, of V and, in particular, $r = n = \dim V$. Relative to this basis θ has the matrix

$$\begin{pmatrix} 0 & 1 & 0 & 0 & \ldots & 0 \\ 0 & 0 & 1 & 0 & \ldots & 0 \\ \cdot & \cdot & \cdot & \cdot & \cdot & \cdot \\ 0 & 0 & \ldots & \ldots & 0 & 1 \\ a_1 & a_2 & \ldots & \ldots & a_{n-1} & a_n \end{pmatrix} \tag{2}$$

Moreover, the minimal polynomial of θ is

$$x^n - a_n x^{n-1} - \cdots - a_1, \tag{3}$$

For if we remember that $v_i = v\theta^{i-1}$, we see that (3) is the polynomial of least degree annihilating v, hence it is a factor of the minimal polynomial of θ, μ say, because μ is the bound of V as θ-space (and so is the LCM of the orders of its elements, cf. **10.6**). Now μ has degree at most n, hence (3) is in fact the minimal polynomial and also coincides with the characteristic polynomial.

Given any monic polynomial of degree n, such as (3), this provides a method of constructing a matrix with (3) as characteristic polynomial, for the matrix (2) has (3) as its characteristic polynomial. The matrix (2) is called the *companion matrix* of the polynomial (3).

Our discussion also leads to a simple criterion for an endomorphism to be cyclic.

THEOREM 1 *An endomorphism of an n-dimensional vector-space is cyclic if and only if its minimal equation has degree n.*

Proof. We have seen that this condition is necessary. Conversely, assume that θ has a minimal polynomial μ of degree $n = \dim V$, then by Prop. 4, **10.6**, there is a vector $v \in V$ whose precise order is μ, and this means that $v, v\theta, \ldots, v\theta^{n-1}$ are linearly independent, for a relation between them would provide an annihilator of lower degree. Thus they form a basis and so V is generated, as a θ-space by v, therefore θ is cyclic. ∎

Let us now consider a cyclic endomorphism whose minimal polynomial is a prime power: $\mu = p^t$. Since every torsion module is a direct sum of primary modules, everything can be built up from this case. Let us take p in the form

$$p = x^s - a_1x^{s-1} - \cdots - a_s \qquad \text{where } st = n.$$

If v is a vector which generates V as θ-space, then, by what has been said, v, $v\theta, \ldots, v\theta^{n-1}$ form a basis of V. We still have a basis if we replace this set by the set $v, v\theta, \ldots, v\theta^{s-1}, vp, vp\theta, \ldots, vp\theta^{s-1}, vp^2, vp^2\theta, \ldots, vp^{t-1}\theta^{s-1}$, where $p = p(\theta)$. For this sequence is obtained from the sequence of vectors $v\theta^i$ by adding to some of the vectors linear combinations of vectors occurring earlier in the sequence. Denote by A the matrix of θ relative to the new basis, then the ith row of A represents the coordinates of the image of the ith basis vector under θ. If $i < s-1$, then $vp^j\theta^i \,.\, \theta = vp_i\theta^{i+1}$, while for $i = s-1$, $j < t-1$, $vp^j\theta^{s-1} \,.\, \theta = vp^{j+1} + \sum_i a_i vp^j\theta^{s-i}$ and for $i = s-1$, $j = t-1$, $vp^{t-1}\theta^{s-1} \,.\, \theta = \sum_i a_i vp^{t-1}\theta^{s-1}$. Thus A takes the form

$$A = \begin{pmatrix} P & N & 0 & 0 & \ldots & 0 \\ 0 & P & N & 0 & \ldots & 0 \\ \cdot & \cdot & \cdot & \cdot & \cdot & \cdot \\ 0 & 0 & 0 & \ldots & P & N \\ 0 & 0 & 0 & \ldots & 0 & P \end{pmatrix}$$

where P is the companion matrix of p and $N = e_{s1}$, i.e.

$$N = \begin{pmatrix} 0 & 0 & \ldots & 0 \\ 0 & 0 & \ldots & 0 \\ \cdot & \cdot & \cdot & \cdot \\ 1 & 0 & \ldots & 0 \end{pmatrix}$$

An important case is that where p is a linear polynomial. This is always so if the field is algebraically closed, e.g. for the complex numbers. Then $s = 1$, $p = x - \alpha$ and A takes the form

$$A = \begin{pmatrix} \alpha & 1 & 0 & 0 & \ldots & 0 \\ 0 & \alpha & 1 & 0 & \ldots & 0 \\ 0 & 0 & \alpha & 1 & \ldots & 0 \\ \cdot & \cdot & \cdot & \cdot & \cdot & \cdot \\ 0 & 0 & 0 & \ldots & \alpha & 1 \\ 0 & 0 & 0 & \ldots & 0 & \alpha \end{pmatrix}$$

E.g., if $\mu = p^3$, where $p = x-2$, then

$$A = \begin{pmatrix} 2 & 1 & 0 \\ 0 & 2 & 1 \\ 0 & 0 & 2 \end{pmatrix}$$

As another example, if $\mu = p^2$, where $p = x^2-3x+7$, then

$$A = \begin{pmatrix} 0 & 1 & 0 & 0 \\ -7 & 3 & 1 & 0 \\ 0 & 0 & 0 & 1 \\ 0 & 0 & -7 & 3 \end{pmatrix}$$

Exercises

(1) Let p be a polynomial with distinct zeros $\alpha_1, \ldots, \alpha_n$, B its companion matrix and $D = \operatorname{diag}(\alpha_1, \ldots, \alpha_n)$. Show that for some vector u, the set $u, uD, \ldots, uD^{n-1}$ is a basis; deduce that $B = PDP^{-1}$, where $P = (p_{ij})$ is the alternant matrix $p_{ij} = \alpha_j^{i-1}$.

(2) Let θ be an endomorphism of a vector space V with characteristic polynomial φ. Given $u \in V$, show that the monic polynomial $f(x)$ of least degree such that $uf(\theta) = 0$ satisfies $f \mid \varphi$; deduce that the zeros of f are eigenvalues of θ. Under what conditions can all eigenvalues of θ (not necessarily with the right multiplicity) be obtained from a single vector u?

(3) Prove the equivalence of the following three conditions on a matrix A: (i) A is cyclic, (ii) A is similar to the companion matrix of its characteristic polynomial, (iii) the minimal polynomial of A equals its characteristic polynomial.

(4) Show that $\begin{pmatrix} 3 & -1 \\ 1 & 1 \end{pmatrix}$ is a cyclic matrix and find a matrix which transforms it to the companion matrix of its minimal polynomial.

(5) Let $f = 0$ be an equation over $\mathbf{C}$ with companion matrix B. If g is a rational function with no poles at the zeros of f, show that the equation with roots $g(\lambda)$, where λ ranges over the roots of $f = 0$, is $\det(xI - g(B)) = 0$. (A. Châtelet)

(6) Show that any matrix commuting with a cyclic matrix A is a polynomial in A. (Hint. Apply the matrix to a generating vector.) Give an example to show that the restriction on A cannot be omitted.

11.4 The Jordan normal form

We now consider the normal form for a general endomorphism. Let θ be an endomorphism of a vector space V over a field k and regard V as a θ-space. Since θ satisfies an equation over k, V is a torsion module and so we have a decomposition

$$V = V_1 \oplus \cdots \oplus V_r,$$

where V_i is a cyclic θ-space, with minimal polynomial μ_i, say, where $\mu_i \mid \mu_{i+1}$. In fact the polynomials $\mu_1, \ldots, \mu_r$ are just the invariant factors of $xI-A$, for any matrix representing θ; they are also called the *invariant factors* of A or of θ. Since V_i is cyclic, the endomorphism θ is represented on V_i by the companion matrix of μ_i, for a suitable basis (chosen as in **11.3**). Denoting this matrix by B_i, we obtain for θ the matrix

$$A = \begin{pmatrix} B_1 & 0 & 0 \ldots 0 \\ 0 & B_2 & 0 \ldots 0 \\ \cdot & \cdot & \cdot \quad \cdot \quad \cdot \\ 0 & 0 & 0 \ldots B_r \end{pmatrix}$$

This is known as the *Frobenius* or *rational canonical form* of A.

Secondly we may decompose V into its primary components according to Th. 3, **10.6**, and then decompose each primary component into its primary constituents. Let us for simplicity take the field k algebraically closed, so that each irreducible polynomial is linear. Then the characteristic polynomial of θ can be split into linear factors:

$$\varphi = (x-\alpha_1)^{r_1} \ldots (x-\alpha_h)^{r_h}, \tag{1}$$

where $\alpha_1, \ldots, \alpha_h$ are distinct elements of k. Now the decomposition into primary components leads to the expression

$$A = \begin{pmatrix} A_1 & 0 & 0 & \ldots & 0 \\ 0 & A_2 & 0 & \ldots & 0 \\ \cdot & \cdot & \cdot & \cdot & \cdot \\ 0 & 0 & 0 & \ldots & A_h \end{pmatrix} \tag{2}$$

where A_i has the characteristic polynomial $(x-\alpha_i)^{r_i}$. If we now apply the decomposition into cyclic spaces (Th. 2, **10.6**), we find

$$A_i = \begin{pmatrix} B_{i1} & 0 & 0 & \ldots & 0 \\ 0 & B_{i2} & 0 & \ldots & 0 \\ \cdot & \cdot & \cdot & \cdot & \cdot \\ 0 & 0 & 0 & \ldots & B_{i\rho_i} \end{pmatrix}, \tag{3}$$

where

$$B_{ij} = \begin{pmatrix} \alpha_i & 1 & 0 & 0 & \ldots & 0 \\ 0 & \alpha_i & 1 & 0 & \ldots & 0 \\ 0 & 0 & \alpha_i & 1 & \ldots & 0 \\ \cdot & \cdot & \cdot & \cdot & \cdot & \cdot \\ 0 & 0 & \ldots & & \alpha_i & 1 \\ 0 & 0 & \ldots & & 0 & \alpha_i \end{pmatrix}. \tag{4}$$

The resulting expression for A is called the *Jordan normal form*; we observe that it is an expression in upper triangular form. If the elementary divisors of A are $(x-\alpha_i)^{t_{ij}}$ where $i = 1, \ldots, h$, $j = 1, \ldots, \rho_i$, then B_{ij} has order t_{ij}.

The type of this matrix may be described by the exponents of the elementary divisors; it is represented by a symbol of the form

$$[(t_{11}t_{12}\dots t_{1\rho_1})(t_{21}\dots t_{2\rho_2})\dots(t_{h1}\dots t_{h\rho_h})], \tag{5}$$

where each pair of parentheses contains the exponents corresponding to a given linear factor $x-\alpha_i$, arranged in descending order of magnitude†. This symbol (5) is called the *Segre characteristic* of the matrix; it presupposes that the characteristic polynomial splits into linear factors as in (1).

Clearly the Jordan normal form of a matrix is completely determined by the Segre characteristic and the eigenvalues, in the order corresponding to that of the symbol (5). To give an example, if A has eigenvalues 2 and 3 and its Segre characteristic is [(211)(3)], then the Jordan normal form of A is

$$\begin{pmatrix} A_1 & 0 \\ 0 & A_2 \end{pmatrix}, \quad \text{where } A_1 = \begin{pmatrix} 2 & 1 & 0 & 0 \\ 0 & 2 & 0 & 0 \\ 0 & 0 & 2 & 0 \\ 0 & 0 & 0 & 2 \end{pmatrix} \text{ and } A_2 = \begin{pmatrix} 3 & 1 & 0 \\ 0 & 3 & 1 \\ 0 & 0 & 3 \end{pmatrix}.$$

This normal form was described by C. Jordan in his 'Traité des substitutions' in 1870. As an application we have

THEOREM 1 *Let A be a matrix whose characteristic polynomial splits into linear factors. Then $A = N+D$, where $ND = DN$ and N is nilpotent, while D is diagonalizable.*

Proof. Let $A' = PAP^{-1}$ be in Jordan normal form. If we can prove the theorem for A', the result will follow for A, for if $A' = N'+D'$, where N' is nilpotent, D' diagonalizable and $N'D' = D'N'$, then if we put $N = P^{-1}N'P$ $D = P^{-1}D'P$, N and D are matrices of the kind required, as is easily checked. Now we see from (2), (3) and (4) that a matrix in Jordan normal form can be written as a diagonal sum of matrices $A_i = \alpha_i I+N_i$, where N_i has zeros on and below the main diagonal and so is nilpotent. Putting these matrices together, we obtain A' as a sum of a diagonal and a nilpotent matrix, which commute, because this is so for each diagonal block. ■

We can use the Jordan normal form to answer other questions, e.g. under what conditions is A diagonalizable? From (4) we see that this is so iff each B_{ij} has order 1, i.e. $t_{ij} = 1$ for all i, j. Thus assuming the characteristic polynomial to split into linear factors, the condition is that each elementary divisor be linear. In any case this is necessary and, when it holds, the characteristic polynomial splits into linear factors, so that the condition is then also sufficient. Hence we have

PROPOSITION 2 *A matrix is diagonalizable if and only if each elementary divisor is linear.* ■

† The order of the exponents is of course merely a matter of convention; it is customary to write the exponents in (5) in descending order, although for the invariant factors in **10.6** the ascending order is more usual.

Now we can also find an expression for the minimal polynomial μ of A. If the invariant factors of $xI-A$ are $d_1, \ldots, d_h$, where $d_i \mid d_{i+1}$ $(i = 1, \ldots, h-1)$, then clearly μ is the last invariant factor: $\mu = d_h$, while the characteristic polynomial is $\varphi = d_1 \ldots d_h$ (cf. (11), **10.5**). This shows not only that $\mu \mid \varphi$, but also that $\varphi \mid \mu^h$, where of course $h \leqslant n$. Thus we obtain

THEOREM 3 *Let A be any square matrix of order n, with minimal polynomial μ and characteristic polynomial φ. Then $\mu \mid \varphi \mid \mu^n$ and moreover, $\mu = \varphi/\Delta_{n-1}$, where Δ_{n-1} is the HCF of all the $(n-1)$th order minors of $xI-A$.* ∎

Finally we ask under what conditions two matrices are similar. Since the Jordan normal is entirely determined by the invariant factors and these are similarity invariants (i.e. they are unchanged when we pass to a similar matrix), we obtain (at least over an algebraically closed field),

THEOREM 4 *Two matrices A and B are similar if and only if they have the same invariant factors or, equivalently, the same elementary divisors.* ∎

If we are willing to use the fact (proved in Vol. 2), that the field can always be extended to one in which the characteristic polynomial splits into linear factors, we deduce that Th. 4 holds generally. But this can also be proved directly, using the

LEMMA 5 *Let R be any ring and $R[x]$ the polynomial ring in an indeterminate x over R. Given $a, b \in R$, there exists a unit u in R such that*

$$ua = bu \tag{6}$$

if and only if there are units f, g in $R[x]$ such that

$$f\,.\,(x-a) = (x-b)\,.\,g. \tag{7}$$

Proof. If (6) holds, then clearly $u(x-a) = (x-b)u$. Conversely, assume (7) and subtract $(x-b)h(x-a)$ from both sides, where h is an element of $R[x]$ to be fixed later. We obtain

$$u(x-a) = (x-b)v, \tag{8}$$

where $u = f-(x-b)h$, $v = g-h(x-a)$. By the division algorithm with monic divisor, we can choose h so that u has degree 0 in x, i.e. $u \in R$. Comparing degrees in (8), we find that $v \in R$; now a comparison of highest terms shows that $v = u$ and so

$$ua = bu.$$

It only remains to show that u is a unit. By hypothesis, f is a unit, say $ff' = 1$, hence $uf'+(x-b)hf' = 1$. On dividing f' by $x-a$ we have $f' = (x-a)q+r$, where $r \in R$ and hence

$$\begin{aligned} 1 = uf'+(x-b)hf' &= u(x-a)q+ur+(x-b)hf' \\ &= (x-b)uq+ur+(x-b)hf', \end{aligned}$$

i.e.

$$(x-b)(uq+hf') = 1-ur. \tag{9}$$

Here the right-hand side is in R; if $uq+hf' \neq 0$, let its degree be m, then the left-hand side has degree $m+1$, a contradiction. Hence both sides of (9) are 0, and $ur = 1$. This shows that u has a right inverse. By symmetry it has a left inverse and so is a unit. ■

Now Th. 4 follows easily: if A, B are $n \times n$ matrices over k, we take $R = k_n$, the ring of all $n \times n$ matrices over k. We know that A and B have the same invariant factors iff $xI-A$, $xI-B$ are associated in $k_n[x]$ and, by the lemma, this holds iff $B = PAP^{-1}$ for an invertible matrix P over k. ■

Exercises

(1) Find the Jordan normal form and the matrix of transformation, for each of the following:

$$\begin{pmatrix} -2 & 0 \\ -14 & 5 \end{pmatrix}, \quad \begin{pmatrix} 12 & -7 \\ 14 & -9 \end{pmatrix}, \quad \begin{pmatrix} 3 & -2 & 0 \\ -1 & 0 & -1 \\ -1 & 3 & 2 \end{pmatrix}.$$

(2) Find a basis consisting of common eigenvectors for the matrices

$$\begin{pmatrix} 0 & -1 & 0 \\ 2 & 3 & 0 \\ 2 & 1 & 2 \end{pmatrix} \quad \text{and} \quad \begin{pmatrix} 0 & 0 & -1 \\ 2 & 1 & 2 \\ 0 & 0 & 1 \end{pmatrix}.$$

(3) Show that any square matrix A is similar to $\begin{pmatrix} B & 0 \\ 0 & C \end{pmatrix}$, where B is the companion matrix of the minimal polynomial of A.

(4) Let P be an orthogonal matrix. Show that the elementary divisors of $xI-P$ occur in reciprocal pairs, except for $x \pm 1$. (Frobenius)

(5) Show that a matrix A is symmetric about the second diagonal (i.e. the SW–NE diagonal) iff $(PAP)^T = A$, where $P = (p_{ij})$, $p_{ij} = \delta_{i+j,n+1}$.

(6) If the block B in Jordan normal form is $B = \alpha I+N$, where $N = \sum E_{i,i+1}$, show that for any polynomial $f(x)$, the matrix $f(B)$ can be obtained by Taylor's formula

$$f(B) = f(\alpha) \,.\, I+f'(\alpha)N+\frac{1}{2!}f''(\alpha)N^2+ \cdots +\frac{1}{n!}f^{(n)}(\alpha)N^n.$$

Verify that $f(B)$ is symmetric about the second diagonal (i.e. running SW to NE).

(7) Let A be a square matrix of order n. Then a complete set of conjugates of A is a set of n matrices $A_1 = A, A_2, \ldots, A_n$ such that (i) the A_i commute pairwise, (ii) all the A_i have the same characteristic equation, (iii) if φ is the characteristic polynomial, then $\prod(xI-A_i) = \varphi(x)I$. Show that any diagonalizable matrix has a complete set of conjugates. By looking at the cases $n = 2, 3$, show that some condition on A is necessary.

(8) Two pairs of matrices (A_1, A_2) and (B_1, B_2) are said to be associated if there exist invertible matrices P, Q such that $PA_1Q = B_1$, $PA_2Q = B_2$. If A_1, B_1 are

non-singular, show that (A_1, A_2) is associated to (B_1, B_2) iff A_1x+A_2 and B_1x+B_2 have the same invariant factors in the ring $k[x]$.

(9) Show that two skew symmetric matrices over a principal ideal domain are congruent iff they have the same invariant factors.

(10) Let N be a nilpotent matrix over a field of characteristic 0 and $r \geqslant 1$. Show that $(I+N)^{1/r}$ is a polynomial in N. If A is a non-singular matrix over **C** show that there exists B such that $A = B^r$.

(11) Let $A = PBQ$, where A, B are both symmetric or both skew symmetric matrices over **C** and P, Q are non-singular. Find R such that $A = RBR^T$. (Hint. Find V with $V^2 = (Q^T)^{-1}P$ and put $R = Q^T V$.)

(12) A matrix whose square is I is called an *involution*. Show that a matrix T over **C** can be written as a product of two involutions iff there is a non-singular matrix A such that $TAT = A$. (Hint. If $TAT = A$, show that A^2 commutes with T; now choose a polynomial B in A^2 such that $B^2 = A^2$ and verify that AB^{-1} is an involution.) Deduce that a matrix T is similar to its inverse iff it is a product of two involutions.

(13) Let A be a non-singular matrix over **C** and P such that $PAP = A$. If $P = U^{-1}P^TU$, show that $B = AU^{-1}$ satisfies $PBP^T = B$. Conversely, if $PBP^T = B$, find A such that $PAP = A$.

11.5 Normal matrices

Let A and B be two diagonalizable matrices, then it is not necessarily the case that A, B are simultaneously diagonalizable, i.e. that an invertible matrix P exists such that $A_1 = PAP^{-1}$ and $B_1 = PBP^{-1}$ are diagonal. For if this holds, then $A_1B_1 = B_1A_1$ and hence $AB = BA$, thus A and B must commute. This necessary condition is also sufficient:

THEOREM 1 *A finite set of endomorphisms $\theta_1, \ldots, \theta_r$ of a vector space V is simultaneously diagonalizable if and only if each θ_i is diagonalizable and the θ_i commute pairwise. If the θ_i are hermitian endomorphisms on a unitary space, they are simultaneously diagonalizable if and only if they commute pairwise.*

Proof. The necessity of the conditions is clear, so assume that they are satisfied. If any one of $\theta_1, \ldots, \theta_r$ is a scalar transformation $u \mapsto u\lambda$, then it does not change under coordinate transformations and may be ignored. Hence we may assume that none of the θ_i is a scalar. We shall use induction on dim V, the case of dim $V = 1$ being trivial (because then all the endomorphisms are scalar). We begin by diagonalizing θ_1—by a unitary transformation in case V is unitary—then V can be expressed as a direct sum of the different eigenspaces of θ_1:

$$V = V_1 \oplus \cdots \oplus V_t,$$

where for $u \in V_j$, $u\theta_1 = u\lambda_j$ and $\lambda_1, \ldots, \lambda_t$ are all different. In the unitary case it follows (by Lemma 2, **8.4**) that the V_j are pairwise orthogonal. Moreover, since θ_1 is not a scalar, $t > 1$, so that $\dim V_j < \dim V$. We claim that each θ_i maps each V_j into itself. For if $u \in V_{j_0}$, let

$$u\theta_i = u_1 + \cdots + u_t, \qquad \text{where } u_j \in V_j,$$

then $u\theta_i\theta_1 = u\theta_1\theta_i = u\theta_i\lambda_{j_0}$, i.e.

$$u_1\lambda_1 + \cdots + u_t\lambda_t = u_1\lambda_{j_0} + \cdots + u_t\lambda_{j_0},$$

hence $u_j = 0$ for $j \neq j_0$ and so $u\theta_i \in V_{j_0}$ as claimed. Now we can apply the induction hypothesis to diagonalize the θs in each V_j. ∎

In Ch. **8** we saw that over the complex numbers, a hermitian matrix can always be diagonalized and moreover, the matrix which accomplishes the diagonalization can be taken to be unitary (a result which was used in the proof of Th. 1). We now ask whether there are other matrices, not necessarily hermitian, which have the same property. The answer is certainly 'yes': If A is hermitian and not zero, then iA is not hermitian, but there is a unitary matrix U such that $UiAU^{-1}$ is diagonal. To answer the question generally we need a definition. A matrix A (over **C**) is said to be *normal* if it commutes with its hermitian transpose:

$$AA^H = A^H A.$$

In particular, when A is real, this means that $AA^T = A^T A$. Correspondingly an endomorphism of a unitary space is called *normal* if it commutes with its adjoint.

THEOREM 2 *Let A be a square matrix over* **C**. *Then there is a unitary matrix U such that UAU^* is diagonal if and only if A is normal.*

Proof. For any matrix A, let us write

$$A_+ = \frac{1}{2}(A + A^{\mathrm{H}}), \qquad A_- = \frac{1}{2i}(A - A^{\mathrm{H}}),$$

then it is clear that A_+, A_- are hermitian, and they commute iff A is normal. Moreover, normality is preserved by unitary transformations and any diagonal matrix is normal. This shows that the condition is necessary; when it is satisfied, we can by Th. 1 transform A_+ and A_- simultaneously to diagonal form and then $A = A_+ + iA_-$ has been diagonalized, as required. ∎

Exercises

(1) If x is an eigenvector of a normal matrix A corresponding to the eigenvalue λ, show that x is also an eigenvector of A^H corresponding to $\bar{\lambda}$.

(2) Give a direct proof that the minimal equation of a normal matrix has distinct roots.

(3) Show that a triangular matrix is normal iff it is diagonal. Using the reduction to triangular form (and the fact that normality is a similarity invariant) obtain another proof of Th. 2.

11.6 Linear algebras

Let k be a field. By a *linear algebra* over k, or briefly, a *k-algebra*, we understand a ring A which is a vector space over k in which the multiplication is bilinear, i.e. we have

$$\alpha(ab) = (\alpha a)b = a(\alpha b) \qquad \text{for } \alpha \in k,\ a, b \in A. \tag{1}$$

This notion may be generalized in various ways, by either (i) not assuming the existence of a unit-element or the associativity of A, or (ii) taking k to be an arbitrary commutative ring instead of a field. But here we shall keep to the case defined above; moreover, A will usually be finite-dimensional as vector space over k. As usual, a *subalgebra* of A is a subring which is also a subspace, and a *homomorphism* between algebras is a ring homomorphism which is also k-linear. In particular, a bijective homomorphism is again called an *isomorphism*.

As an example of a k-algebra, let F be a field containing k as a subfield. Then F may be regarded as a k-algebra; we note that this remains true even if F is a skew field, provided that k is contained in the centre of F. The commutative case will be discussed in Vol. 2. Another important example is the ring k_n of all $n \times n$ matrices over k; this may be regarded as a k-algebra in a natural way. Below we shall see that every finite-dimensional k-algebra is isomorphic to a subalgebra of k_n, for suitable n.

Given a k-algebra A and an element $a \in A$, we have two linear mappings of A into itself, namely the *left multiplication* by a:

$$a_L\colon x \mapsto ax,$$

and *right multiplication* by a:

$$a_R\colon x \mapsto xa.$$

They are analogous to left and right multiplication in a group and we can use them to prove an analogue of Cayley's theorem.

THEOREM 1 *Let A be an n-dimensional k-algebra; then A is isomorphic to a subalgebra of k_n.*

Proof. We show that the mapping $a \mapsto a_R$ is an embedding of A in $\mathrm{End}_k(A)$; since the latter is the full $n \times n$ matrix algebra over k, this will establish the conclusion. It is clear that $(a+b)_R = a_R + b_R$ and by (1), $(\alpha a)_R = \alpha a_R$. Moreover, $(ab)_R = a_R b_R$ and $1_R = 1$, as is easily verified. It remains to show that the mapping is injective. Suppose $a_R = 0$, then $a = 1 \,.\, a_R = 0$, hence the kernel of the mapping is 0, i.e. it is injective. ■

In the proof our assumption, that A has a unit-element, is essential. But even if B is a linear algebra without a unit-element, we can embed B in an algebra with 1, by taking the vector space $A = k \oplus B$ and defining the multiplication in A by

$$(\alpha, a)(\beta, b) = (\alpha\beta,\ ab + \alpha b + a\beta).$$

It is easily seen that with this law A becomes a k-algebra with $(1, 0)$ as unit-element and B as subalgebra. Thus Th. 1 extends to algebras without 1 in the following form:

COROLLARY *Any n-dimensional algebra without 1 is isomorphic to a subalgebra of* k_{n+1}. ∎

When A is finite-dimensional we can apply the theory of this chapter to choose a basis in A such that some a_L or a_R takes on a canonical form. We shall not carry this out in detail, but note one simple consequence that will be of use to us later.

THEOREM 2 *Let A be an n-dimensional k-algebra. Then every element of A satisfies an equation of degree at most n and an element c of A is a unit if and only if it satisfies an equation with non-zero constant term.*

This is almost obvious. Since A is n-dimensional, the elements $1, c, c^2, \ldots, c^n$ are linearly dependent and so c satisfies an equation of degree at most n. Let us take such an equation,

$$\alpha_0 c^n + \alpha_1 c^{n-1} + \cdots + \alpha_n = 0, \tag{2}$$

where $\alpha_i \in k$ and not all the αs vanish. If $\alpha_n \neq 0$, then on dividing by α_n we may take the constant term to be 1 and then $-(\alpha_0 c^{n-1} + \cdots + \alpha_{n-1})$ is an inverse for c. Conversely, if c is invertible, take an equation (2) of least degree; if its constant term is 0, then on multiplying by c^{-1}, we get an equation of lower degree, which is a contradiction. Hence in particular, the minimal equation for c has non-zero constant term. ∎

The last conclusion remains true even if c is only known to be a non-zerodivisor. For in that case, if an equation of least degree has a zero constant term,

$$\alpha_0 c^r + \alpha_1 c^{r-1} + \cdots + \alpha_{r-s} c^s = 0 \qquad \text{where } s > 0,$$

then $c^s(\alpha_0 c^{r-s} + \cdots + \alpha_{r-s}) = 0$ and hence

$$\alpha_0 c^{r-s} + \cdots + \alpha_{r-s} = 0,$$

which contradicts the minimality of r. Thus if c is a non-zerodivisor, it must be invertible and we have proved

PROPOSITION 3 *In a finite-dimensional k-algebra, every non-zerodivisor is a unit. In particular, if a finite-dimensional k-algebra is entire, it must be a (skew) field.* ∎

As in the case of matrices we see that every element in a finite-dimensional algebra has a uniquely determined minimal equation. However, if we apply the proof of Th. 2 to an $n \times n$ matrix, we obtain the bound n^2 for the degree of the minimal equation, rather than the true bound n. The reason for this is that a single matrix generates a *commutative* subalgebra of k_n; and a commutative subalgebra of k_n, at least of diagonalizable matrices, is at most n-dimensional (by Th. 1, **11.5**).

Exercises

(1) Find all algebras up to dimension 4 (with and without 1).

(2) Give an example of an n-dimensional algebra without 1 which is not embeddable in k_n.

(3) If a linear algebra ($\neq 0$) is defined over a skew field k, show that k must be commutative.

(4) Show that a simple ring may be regarded as an algebra (possibly infinite-dimensional) over $\mathbf{Q}$ or $\mathbf{F}_p$, in a natural way.

(5) Let A be a k-algebra with basis $v_1, \ldots, v_n$. Show that the multiplication in A is completely determined by the equations

$$v_r v_s = \sum v_t \gamma_{rs}^t \qquad (\gamma_{rs}^t \in k).$$

Write down the conditions satisfied by the multiplication constants γ_{rs}^t to ensure associativity, and likewise for commutativity. Find the multiplication constants (relative to some basis) for (i) the algebra generated by a matrix with minimal polynomial p and (ii) the full matrix algebra over k.

(6) Show that a matrix A over a field is invertible iff its minimal polynomial has a non-zero constant term. If this is so and the minimal polynomial has degree m, show that A^{-1} can be written as a polynomial of degree $m-1$ in A.

Further exercises on Chapter 11

(1) Let A be a real matrix in which all entries are $\geqslant 0$ and all except all those on the main diagonal and the two diagonals bordering it are 0 (i.e. $a_{ij} = 0$ iff $|i-j| > 1$). Show that A has the form $D^{-1}BD$, where D is diagonal and B symmetric.

(2) For any square matrix A, show that $xI-A$ and $xI-A^T$ have the same invariant factors. Deduce that every square matrix is similar to its transpose.

(3) Let $t(A)$ be a linear function of the entries of a square matrix A such that $t(AB) = t(BA)$. Show that $t(A) = c \,.\, \mathrm{tr}\,(A)$, where c is a constant.

(4) Let A be a square matrix of order n, M a matrix with n columns and N a matrix with n rows. If $MA^\nu N = 0$ for $\nu = 0, 1, \ldots, n-1$, show that $MA^\nu N = 0$ for all ν. (Hint. Let A^ν act on the space spanned by the columns of N.)

(5) Find a basis of eigenvectors for each of the following matrices:

$$\text{(i)} \begin{pmatrix} 3 & -2 & 2 \\ 2 & -1 & 2 \\ 2 & -2 & 3 \end{pmatrix}, \quad \text{(ii)} \begin{pmatrix} 15 & 8 & 22 \\ 11 & 10 & 20 \\ 8 & 13 & 20 \end{pmatrix}, \quad \text{(iii)} \begin{pmatrix} 0 & 2 & 2 \\ -2 & 3 & 0 \\ -2 & 0 & -3 \end{pmatrix}.$$

(6) Let A be an invertible matrix. If for some P, Q, $PAQ = A$ and $P+I$, $Q+I$ are both invertible, show that

$$A(I-Q)(I+Q)^{-1} = (P+I)^{-1}(P-I)A,$$

and if the common value is denoted by M, then $P = (A+M)(A-M)^{-1}$, $Q = (A+M)^{-1}(A-M)$. Conversely, if $A-M$, $A+M$ are invertible and P, Q are defined by these equations, then $PAQ = A$. (Cayley)

(7) Show that the companion matrix of a polynomial f is similar to a diagonal matrix iff f splits into distinct linear factors.

(8) An endomorphism θ on a 5-dimensional space has minimal polynomial $(x^2-1)(x+2)$. Show that V is a direct sum of three primary θ-subspaces. Find all possible normal forms.

(9) The correspondence $f\colon x \mapsto \dfrac{ax+b}{cx+d}$ $(a, b, c, d \in k,\ ad-bc \neq 0)$ on the projective line over k is called a *homography*. Given a homography f over $\mathbf{C}$, show that there is another homography g such that $g^{-1}fg$ has one of the following forms: (i) $x \mapsto mx$ $(m \neq 0)$, (*ii*) $x \mapsto x+c$.

(10) Write $\Delta(t_1, \ldots, t_r) = \prod_{i>j} (t_i - t_j)$ for the product of the differences. If A is a matrix with distinct eigenvalues $\alpha_1, \ldots, \alpha_n$, show that A satisfies the equation $\Delta(\alpha_1, \ldots, \alpha_n, x) = 0$. If further, $\Delta = \Delta(\alpha_1, \ldots, \alpha_n)$, show that for any polynomial f,

$$f(A) = \sum \frac{f(\alpha_i)}{\Delta} \Delta_i (A-\alpha_1 I) \ldots (A-\alpha_{i-1}I)(A-\alpha_{i+1}I) \ldots (A-\alpha_n I).$$

where $\Delta_i = (-1)^{n-i}\Delta(\alpha_1, \ldots, \alpha_{i-1}, \alpha_{i+1}, \ldots, \alpha_n)$. (Sylvester's interpolation formula). More generally, show that this holds when f is any power series convergent for all α_i.

(11) Show that a power series $\sum c_r A^r$ converges whenever all eigenvalues of A lie in the circle of convergence of the series $\sum c_r z^r$. Show that any such power series may be expressed as a polynomial in A. Deduce that $\exp A$ exists for all A; if $AB = BA$, show that $\exp(A+B) = \exp A . \exp B$. (Hint. Use Ex. (10) or the normal form and Ex. (6), **10.4**.)

(12) Show that every orthogonal matrix of determinant 1 has the form $\exp S$, where S is skew-symmetric. (Hint. Take the case of order 2 first, and express the general case as a diagonal sum of the cases $n = 1, 2$.)

(13)* Show that the number of linearly independent solutions of the equation $AX = XA$, where A has Segre characteristic $[(t_1 \ldots t_r)]$ is $t_r + 3t_{r-1} + 5t_{r-2} + \ldots$. Find the number in the general case. (Frobenius)

(14) Find a commutative algebra of $2n \times 2n$ matrices, of dimension n^2+1. (Hint. Consider $n \times n$ blocks.)

Appendices

1 Further reading

General

N. Bourbaki, *Algèbre*, Vol. 1, new ed. 1970, Hermann, Paris.
R. Godement, *Cours d'algèbre*, 1963, Hermann, Paris.
N. Jacobson, *Lectures in Abstract Algebra*, Vols. I–III, 1951, 1953, 1964, van Nostrand.
S. Lang, *Algebra*, 1965, Addison-Wesley.
S. MacLane and G. Birkhoff, *Algebra*, 1967, Macmillan.
B. L. v.d. Waerden, *Algebra*, Vols. I and II, 1971, 1967, Springer.
H. Weber, *Lehrbuch der Algebra*, Vols. I–III, 1894–1908, Chelsea reprint, 1963.

Logic and set theory

P. Halmos, *Naive set theory*, 1960, van Nostrand.
A. Tarski, *Introduction to Logic*, 1941, Oxford University Press.
R. L. Wilder, *Introduction to the Foundations of Mathematics*, 1965, Wiley.

Number theory

G. H. Hardy and E. M. Wright, *Introduction to the Theory of Numbers*, 1945, Oxford University Press.
S. Lang, *Algebraic Number Theory*, 1971, Addison-Wesley.
J.-P. Serre, *Cours d'Arithmétique*, 1970, Presses Univ. France.

Group theory and symmetry

H. S. M. Coxeter, *Introduction to Geometry*, 1961, Wiley.
A. G. Kuroš, *Theory of Groups*, Vols. I, II, 1956, Chelsea.
W. Ledermann, *Introduction to group theory*, 1973, Oliver and Boyd.
J. J. Rotman, *The Theory of Groups*, 1973, Allyn and Bacon.
A. Speiser, *Theorie d. Gruppen v. endl. Ordnung*, 1945, Dover reprint.
H. Weyl, *Symmetry*, 1952, Princeton University Press.

Ring theory

C. W. Curtis and I. Reiner, *Representation Theory of Finite Groups and Associative Algebras*, 1962, Interscience.
N. Jacobson, *Structure of Rings*, 1964, Providence.
J. Lambek, *Rings and Modules*, 1966, Blaisdell.
O. Zariski and P. Samuel, *Commutative Algebra*, Vols. I, II, 1958, 1960, van Nostrand.

Categories

S. MacLane, *Categories for the Working Mathematician*, 1972, Springer.
B. Mitchell, *Theory of Categories*, 1965, Academic Press.

2 Some frequently used notations, with a reference to the page where they are defined or first used

$\mathbf{N}$	the natural numbers	21
$\mathbf{Z}$	the integers	21
$\mathbf{Q}$	the rational numbers	35
$\mathbf{R}$	the real numbers	35
$\mathbf{C}$	the complex numbers	35
$\mathbf{Z}/m$	the integers mod m	30
$\mathbf{F}_p$	the field of p elements	36
C_n	the cyclic group of order n	52
δ_{ij}	the Kronecker delta	81
Sym_n	the symmetric group of degree n	54
Alt_n	the alternating group of degree n	57
$\mathrm{Hom}\,(A, B)$	the group of homomorphisms from A to B	79
$\mathrm{End}\,(A)$	the endomorphism ring of A	82
$\mathrm{Aut}\,(G)$	the automorphism group of G	227
G'	the derived group	231
G^{ab}	the group made abelian	231
$\mathfrak{M}_n(R) = R_n$	the ring of all $n \times n$ matrices over R	83
$\mathbf{GL}_n(k)$	the general linear group	99
$\mathbf{Af}_n(k)$	the affine group	46, 62
$\mathbf{O}_n(k)$	the orthogonal group	200
$\mathbf{U}_n(\mathbf{C})$	the unitary group	200
$\mathbf{Sp}_{2m}(k)$	the symplectic group	209

Index